GEOLOGY: THE SCIENCE OF A CHANGING EARTH

GEOLOGY:

THE SCIENCE OF A CHANGING EARTH

Sixth Edition

Ira S. Allison
Professor Emeritus of Geology
Oregon State University, Corvallis

Robert F. Black
Professor of Geology
University of Connecticut, Storrs

John M. Dennison
Professor of Geology
University of North Carolina, Chapel Hill

Robert K. Fahnestock
Associate Professor of Geology
State University of New York, Fredonia

Stan M. White
Associate Professor of Geology
California State University, Fresno

McGraw-Hill Book Company

New York   St. Louis   San Francisco
Düsseldorf   Johannesburg
Kuala Lumpur   London   Mexico
Montreal   New Delhi   Panama
Paris   São Paulo   Singapore
Sydney   Tokyo   Toronto

GEOLOGY: THE SCIENCE OF A CHANGING EARTH

Copyright © 1955, 1960, 1974 by McGraw-Hill, Inc. All rights reserved.
Copyright 1932, 1939, 1949 by McGraw-Hill, Inc. All rights reserved. Copyright renewed
1960 by Virginia C. Emmons, George A. Thiel, Clinton R. Stauffer, and Ira S. Allison.
Copyright renewed by 1967 by Virginia C. Emmons, George A. Thiel, Eva W. Stauffer, and
Ira S. Allison. All rights reserved. Printed in the United States of America. No part of this
publication may be reproduced, stored in a retrieval system, or transmitted, in any form or
by any means, electronic, mechanical, photocopying, recording, or otherwise, without the
prior written permission of the publisher.

2 3 4 5 6 7 8 9 0 M U B P 7 9 8 7 6 5 4

This book was set in Times Roman by Textbook Services, Inc. The editors were Robert H.
Summersgill and James W. Bradley; the designer was John Horton; and the production
supervisor was Joe Campanella. The drawings were done by B. Handelman Associates,
Inc.
The printer was The Murray Printing Company; the binder, The Book Press, Inc.

Library of Congress Cataloging in Publication Data
Main entry under title

Geology: the science of a changing earth.

   First ed. by W. H. Emmons published in 1932 under
title: Geology.
   Includes bibliographies.
   1. Physical geology. I. Emmons, William Harvey,
date Geology. II. Allison, Ira Shimmin, date
QE28.2.G46 1974   550   73-18281
ISBN 0-07-001118-4
ISBN 0-07-001119-2 (pbk.)

CONTENTS

PREFACE

The sixth edition of *Geology: The Science of a Changing Earth*, provides a sound, basic introduction to physical geology. It is designed for that student seeking an awareness of the physical earth he lives on.

The need for an increased awareness of earth, its materials and its processes, has increased tenfold since the late 1960s. It has come about because of the increased concern for our environment: for ecology and ecological balances; for population increases and dwindling natural resources; and for geological hazards, such as earthquakes. These concerns exist because the earth exists, and they are inevitably related to geology. Understanding geology or developing an awareness of geology will not alleviate these concerns, but it will make it easier to understand why they exist.

Hardly a day passes during which some geological event of our earth is not reported by the various news media, magazines, or television. Consider a few examples:

The earthquake in Managua, Nicaragua, during December 1972

The comprehensive article on earthquakes in the January 1973 *National Geographic*

The eruption on January 22, 1973, of the volcano Helgafjell on Heimany Island near Iceland and Surtsey

The extensive flooding in coastal California communities or the renewed earthquakes near the San Fernando Valley, California, in January 1973

The continuing reports from Scripps Institution of Oceanography on the progress and results of the deep sea drilling vessel *Glomar Challenger*

The above examples occurred during a 1-month period from December 1972 to January 1973. Imagine, then, what has happened upon our earth since its conception nearly 4.5 billion years ago. What will happen to our earth this year?

This text will provide the means to understand past, present, and future geologic processes on our earth.

This edition is divided into four sections, each presenting a set of centralized concepts. Section One presents general concepts of geology and related sciences in order to provide a basic scientific foundation. Section Two considers earth materials, their origin, occurrences, and use as natural resources. Section Three provides an in-depth study of the earth, its structure, and the internal processes that create mountains, cause earthquakes, or shift continents. Section Four turns attention to the earth's surface, its landscapes and their origin and description. The appendices on mineral properties and the metric system can be utilized as the need arises.

Previous editions of this work were published in 1932, 1939, 1949, 1955, and 1960. Of the former authors, Dr. Emmons and Dr. Stauffer are deceased, and Dr.

Thiel and Dr. Allison are retired. For the sake of a broader coverage, fresher viewpoints, and future continuity, other authors were invited to share in the preparation of this sixth edition: Professors Black, Dennison, Fahnestock, and White. Each author has contributed his specialized knowledge toward the development of individual chapters, as well as contributing to the overall design. Stan White collated and integrated the various chapters, arranged them into related sections, provided introductions to the sections, shortened the original text into a more manageable form, and wrote new pages and paragraphs to establish coherence.

The authors express their indebtedness to their colleagues and families for their help, guidance, and understanding in connection with this edition. Robert L. Heller, Consulting Editor for the McGraw-Hill Series in Undergraduate Geology, is due a special note of thanks for his valuable suggestions and contributions of time and effort in the preparation of this text.

Ira S. Allison
Robert F. Black
John M. Dennison
Robert K. Fahnestock
Stan M. White

GEOLOGY: THE SCIENCE OF A CHANGING EARTH

GENERAL CONCEPTS

CHAPTER 1
GEOLOGY: THE SCIENCE OF A EARTH

Most tourists standing at the overlook of the Grand Canyon of the Colorado River in Arizona are able to grasp something of the significance of the panorama before them. The Colorado River in the course of time has carved a canyon 1.6 kilometers (1 mile) deep and 20 to 25 kilometers (12 to 15 miles) wide from rim to rim. It has exposed to view a thick section of nearly horizontal beds, which were laid down layer by layer hundreds of millions of years ago (Fig. 1.1). However, the full geologic story of these rocks is virtually unknown to the tourist, as are other implications of the erosional form, the canyon proper.

To the geologist, or to the student of geology, who is trained to interpret them, the layered rocks tell of conditions prevailing in the Arizona area long ago. They tell him that hundreds of millions of years ago lime (calcium) muds accumulated on an ancient sea floor on which invertebrate animals lived in large numbers; the sun-cracked mud may indicate to him a former river flood plain. The muds of the past have all long since hardened into solid rock, but they still bear telltale signs of their origin. On the one hand, the Grand Canyon is a spectacular river canyon cut in solid rock. On the other, the canyon and the rocks are a "geologic encyclopedia." They tell the geologic history of one part of our country.

GEOLOGY AND MAN

A knowledge of geology can contribute to a deeper appreciation of one's physical environment. Geology demonstrates the transience of apparently permanent mountains, volcanoes, valleys, and lakes, and it illuminates our travels. Visits to such scenic areas as the Grand Canyon (Fig. 1.1), the Black Hills, the Appalachians, Zion Canyon, Crater Lake (Fig. 1.2), or the Rockies are enriched by a knowledge of their origin and history.

Geology and geologic processes, however, have unpleasant aspects. In 1969, some Californians panicked over rumors that their state might collapse into the Pacific Ocean as a result of a disastrous earthquake. The San Fernando earthquake of 1971, or predictions of severe earthquakes to come, did nothing to alleviate these fears. It is encouraging to note that, on December 30, 1972, the "great earthquake predictor," Rueben Greenspan, announced he would no longer predict earthquakes and was withdrawing his

FIGURE 1.1   *Grand Canyon of the Colorado River, as seen looking east toward the Painted Desert. This great canyon, approximately 20 to 25 kilometers wide and 1.6 kilometers deep, has been eroded by the river and its tributaries. (Spence Air Photos.)*

prediction of another disastrous San Francisco earthquake that was to occur at 9:20 A.M. on January 4, 1973. Arizonans are concerned over the cost and difficulty of getting adequate water supplies, while Midwesterners are concerned about too much water. Immense spring floods of the Mississippi River and its tributaries often create havoc. Coastal homes, highways, and beaches of the Atlantic and Gulf Coasts are periodically ravaged by storm waves caused by hurricanes. Seismic sea waves generated by earthquakes cause destruction along the Pacific Ocean shores of Crescent City, California. A new volcanic island (Surtsey) appears suddenly off the coast of Iceland, in 1963; another volcano near Iceland erupts suddenly, in 1973; and on the island of Hawaii, lava flows threa-

ten parts of the island. These volcanic events strongly affect the lives of the people in these areas.

Geology, Man's Needs, and the Environment

Newspapers and magazines continually report catastrophic events that affect man in some part of the world. Unfortunately, few people seem to appreciate the more prosaic, day-to-day, geological processes. Consider, for example, the use of earth materials, which have long affected the activities and welfare of man. He made his first crude weapons and utensils from rock, wood, and bone. When he became more adept in working rock into various shapes, the Stone Age began. With the discovery of methods of isolating metals from their ores, man passed successively into the Copper, Bronze, Iron, and Atomic Ages. At each step, earth materials assumed progressively greater importance.

Today, mineral resources play a role second in importance only to that of agriculture. Man has found thousands of uses for the various materials of the earth, including metals, fuels, fertilizers, structural materials, abrasives, fillers, and chemicals. According to the U.S. Bureau of Mines, the total value of the mineral production of the United States in 1972 was nearly $31 billion. About 70 percent of the total value was supplied by the mineral fuels—petroleum, coal, and natural gas—while metals accounted for 11 percent and nonmetals, other than fuels, 19 percent.

These resources, which are used to support man's standard of living, also create problems. To harassed city officials, the disposal of sewage, garbage, and trash (at reasonable cost and without polluting the local lakes, streams, or groundwater) is at best a difficult problem. What can be done with unsightly and commonly toxic mine wastes or messy gravel pits and rock quarries? These are the same mines, pits, and quarries that provide natural resources for our use. In addition, alternate supplies of mineral resources and new disposal areas steadily become scarcer and more expensive.

Man is beginning to appreciate his dependence on a finite world that geology helped shape. For decades, poor farming practices permitted accelerated erosion and near-overnight destruction of soils that took thousands of years to form by natural geologic processes. This loss has been recognized, and now soil conservation is practiced widely. Exploitation of mineral resources quickly removes the end products of geologic changes that required millions of years. Mining cannot be stopped, but more efficient extraction, rather than "highgrading" or taking only high-grade ore, will permit more ore or oil to be produced.

National interests and global politics increasingly dictate who shall use these resources. No country, particularly the United States, is self-sufficient. Furthermore, at both national and local levels, we must decide whether to permit commercial development of geologic resources or to preserve the land for recreation and conservation purposes. Citizens must directly, and indirectly, involve themselves with geological matters if they are to vote intelligently. Their decisions will determine our standard of living, our way of life, and, ultimately, the survival of mankind.

This book will present the methodology and results of systematic inquiries into the nature of the earth and earth processes and how the earth and life upon it evolved. This is the science of geology.

FIGURE 1.2 *Crater Lake and Wizard Island, as seen from near the park lodge, are set in the crater of an ancient volcano, Mt. Mazama. Crater Lake, in Oregon's southern Cascade Mountains, is about 600 meters deep, 8 kilometers wide, and covers 52 square kilometers. It was formed when Mt. Mazama, more than 3,000 meters high, was destroyed by volcanic action about 6,600 years ago. (Oregon State Highway Department Photo.)*

CONCEPTS OF GEOLOGY

Geology deals primarily with minerals, rocks, fossils, and ores—how they are made, the changes they undergo, and the geologic history they have to tell. Rocks appear to be inert, unchanging things, but each rock carries within it a record of its origin and of the changes it has endured.

Law of Uniform Change

The science of geology is based on change and the proposition that the principles of chemistry, physics, and biology are independent of time. Geologists assume that the forces and processes, or the agents of change, acting upon the earth today operated in the same general way in the past. Winds, rains, rivers, waves, and volcanoes affected the surface of the earth in the past just as they do today.

Everything is subject to change. Granitic rock, the traditional symbol of stability, weathers to loose sand and clay and washes away.

Rocks tumbled on a beach or along a stream bed today are worn to roundness; thus, the rounded rocks present in a layer of solid rock probably have undergone similar rounding in a comparable setting in the past. At the present time, reef-forming corals live in warm, clear, shallow seawater. It is reasonable to assume, therefore, that coral-reef limestones in Indiana are also indicative of warm, clear, shallow seas in the past. The preserved symmetrical ripple marks on a sandstone surface tell of the oscillations of shallow water over loose sand, just as identical ripple marks are observed to form today. The steeply inclined layers of an ancient lava flow in upper Michigan tell of ancient volcanic activity and of an earth movement that later tilted these flows. The present ice sheet on southern Greenland scratches fine lines and grooves on the underlying solid rock. It also leaves in its path jagged or even locally rounded rock surfaces, covered in places by a heterogeneous mixture of boulders, sand, and silt. Similar markings and deposits now found in New England are attributed to glaciation by a former ice sheet. The examples considered thus far illustrate a geologic cause and effect, taking into account time.

Time is a necessary element in effecting geologic change. The minute, slow, day-to-day or even lifetime changes may seem inconsequential. Given thousands, millions, or billions of years, however, geological forces are capable of producing vast results such as the erosion of the Grand Canyon or the leveling of mountains. A former mountain range that extended across upper Michigan, northern Wisconsin, and central Minnesota, for example, has been reduced by erosion over a long period of time to a plain. Periodic outbursts of lava and volcanic cinders eventually build up huge volcanic cones like Mount Rainier, or Parícutin in Mexico, only to have wind, streams, and glaciers go to work tearing it down. Shiprock, New Mexico, is now a relic of just such a former volcano (see Fig. 6.11). An understanding of the meaning of rocks requires accurate and thorough knowledge, both of current earth processes and of the rock record.

Spheres of Geologic Change

The role of geologic change exists in and between several spheres. It is the characteristics of these spheres that initiate the processes and produce the results of geologic change. Only a few characteristics of each sphere are noted, but they will be discussed further in the text.

The Atmosphere   This is the gaseous envelope encircling the earth. The distribution of heat, pressure, and water-vapor content within this envelope gives the earth weather and climate. It is also within this sphere that the starting point for a geologic age-dating system, the carbon-14 system, originates.

The Hydrosphere   This is the water envelope on the solid earth. Water exists in many forms, and it is within this sphere that the hydrologic cycle originates. The interchange, however, is triggered by characteristics of the atmosphere. The elements of this sphere, waves and tides, also change the beaches and coasts.

The Geosphere   This is the solid earth, the geologist's main area of concern (Fig. 1.3). The solid earth consists of several divisions, each with its own special characteristics. These divisions are the thin (5- to 40-kilometer) outer shell, or crust, the inner zone, or mantle, and the center zone, or core. The core itself has two subdivisions: a liquid outer core and a solid inner core. It is on the geosphere where change is most noticeable. Processes within the divisions of the

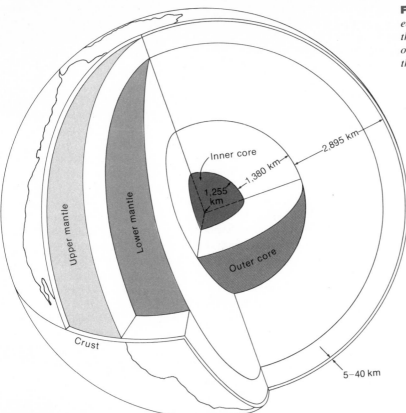

geosphere even affect the distribution of our continents and ocean basins.

The Biosphere This envelope of life is more easily thought of as a zone of influence upon geologic processes. The life within this sphere, from the smallest oceanic plankton to man, exerts a strong influence on the geologic development of the earth, from the formation of sedimentary rocks to their weathering.

Change

Change is the key to geology. One knows that change has occurred by observation. If something appears different, in any way, than it did when first observed, change has occurred. Observation in this case means use of all our senses, plus the extension of our senses through instrumentation.

The random dissipation of energy can cause work

to be done, and this work produces changes in the position of, or state of, matter. In the complex system of the earth, energy is also transferred from one place to another; this is the cause of the changes that constitute the process of geology. This transfer, the cause of change, then, is an attempt to reach equilibrium. Geologic change is no different, only more unfamiliar in the results and the processes that cause the change.

Processes of Change

Gradation The activity of water and ice in breaking and wearing down the high places, and subsequently filling in the low places, is called gradation. The destructive aspect is degradation, and the constructive phase is aggradation. The terms weathering, erosion, and deposition are more commonly used in discussing such changes. When rain falls, the moisture can cause rocks to physically or chemically decay in place, or the runoff may carry with it particles of previously decayed rock. These materials, which are deposited along the streams, in desert basins, in lakes, or in the ocean, erode the surface over which they pass. Waves beat against the coasts and wear away the land, moving away the material to beaches or offshore areas. Winds carry rock particles and roll them along the ground, depositing them as sand dunes or as layers of soil and dust. Ice in motion, called a glacier, carries rock fragments, many of boulder size. When the ice melts, these are deposited.

Change brought about by these geologic agents is generally slow and is usually continuous. Over long periods of time its results are quite significant.

Igneous Activity The processes involving the movement and emplacement of molten material beneath the surface of the earth and the expulsion of lava, cinders, pumice, and ash above the surface of the earth are igneous activities. The subsurface phase is intrusive and the external one is extrusive. Intrusive igneous rocks become exposed to observation by removal of their former cover. Igneous activity, with its high temperatures and pressures, often results in another form of change—metamorphism.

Metamorphism This is a process in which preexisting rocks undergo significant chemical, mineral, or textural change when subjected to high temperatures and pressures or chemically active fluids. Although this process does appear similar to igneous activity, it is different in that pre-existing rocks are subjected to changes while in a solid form.

Diastrophism Rocks of the outer part of the earth undergo differential movement. Rocks formed in a marine environment have been raised thousands of meters above sea level in the Canadian Rockies and in the Colorado Plateau. Other areas, such as the Pacific Ocean floor west of Midway Island, have been depressed. Former volcanic islands, called guyots, which have been truncated by wave action, are now

FIGURE 1.4 *Aerial photograph of Little Dome, a southeast-plunging anticlinal fold in Fremont County, Wyoming. The processes of erosion have sculptured this feature, exposing the folded rock units (Jurassic) and intensifying their appearance on the Wyoming landscape. (U. S. Geological Survey Air Photo.)*

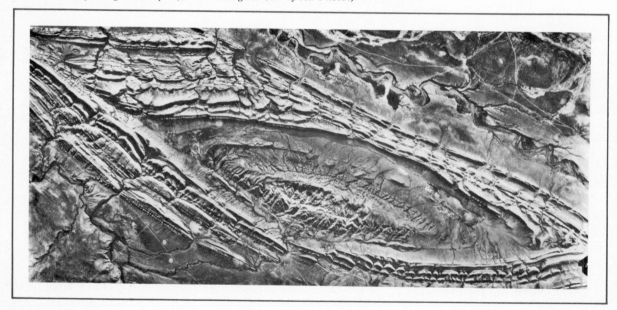

FIGURE 1.5 *Aerial photograph of the San Andreas fault in Southern California. Note that the rock units of the mountain area to the right are tilted and folded. Note also that streams from these mountains have deposited sediments on the lowland to the left. (Spence Air Photos.)*

found thousands of meters below sea level. Large continental areas, such as our Gulf Coast, have been submerged and subsequently elevated above sea level. In other places, such as in Wyoming or the Appalachian Mountains, originally horizontal rock layers have been bent into tight folds (Fig. 1.4). Beneath the Great Plains, rock layers are gently tilted. In still another area, the Great Basin area of Utah and Nevada, large blocks of rock have been displaced vertically thousands of meters. In California, on the other hand, sections of the earth's crust have moved horizontally along fractures called faults (Fig. 1.5).

All these movements, whether uplift, subsidence, folding, tilting, or faulting, are encompassed under the term diastrophism. The regional broad-scale movements, such as general uplift without intense deformation, are epeirogenic; folding and faulting, ordinarily affecting comparatively narrow belts, are orogenic, or mountain-making, earth movements. Without repeated diastrophic changes, all the land would have been eroded practically to sea level long ago.

SUMMARY

The science of a changing earth is geology. Understanding geology means an increased awareness of the beauty of our surroundings and less apprehension toward seemingly damaging geologic changes.

REFERENCES

Earth Science Curriculum Project, 1973, *Investigating the Earth,* Houghton Mifflin Company, Boston.

Holmes, A., 1965, *Principles of Physical Geology,* 2d ed., The Ronald Press Company, New York.

National Academy of Sciences, Committee on Geological Science, 1972, *The Earth and Human Affairs,* Canfield Press, San Francisco.

Shelton, J. S., 1966, *Geology Illustrated,* W. H. Freeman and Company, San Francisco.

Spencer, E. W., 1972, *The Dynamics of the Earth,* Thomas Y. Crowell Company, New York.

Strahler, A.N., 1971, *The Earth Sciences* 2d ed., Harper & Row Publishers, Incorporated, New York.

CHAPTER 2
MATTER AND ENERGY

Geologic change is the result of the physical and chemical behavior of the materials that compose the earth. Since understanding geology requires a certain fundamental background in chemistry and physics, this chapter presents a brief review of energy and matter. It is intended to provide an introduction to the physical and chemical principles sufficient for an introductory approach to geology.

COMPOSITION OF MATTER

Atoms

The smallest particle of an element that retains all the chemical properties of that element is called an atom. An individual atom is much too small to see, even with an electron microscope; yet, physicists have recognized over 100 fundamental particles smaller than atoms. For geologic purposes, we need concern ourselves with only three types of subatomic particles—protons, neutrons, and electrons. Most of the matter in an atom is located at its center, or nucleus. Except for the common form of hydrogen, nuclei of atoms contain protons, which have a positive electrical charge, and neutrons, which have approximately the same mass as protons but have no electrical charge. Common hydrogen has only the one proton in its nucleus. In the space around the nucleus of an atom, electrons move in rapid motion in orbits, or shells. Each electron has a negative charge that exactly equals the positive charge of a proton. The mass of an electron, however, is only about $\frac{1}{2,000}$ that of a proton. According to current theory, electrons are clustered in orbits at definite distances from the atomic center. Only a certain number of electrons can exist in a specific shell, and additional electrons must occur in other shells farther from the nucleus. Figure 2.1 portrays the structure of a potassium atom, the symbol K standing for potassium. The smallest combination of atoms of a compound of matter, which still has all the properties of that compound, is called a molecule. Molecules can be further decomposed into constituent particles of elements.

Elements

The number of protons in the nucleus is the fundamental property that makes that atom a particular element. Chemists have identified 104 elements, and these have 1 to 104 protons in their nuclei. Of the 104 elements, 87 occur naturally. The others have been produced in laboratories. Many of these elements are radioactive; they spontaneously eject subatomic particles from their nuclei, thereby altering the number of protons and neutrons remaining and so changing into new elements. Elements with more than 93 protons are very unstable and decay so readily that they have long since disappeared from this ancient earth. The elements with 43 and 61 protons exhibit rapid radioactive decay and either do not occur in nature or are produced for just an instant during the process of radioactive decay.

Only a few elements are significant in a beginning study of geology. In Table 2.1 a list of only 14 geologically important elements is given. Each element is represented by a chemical symbol that is a scientific type of shorthand.

Atoms with different numbers of neutrons but the same number of protons in their nuclei are the same chemical element, since the number of protons determines chemical identity. These unlike atoms with a different number of neutrons are called isotopes. Many elements have more than one isotopic form. Hydrogen, for example, has three isotopes with 0,1, and 3 neutrons, and carbon, with 6 protons in its nucleus, has five isotopes with 4,5,6,7, and 8 neutrons.

TABLE 2.1   Some Geologically Important Elements

Symbol	Name	Atomic no. (no. of protons)	Atomic weight (protons plus neutrons)
Al	Aluminum	13	26.98
Ca	Calcium	20	40.08
C	Carbon	6	12.01
H	Hydrogen	1	1.008
Fe	Iron	26	55.85
Pb	Lead	82	207.21
Mg	Magnesium	12	24.32
N	Nitrogen	7	14.008
O	Oxygen	8	16.00
K	Potassium	19	39.10
Si	Silicon	14	28.09
Na	Sodium	11	22.99
Ti	Titanium	22	47.90
U	Uranium	92	238.07

COMPONENTS AND STATES OF MATTER

Material consisting of several compounds that can be separated by their differing physical properties is called a mixture. The various compounds in a mixture can be separated, or purified, by using the properties of color, density, different melting temperatures, or different reaction to a magnet. The igneous rock granite, composed of grains of different minerals, is a good example of a mixture. If the component particles are mixed with a liquid, the mixture is a chemical solution. Examples of solutions include salt and water in the sea, a natural gold-silver alloy of two solid metals (a physical solution), oxygen dissolved in water, or the solution of gases known as the atmosphere.

Compounds or molecules may exist in three states: solid, liquid, or gas. The state of a substance depends on its temperature and pressure. Under normal pressure, pure water is a solid (ice) at temperatures colder than 0°C; between 0 and 100°C, it is liquid; and at temperatures above 100°C, it is a gas, water vapor. Water vapor, as in air, can exist in trace amounts at various temperatures. For all substances, progressively higher temperatures are necessary to produce the successive states of solid, liquid, and gas. Physical state is clearly dependent upon heat energy, and so addition of heat is required to melt or evaporate a substance. Conversely, heat is released when a substance

FIGURE 2.1 *Sketch of the atomic structure of potassium. Atomic number 19; 19 protons; 20 neutrons; 19 electrons (2 in shell 1, 8 in shell 2, 8 in shell 3, 1 in shell 4); atomic symbol K.*

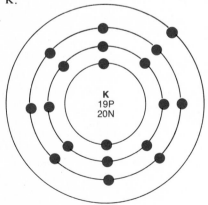

changes from a gas to a liquid, or from a liquid to a solid.

Chemical Combination

The chemical method that produces compounds should now be considered. The two principal mecha-

nisms of chemical combination of elements are ionic bonding and covalent bonding.

Ionic Bonding In ionic bonding, the participating atoms gain or lose electrons to acquire a net positive or negative electrostatic charge. An electrostatic attraction between oppositely charged atoms causes them to adhere to each other. A typical example is provided by sodium chloride (salt). The sodium atom has a nucleus consisting of 11 positively charged protons and 12 neutrons (Fig. 2.2a); 11 negatively charged electrons move about the nucleus: 2 in the first shell, 8 in the second, and 1 in the third electron shell. Therefore, in the total structure of the sodium atom there are 11 positive and 11 negative charges, with a net electrostatic charge of zero. On the other hand, the chlorine atoms in NaCl have 17 protons, 18 neutrons, and 17 electrons, with a net charge of zero.

If an atom gains or loses an electron, an ion (charged particle) is formed. Atoms tend to be more stable if their outermost electron shells have 8 or, in a few examples, 2 electrons. If the single electron in the third shell of a sodium atom moves over to a chlorine atom, then 2 ions form, each having a full outermost shell. The sodium ion has two shells, the innermost with 2 and the outermost with 8 electrons, and the chloride ion has three complete shells with 2, 8, and 8 electrons (Fig. 2.2b). The sodium ion has one more proton than it has electrons, and so it has a +1 charge. The chloride ion has a −1 charge due to an excess of 1 electron over the number of protons.

Unlike charges are attracted to each other, and so the positively charged sodium and the negatively charged chloride ions are attracted to form a compound of sodium chloride (Fig. 2.2c). Rather than form molecules consisting of pairs of specifically linked Na⁺ and Cl⁻ ions, the ions join in a patterned arrangement in which each sodium ion is surrounded by 6 chloride ions, and each chloride ion is surrounded by 6 sodium ions. Note that the compound has electrostatic neutrality. All the plus and minus charges cancel, and the space in the compound is efficiently used.

The difference between numbers of electrons and protons associated with a particular ion determines the valence of that particle. If the number of protons and electrons are equal, the valence is zero and the element exists in atomic form, not ionic form. If the protons exceed the number of electrons, then the ion has a

FIGURE 2.2 *Formation of sodium chloride. (a) Sodium and chlorine atoms (electrons shown as circles in shells). (b) Sodium and chloride ions (electrons shown as circles in shells). (c) Two-dimensional representation of sodium and chloride ions in a crystal. In three dimensions, each chloride ion is surrounded by six sodium ions, and vice versa.*

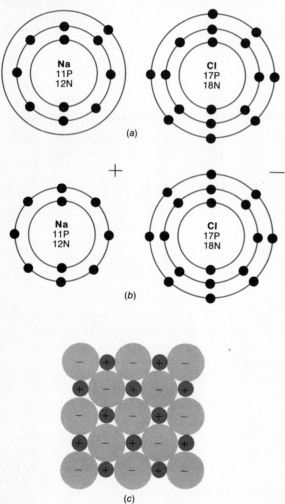

positive valence. If the electrons exceed the number of protons, the valence is negative. Some elements exhibit several valence states. Iron, for example, can have any of three valences: the 0 valence of iron in the familiar metal, the +2 valence of ferrous iron (two electrons missing) in some compounds, or the +3 valence of ferric iron (3 electrons missing) in other compounds.

Covalent Bonding   In covalent bonding, electrons are shared by adjacent atoms so that each atom effectively has a completed shell. In Fig. 2.3a a carbon atom is shown with 4 electrons in its outermost shell, and an

FIGURE 2.3   *Formation of carbon dioxide molecules. (a) Carbon and oxygen atoms (electrons shown as circles in shells). (b) Carbon dioxide molecule (electrons from each original atom are indicated; actually, they are shared equally in all outermost shells and are indistinguishable). (c) Solid carbon dioxide, with crystals made up of an ordered arrangement of discrete molecules.*

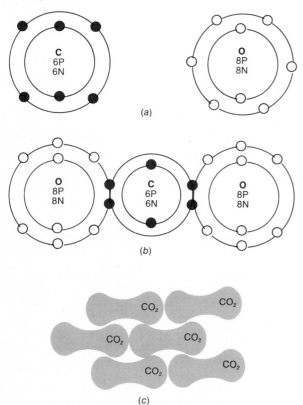

oxygen atom is shown with 6 electrons in its outermost shell. If the carbon atom shares 2 electrons with each of 2 oxygen atoms, then the electron orbits of both carbon and oxygen are filled by sharing. Such an arrangement has greater energy stability than carbon and oxygen in their elemental forms, and so the compound carbon dioxide (CO_2) develops (Fig. 2.3b). This substance occurs as distinct molecules. Crystals of carbon dioxide (dry ice) consist of an ordered arrangement of distinct molecules (Fig. 2.3c). When dry ice changes state to a gaseous form, the molecules abandon this ordered arrangement and move randomly about in the air. The earth's temperature is too warm for crystals of CO_2 to occur naturally, but the gas is present in the air, and it plays an important role in many geological phenomena.

A far more important example is the compound dihydrogen oxide, more familiarly called water. This is the most abundant liquid on the earth's surface and is very important geologically. A molecule of H_2O forms by covalent bonding (Fig. 2.4a). Each of two hydrogen atoms shares its single electron with the oxygen, effectively filling the outermost shell of the oxygen atom as well as the outermost shells of the hydrogen atoms. The hydrogen atoms are asymmetrically placed about the oxygen. The entire molecule is electrostatically balanced (the total number of protons equal the total number of electrons), but their arrangement in space is such that one end of the molecule has a slight negative charge and the other end has a corresponding positive charge (Fig. 2.4b). Such a molecule is electrostatically polar, and this polarity is very critical in the behavior of water as a solvent in geologic processes.

PHYSICAL NATURE OF MATTER

Mass, Density, and Specific Gravity

Matter is anything that has mass (Fig. 2.5). Every piece of matter on the earth has weight, which means that it is physically attracted by the mass of the earth so that this matter can be made to deflect a spring balance. The masses of different objects at the earth's surface are proportional to the relative amounts of deflection that they produce on a balance. Matter also displays the property of inertia. The greater the mass of an object, the greater the force that must be exerted

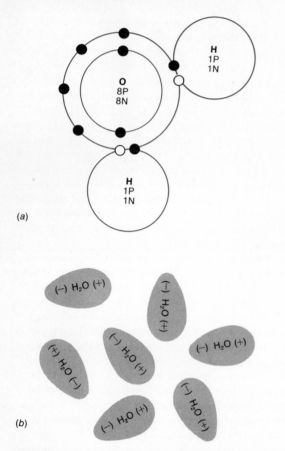

(a)

(b)

FIGURE 2.4 *Structure of water molecules. (a) Electron pattern in water molecule (electrons shown as circles in shells). (b) Schematic representation of molecules in liquid water, showing slightly unbalanced electrical charges at each end.*

to move it and the greater the force required to stop or deflect it. So the quantity of matter, or mass, of an object can be measured by the amount of force necessary to move that object a specified distance.

In general, large objects have greater mass than small ones. However, some objects have more mass in a given volume than others, and vice versa. Density is the measure of mass per unit of volume. In most scientific work, mass is measured in grams and volume in cubic centimeters. Water has a density of 1.00 gram per cubic centimeter. The mineral quartz has a density of 2.65 grams per cubic centimeter. Specific gravity is

the mass of an object divided by the mass of an equal volume of water; specific gravity is expressed as a simple number without any units of measurement. In the metric system, it is equivalent to density. The specific gravity of quartz is 2.65, diamond 3.52, and native copper 8.94. Simply expressed, quartz is 2.65 times heavier than an equal volume of water.

Energy

Unlike matter, energy does not have mass; instead, it is measured by the change it can produce in the motion of an object having a certain mass. In other words, energy is the ability to do work. The movement of matter from one place to another is a direct result of energy acting on matter. Energy occurs in a variety of interchangeable forms, and all these forms can ultimately be made to do physical work, which means changing the velocity of some object having mass. The magnitude of a quantity of energy is measured by the amount of work it can do.

Work is a measure of the amount of energy that has been converted to provide the force to move an object (work = force × distance). The different forms of energy can be classified as shown in Fig. 2.5. Geologic changes in the earth result from energy transfer, and so at least a modest understanding of the nature of energy is necessary for beginning geology.

Kinetic Energy The energy of motion is kinetic energy. A boulder acquires kinetic energy when it falls from a cliff. Kinetic energy can do physical work, such as crush another rock that the boulder might hit at the base of the cliff. The amount of kinetic energy is obtained by the formula

$$K.E. = \tfrac{1}{2} mv^2$$

where m = mass and v = velocity.

Because of its greater mass, a truck moving at 55 kilometers per hour has more kinetic energy than a Volkswagen moving at the same speed. Doubling the speed of either vehicle from 55 to 110 kilometers an hour quadruples its kinetic energy. Kinetic energy is present in the waves on the sea, in wind, in running water, in a mudslide, and in the rotation of the earth. There are some special types of kinetic energy.
(a) Sound Sound is a special type of kinetic energy whereby a concussion, or shock wave, is transmitted through an object as vibration. Molecules impact

against one another to disperse the energy of the sound wave through the mass. Sound travels through gases, liquids, and solids. It cannot cross a vacuum. A vacuum exists because there is no matter within, and therefore, no matter to transmit the kinetic energy of the shock wave. Many sounds are insignificant geologically, but the noise of an exploding volcano may dislodge soil or rocks. Earthquakes are a series of intense vibrational waves passing through the earth, produced chiefly by rock masses moving or breaking inside the earth. Most earthquake vibrations are below audible frequencies but on occasion they can be heard as sounds.

(b) Heat Heat consists of random vibrations of molecule-size particles, or ions, in a crystal. Temperature is directly proportional to the average kinetic energy of the particles within the mass. In a crystalline solid, such as ice, the constituent molecules in a low-energy state occupy a fixed patterned arrangement. As the temperature rises, the molecules vibrate or oscillate within this geometric framework with increased intensity. When the temperature reaches 0°C, the

FIGURE 2.5 *Classification of physical phenomena.*

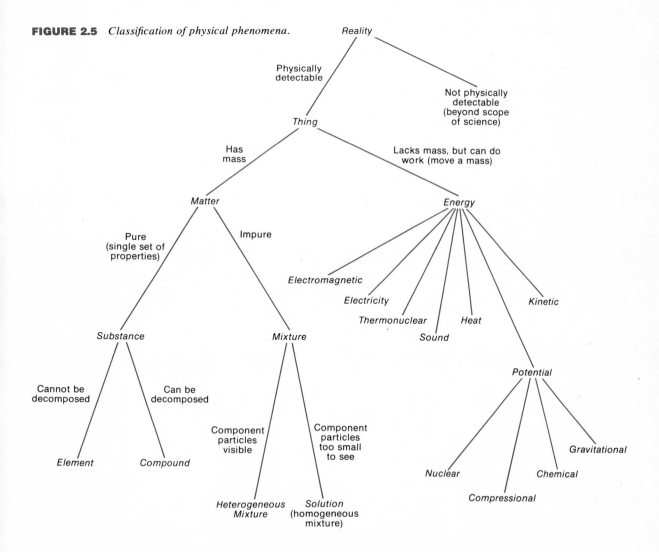

vibration becomes so strong that it overcomes the electrostatic attraction which binds the molecules together and the ice crystal melts. After the ice is converted into a liquid, further addition of energy causes the temperature to rise. With this increased kinetic energy of the vibrating molecules, some of the water molecules escape from the liquid to evaporate, or form water vapor. With continued heating at the boiling temperature, all the remaining liquid changes to water vapor. Under confining pressure, the temperature of water vapor can be raised far above the ordinary boiling temperature of water, such as occurs with a geyser. Because of their stronger bonding, ionic compounds generally have higher melting, freezing, and boiling temperatures than covalent compounds.

Heat travels by conduction, convection, and radiation. Conduction operates through solids, liquids, and gases by the gradual transfer of kinetic energy from the molecules of one part to another part of the substance, as in the heating of the handle of a silver spoon whose other end is in hot soup. Heat from inside the earth is slowly conducted toward its surface, which is cooler than its interior. Convection is the transfer of heat in liquids and gases by physical movement of the material from one place to another. It is usually in response to a decrease in density accompanying a rise in temperature. Convection of light-warm and heavy-cold air causes the geologically important phenomena of wind and vertical air currents. Radiation of heat is a transfer of energy by a certain frequency of electromagnetic

waves that can even travel across a vacuum or through a gas to produce warming of an object. This is how the sun's energy reaches the earth.

(c) Electromagnetic Energy Electromagnetic waves are fast pulses of energy that can cross vast distances in a vacuum or penetrate certain types of matter. The faster the pulsing of energy, the greater its frequency and the shorter its wavelength. Figure 2.6 shows the variation in types of electromagnetic energy in a presentation known as the electromagnetic spectrum. Electromagnetic energy is significant to geology in the arrival of heat and light from the sun and the radiation back into space of heat and light from the surface of the planet earth.

The visible part of the electromagnetic spectrum, called light, is a very limited segment of the spectral range. Different colors of light have different frequencies: the red end of the spectrum has a lower frequency (longer wavelength) than the violet end. Certain chemical changes release electromagnetic energy in the form of heat or light. Fire is a dramatic example of both. Slow oxidation, such as in the decay of grass, releases heat but no light.

(d) Thermonuclear Energy Thermonuclear energy is released by modification of the nucleus of an atom. When nuclear transformations occur, a fraction of the mass is converted to energy, either electromagnetic energy (including radiant heat) or kinetic energy of ejected subatomic particles. The conversion of mass to energy is in accord with the famous formula devel-

FIGURE 2.6 *The electromagnetic spectrum.*

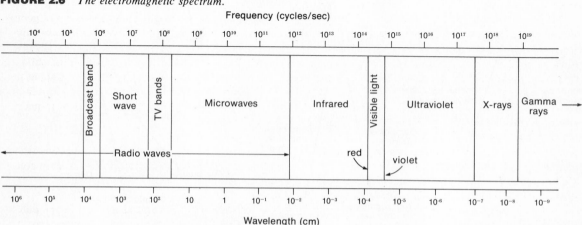

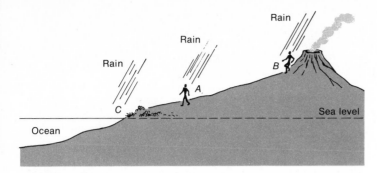

FIGURE 2.7 *Increase of potential energy with increasing altitude. A hiker gains potential energy as he climbs. The energy consumed while climbing is partly stored as potential energy to aid his descent. Rain that falls to the earth at B has greater potential energy for doing geological erosion than rain falling at A. Rain falling on the beach at C has no potential energy for stream erosion.*

oped by Albert Einstein in 1906, $E = mc^2$. This formula relates the total energy to the mass consumed (m) times a constant (c^2, the speed of light raised to the second power). The energy potential by this change is millions of times greater than the energy of any chemical explosion. Usually only a fraction of the total mass is converted to energy in a nuclear reaction. Some nuclear reactions are of the atom-splitting, or nuclear-fission, type. Examples include the slow-release energy by certain naturally radioactive elements in the earth or the sudden explosion of an atomic bomb. Another type of reaction is nuclear fusion, whereby heavier atoms are built up from lighter ones. This is what happens in a hydrogen bomb explosion. Astronomers consider the "burning" of the sun and other stars to be a sort of continuing hydrogen bomb reaction of many billions of years duration.

(*e*) *Electricity* An electric current results from the flow of electrons, protons, or ions through matter or across a vacuum. Chemical, and some physical, change releases electrons, and these produce currents as they move. Electric currents occur inside the earth, but their geological significance is not well understood. Chemical reactions in geological processes sometimes produce faint electric currents.

Potential Energy Potential energy is energy stored in matter by virtue of its position. It may appear as kinetic energy under a following motion. If a mountain climber (Fig. 2.7) ascends a peak, he will, in the process, acquire potential energy by virtue of his newly climbed position. If he should fall, his potential energy will become kinetic energy. Similarly, a rock lying on a mountain top has potential energy proportional to its distance above sea level. This potential

energy equals the work that would be necessary to move that rock from sea level up to the elevation of the mountain top. This sort of energy is gravitational potential energy, stored up because of position relative to the force of gravity, which is directed downward toward the center of the earth. Water occurring at the surface at high elevations on the earth possesses great gravitational potential energy.

The concept of the force of gravity needs an introduction at this point. All material objects are attracted to all other matter by the force of gravity. The gravitational attraction of two bodies is proportional to their masses, as determined by inertial studies, and inversely proportional to the distance between them. Consequently, the attraction of the earth exerts an attractive force, called weight, on an object, and this force is proportional to the mass of that object as measured by its inertia. The earth's gravity causes all eroded geologic materials to move inexorably downhill.

An excellent example of potential energy in geology is compressional energy, as a confined gas pent up inside a volcano that may be released explosively in a volcanic eruption. Similarly, a distorted elastic solid also stores compressional potential energy, such as the wound spring of a clock or the bent rocks along a fault that are ready to snap loose to create an earthquake.

Force

Earlier in this chapter it was noted that work is a measure of the amount of energy which has been converted to provide the force to move an object ($W = F \times D$).

Forces allow one object to exert a push or a pull on another object. When this exertion of force produces a change of velocity or direction of travel of a moving object, work is accomplished. Consider some types of forces or force fields.

Directed Forces The force of one object physically in contact with another may be directed with a specific orientation. Horizontal pushing on a cart in a supermarket overcomes its inertia and friction, and so the cart moves across the floor. Geologic forces directed to the northwest folded an area extending from western Virginia to Pennsylvania some 200 million years ago, resulting in the folded rocks of the Appalachian Mountains.

Pressure Confined liquids and gases exert a pressure in all directions from a specific point in the fluid. In liquids the pressure increases with depth, such as the progressively greater pressures on a research submarine as it descends deeper along the sea floor or the increasing pressure with water depth in a well. Gas pressures of hundreds of kilograms per square centimeter are common in deep wells drilled in search of natural gas. In a container of gas, the pressure is essentially uniform throughout.

Force Fields By definition, a field is a portion of space where there is a value for a measured quantity at every point in that space. A force field is a field in which the measured quantity is a type of force. Three of these force fields occur in nature: gravitational, electrostatic, and magnetic. In all three types, the force is directly proportional to some property inherent in two objects and inversely proportional to the square of the distance between them.

Magnetic force is familiar to everyone who has worked with a magnet. Like magnetic poles (i.e., north to north) repel, and unlike poles (i. e., north to south) attract. The earth's magnetic field attracts a compass magnet to make it point northward. Some bodies in the solar system, Venus, Mars, and the Moon, lack a significant magnetic field, while others, the Sun, Earth, and Jupiter, have strong magnetic fields.

Magnetic and electrical phenomena are often associated. The aurora borealis, "northern lights" in the sky, is localized near the North Pole because protons ejected from the sun are attracted to the magnetic

poles of the earth. More common in everyday experience is an electric motor. At a generator station, mechanical motion of a magnet causes a dynamo to generate an electric current in nearby wires. In the electric motor at the other end of the wires, perhaps hundreds of kilometers away, the energy reaction of the electric current with a magnet produces the kinetic energy of the motor. The magnetic field of the earth will be considered further in Chap. 14. Before leaving the discussion of matter, energy, and forces, an attempt should be made to interrelate them with change, the keystone to the study of geology. To do this, consider another concept, equilibrium.

EQUILIBRIUM

Things are inert because the energy available to do work is not enough to produce change. If there is an energy imbalance, then matter must alter its position, or state, to restore energy balance or equilibrium. In all situations, including geologic ones, if the environment changes, then the system tends to respond so as to adapt to the changed circumstances.

When a surplus of heat energy is taken up by ice, melting results. If the strength of rocky areas on a mountain or steep slope is weakened (by an excess of water) so that they cannot withstand the force of gravity, then a rockslide may occur to restore equilibrium. Thus changes in the physical and chemical environment of rocks tend to produce an equilibrium shift. Another illustration is the weathering of rocks. Solid granitic rock is formed at great depths, and when it is exposed at the surface of the earth it will weather and develop into a soil. The weathered products are chemically more stable than the granite at the surface. All these changes are examples of equilibrium adjustments.

Equilibrium situations are two types, static and dynamic. Static equilibrium is a situation in which no change occurs. A rock sitting on a flat surface does not move because the downward force of gravity is counterbalanced by the supporting ground underneath the rock. In dynamic equilibrium, changes are occurring, but opposing changes occur at approximately balanced rates. For example, heat is added to the earth in the form of radiant energy from the sun, and the earth radiates an equal amount of heat outward into the cooler space surrounding it. The temperature of the air on a particular day is the result of temporary local shifts in a dynamic equilibrium between heat income and heat outgo. Thus, the interplay of matter, energy, forces, equilibrium, and change highlights the complexity of the earth.

Through careful use of the laboratory, scientists have studied the behavior of matter and energy and have generalized their findings into the laws of chemis-try, physics, and biology. But remember, the carefully controlled and simplified environment of the laboratory does not occur in nature. The natural world is a composite of varied environments where many uncontrolled and partially interfering processes go on simultaneously. The challenge for a geologist is to attempt to isolate the specific chemical and physical changes that occur.

REFERENCES

Ahrens, L. H., 1965, *Distribution of the Elements in Our Planet,* McGraw-Hill Book Company, New York.

Akasofu, S., 1965, *The aurora,* Scientific American, vol. 213, no. 6, pp. 54-62.

Asimov, I., 1961, *Building Blocks of the Universe,* 2d ed., Abelard-Schuman, Limited, New York.

Freeman, I. M., 1973, *Physics: Principles and Insights,* 2d ed., pp. 267-269, McGraw-Hill Book Company, New York.

Hampel, C.A. (ed.), 1968, *The Encyclopedia of the Chemical Elements,* Reinhold Book Corporation, New York.

Nechamkin, H., 1968, *The Chemistry of the Elements,* McGraw-Hill Book Company, New York.

Orear, J., 1967, *Fundamental Physics,* 2d ed., John Wiley & Sons, Inc., New York.

Physical Science Study Committee, 1965, *Physics,* 2d ed., D. C. Heath and Company, Boston.

Rosen, S., Siegfried, R., and Dennison, J. M., 1965, *Concepts in Physical Science,* Harper & Row Publishers, Incorporated, New York.

Shortley, G., and Williams, D., 1965, *Elements of Physics,* 4th ed., Prentice-Hall, Inc., Englewood Cliffs, N. J.

Sienko, M. J., and Plane, R. A., 1974, *Chemistry: Principles and Properties,* 2d ed., McGraw-Hill Book Company, New York.

CHAPTER 3

THE EARTH
IN SPACE

Geologists primarily study the solid earth on which we live, but certain geological phenomena are related to the earth's neighbors in space. Therefore, we must consider the astronomic setting of our planet in order to grasp the effect of the universe on geology. The earth is one of nine planets orbiting the sun. The sun is just one of billions of stars that comprise the Milky Way galaxy, and so other stars are likely to have planets also. Understanding the earth's place in the vastness of space helps us deduce the earliest history of our planet and sets the stage for its more recent geologic history, after the earth's solid surface was formed.

THE DESIGN OF SPACE

The Milky Way Galaxy

The Milky Way spiral galaxy is one of a host of galaxies of various sizes and shapes that exist in the universe. Based on information obtained by optical and radio telescopes, some of these galaxies appear to be as much as 10 billion (10×10^9) light years,* or 96×10^{21} kilometers, away (Fig. 3.1). The Milky Way's stars and other matter are held together by mutual gravitational attraction. Because the sun and the earth are located within it and we view it edge-on from the inside looking out, our galaxy does not appear spiral to us. It appears as a band of stars across the sky. This faint band of light arcing across the night sky was first shown by Galileo in 1610 to consist of millions of stars too far away to be seen individually without a telescope. About 6,000 stars are visible to the unaided eye. The nearest (except the sun) is Alpha Centauri, about 4 light years away.

Figure 3.2 is a sketched oblique view of the Milky Way showing how it would appear as seen from some distant galaxy. The spiral is about 9.67×10^{17} kilometers in diameter (100,000 light years across), with the sun located about 30,000 light years from the center and within the cloud of interstellar dust near the galac-

*Astronomers generally measure distances in light years, the unit of distance traveled by light in one year, equal to about 6×10^{12} miles (9.6×10^{12} kilometers). Light-year distances are convenient because the numbers are smaller than those expressing distances in miles or kilometers. If an astronomic body is 1 million light years away, what we observe now is an event which occurred 1 million years ago.

FIGURE 3.1 *A spiral galaxy in the constellation Ursa Major. It appears to be rotating counterclockwise. (Photography courtesy of the Hale Observatories.)*

tic equator. The galactic center is usually hidden by interstellar dust but is located toward the constellation Sagittarius in the southern sky. The center of the galaxy is surrounded by some 120 visible globular clusters of stars, each made up of 10,000 to 1 million stars. The other stars in the galaxy revolve about the galactic center, and the sun, accompanied by the earth, makes one revolution every 200 million years, moving with a speed of 240 kilometers per second. This revolving motion of the stars produces the flattened spiral shape and localizes the cloud of interstellar dust along the galactic equatorial plane.

None of the stars, except the sun, has directly observable planets. However, in 1963, Peter van de Kamp announced that Barnard's star has an invisible planetary companion. This star, 6 light years away from the earth, exhibits an oscillatory or rhythmic motion. This indicates that it is affected by a companion body because the two revolve about a common gravitational center.

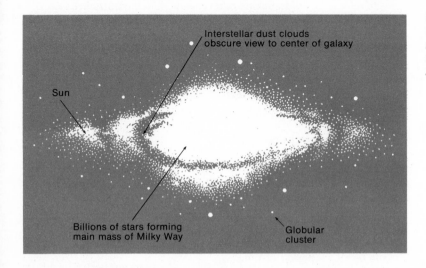

FIGURE 3.2 *Diagram of the Milky Way galaxy, showing the position of the sun. On this scale the earth and sun are at the same point in the sketch. (After Rosen, Siegfried, and Dennison, 1965.)*

Interstellar dust clouds obscure view to center of galaxy

Sun

Billions of stars forming main mass of Milky Way

Globular cluster

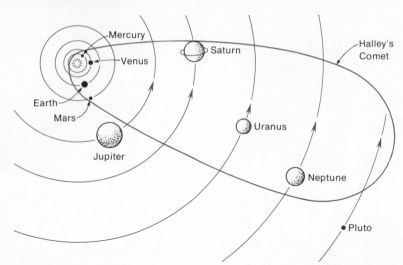

FIGURE 3.3 *Diagram showing the components of the solar system in order of their distance from the sun. The distance from earth to sun is approximately 150 million kilometers. All the planets revolve about the sun in elliptical orbits that are nearly in the same plane. (Not to scale.)*

The Solar System

The sun is the gravitational center of a variety of orbiting planets, satellites, and asteroids known collectively as the solar system. The names of the nine planets are familiar, but in addition, the system contains 32 planetary satellites, as many as 50,000 tiny solid asteroids revolving about the sun, and several hundred comets.

Over 99.8 percent of the total mass of the solar system is in the sun itself. Most of the components of the solar system revolve in orbits inclined no more than a few degrees from the plane of the earth's orbit, and so the whole system can be illustrated in a single sketch like Fig. 3.3. The solar system is about 5.9×10^9 kilometers in diameter out to the orbit of Pluto, yet it is quite isolated in the galaxy.

The Sun   The sun is a fairly typical star within the Milky Way. Because it is so close to us, it appears to be by far the brightest celestial object. About half the stars in the Milky Way occur as isolated bodies like the sun; others are double or triple stars, moving about each other and bound by gravity. Most stars have temperatures approximately half of the sun (6000°C), but some reach 50,000°C. The sun is a middle-size star 1.39×10^6 kilometers in diameter. Other stars range from about $\frac{1}{300}$ the size of the sun, less than half the diameter of the earth, to over 1,500 times the diameter of the sun, and from $\frac{1}{25}$ the mass of the sun to 30 times its mass.

(a) Solar Energy   The heat and light of the sun are an important energy source for many geologic processes; yet, it is the origin of the heat and light of the sun that

is of prime consideration. In recent decades, nuclear physicists have discovered an adequate source of energy—nuclear fusion. Hydrogen nuclei combine to form helium nuclei, releasing energy as some of the mass is transformed into energy in accord with Einstein's equation, $E = mc^2$. The sun retains its relatively constant diameter, however, because the tendency to explosively expand out into space is counterbalanced by the gravitational attraction of its constituent matter, which pulls the exploding gases into a compact spherical shape. The surface temperature of the sun is about 6000°C; but toward the center, where nuclear fusion occurs, the temperature increases to millions of degrees.

Elements other than helium have also been formed by the fusion process, but in sparse amounts. So far, astronomers have detected 67 elements in the sun; other earth elements probably occur, but in amounts too small to be detected. The other stars have compositions similar to the sun, mostly hydrogen and helium, with lesser amounts of other elements. Probably all the elements occurring naturally on the earth or elsewhere in the universe (Table 5.1) originated by nuclear fusion processes in stars, starting with the simplest atom, hydrogen.

The sun's energy output remains remarkably constant. Life on the earth can tolerate a temperature range of only about 100°C. The occurrence of a continuous evolutionary sequence of life traceable over 2 billion years in the fossil record is clear testimony to the constancy of the sun's temperature over a long period of time.

(b) Age of the Sun Calculations from theoretical nuclear physics suggest that the sun is about 5 billion years old. Independent confirmation of this age comes from other parts of the solar system. Use of radioactive-decay measurements permits age calculations which have determined that the oldest rocks on earth are 3.5 to 4.0 billion years old. Rocks of 4.7 billion years in age have been found on the moon, and meteorites yield age dates of as much as 4.6 million years. These ages tend to support an age of approximately 5 billion years for the sun and its satellites.

Computations indicate that the sun contains sufficient fuel to last another 5 billion years, and so it is truly a middle-age star. As stars enter old age, they become explosively unstable and eventually run out of fuel, remaining as cold, dark matter in space—a sort of stellar ash heap.

Planets Planets are bodies that vary greatly in size and that orbit about the sun (Table 3.1). They may be composed of solid material and surrounded partially or completely in concentric envelopes of liquid, solid (frozen), and gaseous atmospheres. The four planets nearest the sun are rather small and are probably composed of rocky or metallic materials similar to those of the earth. Jupiter, Saturn, Uranus, and Neptune are much larger and probably contain substantial amounts of frozen ammonia, carbon dioxide, and methane, surrounding a dense rocky core. Pluto is so distant that little is known other than its diameter.

(a) The Earth The earth is the largest of the four rocky planets near the sun. It orbits in a nearly circular orbit averaging 1.5×10^8 kilometers from the sun. It moves around the sun at the rate of 29.7 kilometers per second, making a complete orbit every 365.25 days. The earth rotates on its axis once every 23 hours 56 minutes. The rotation produces a slight equatorial bulge and polar flattening, and so the earth's diameter is 43 kilometers greater through the equator than through the poles of rotation. The total mass of the earth is 5.976×10^{27} grams. It cannot be measured directly, but the calculations are relatively simple when using the formula for gravitational force:

$$F = k_1 \frac{m_1 m_2}{d^2}$$

where k_1 is the gravitational constant 6.67×10^{-8} (metric system)
$m_1 m_2$ is the mass of the bodies (grams)
d is the distance between centers of the bodies (centimeters)

The volume of the essentially spherical earth is also readily calculated since its radius is known precisely from surveying measurements. The volume is 1.083×10^{27} cubic centimeters. By knowing the mass and volume, the density of the earth is found to be 5.52 grams per cubic centimeter, or 5.52 times heavier than water.

The outermost rocky parts of the earth's crust are known to have an average density of 2.8, and so the in-

TABLE 3.1   The Principal Planets

Planet	Diameter (km)	Density ($H_2O=1$)	Rotation period	Distance from sun ($\times 10^6$ km)	Revolution period	Number of satellites	Surface temperature (°C)
Inner planets							
Mercury	4,900	5.0	88d	58	176d	1	430
Venus	12,258	4.9	243	108	225d	0	300
Earth	12,742	5.5	23h,56m	150	365¼d	1	20
Mars	6,774	3.8	24h,37m	229	687d	2	−10
Outer planets							
Jupiter	140,322	1.3	9h,50m	779	12y	12	−100
Saturn	116,129	0.7	10h,14m	1,429	30y	9	−117
Uranus	50,000	1.7	10h,42m	2,871	84y	5	−114
Neptune	53,226	1.6	15h,48m	4,500	165y	2	−120
Pluto	5,806	?	6.4d	5,919	248y	0	−205

terior portions must be much denser than the surface. Figure 1.3 shows the probable nature of the earth's interior. The increase in density with depth can be explained by differences in composition and by the enormous pressure that the outer parts exert on the inner portions. The inner core is thought to have a specific gravity of about 13, which apparently corresponds to metallic iron at these pressures.

(b) *Planetary Geology*   Direct exploration of the solar system by spacecraft has resulted in the new field of planetary geology, the interpretation and comparison of the solid surfaces of the other planets. Already Mars has been reached by the unmanned American Mariner spacecraft and the Russian Venera. A brief comment on the geology of two planets is appropriate to appreciate their similarities to and differences from the earth.

Mars. This planet has half the diameter of the earth, but its average density of 4.0 suggests a similar composition. Mars has two satellites, Phobos and Deimos, each about 16 kilometers in diameter. The satellites are synchronous with Mars and are heavily cratered. The Martian atmosphere is quite different from our own. Its density is only about 1 percent as great as the earth's, and so Martian winds may be ineffective erosion agents. However, the Mariner 9 spacecraft that reached Mars on November 13, 1971, did record a planetwide dust storm. It filled the entire troposphere layer (11 kilometers), reaching altitudes of 50 to 60 kilometers, and lasting for 6 weeks. Carbon dioxide is 14 times more abundant than on the earth, and water vapor is virtually absent (so rain and streams are lacking), as are methane, nitrous oxide, ammonia, and carbon monoxide. Oxygen probably is missing as a free gas in the atmosphere. Throughout the year, Martian temperatures range from a high of 21 to −54°C, with a daily mean of −4°C. The changing polar "ice caps" may actually be a thin snow cover of frozen carbon dioxide. The composition and temperature of the atmosphere suggest that the possibility of life there is very remote.

The Mariner photographs show the surface of Mars to be similar to that of the moon, dominated by impact craters and large flat surfaces. Martian craters are generally rimless or have low rims, and they have smoother and flatter-bottomed floors than those on the moon, suggesting sedimentary filling of low areas. Other craters, such as South Spot, are flat with ter-

raced walls and a surrounding zone of vertical fractures. Some areas are characterized by a complex group of short ridges and sinuous, linear grooves. An equatorial-centered crevass was photographed that has a length exceeding 4,000 kilometers and a depth three to four times deeper than the Grand Canyon. Other crevasslike features may have been formed by free-flowing water in the past.

The composition of Martian rocks has not been determined in any detail. Some data indicate a silicate (silicon-oxygen compounds) content of 55 to 65 percent. The brownish to reddish color of Mars may be a result of a surface high in the iron oxide minerals such as limonite or hematite, which help to produce similar soil colors on earth. This "soil surface" may contribute material for the extensive dust storms. There is no evidence on Mars of mountain systems such as those formed by compressive forces in the earth's crust, but evidence of active crustal deformation is present in volcanic mountains and calderas. Mars has a gravity field that is apparently more variable than the earth's or the moon's field.

The Mariner 9 studies reveal at least four distinct geological provinces: the Nix Olympica /Tharsis volcanic province, the Ophir-Eos equatorial plateau region, the cratered and smooth terrain in the southern and northern hemispheres (the circular basins here, such as Hellas and Argyre I, resemble lunar impact basins), and the Polar cratered terrains with 100-meter-thick glacial sediment blankets.

Venus. The geologic characteristics of Venus are less well known than Mars since a heavy cloud cover completely obscures its solid surface. Venus makes one rotation every 225 days. Its atmosphere contains 1,000 times as much carbon dioxide as the earth. This gas effectively deters radiation of heat back into space (the "greenhouse" effect). An instrument package dropped from the Russian Venera 4 spacecraft recorded a temperature of 425°C, about 100°C above the melting point of lead. The barometric pressure is about 100 times greater than on the earth, and the atmospheric composition is about 90 percent carbon dioxide, 7 percent nitrogen, less than 1 percent water vapor, but no free oxygen. There are traces of hydrogen chloride and hydrogen fluoride.

Radar reflection studies through the thick cloud cover indicate a much-less-rugged topography. By earth standards, this is an indication of smoothing of the terrain by erosional and depositional processes. One long mountain chain rising a few kilometers above the surrounding plains has been identified, indicating deformation processes. Erosion by streams, ice, and waves obviously is impossible on Venus, but wind and gravity transfer may wear down highlands and create sediments. The lack of a magnetic field on Venus suggests that its interior is rigid, in marked contrast to the liquid outer core of the earth.

The Russian Venera 8 spacecraft examined the soil on Venus and noted that the density is 1.5. The sample contained 4 percent potassium, 0.0002 percent uranium, and 0.00065 percent thorium, similar to terrestrial granite.

(c) The Moon The earth is the only planet with one known satellite. The moon is 3,475 kilometers in diameter and revolves about the earth at a mean distance of some 400,000 kilometers. It takes $29\frac{1}{2}$ days to complete its monthly cycle from full moon to full moon. The moon's orbit is tilted 5° from the earth's orbit, and so the full moon is seldom obscured by the earth's shadow. Solar and lunar eclipses attest to the fact that intersecting positions of the moon, the sun, and the earth do occur. Because the rates of lunar rotation and revolution are the same, one side of the moon always faces the earth. The moon is slightly distorted in shape and points toward the earth because of gravitational attraction on its irregular form. The lunar surface is very bright and hot during the day (116°C), but it is exceedingly cold at night(−136°C) because of the absence of an insulating atmosphere. The average density of the moon is 3.3 grams per cubic centimeter, a little denser than the average of most rocks or the earth's crust.

The moon's surface has been studied and recorded in great detail by the astronauts and remote-guided spacecraft (Figs. 3.4 to 3.8). It is marked by craters that range from a few meters to 241 kilometers in diameter. The rim of the largest crater is approximately 4,000 meters high. These craters may be in part volcanic, but a great many are thought to be meteorite-impact scars. The even larger, smooth dark areas are called mares, or maria. They are thought to be the relatively smooth surfaces of lava flows. Also present are rilles, rays, wrinkle ridges, and crater chains, The rilles are sinuous channels, originating in craters or depressions and trending downslope to maria surfaces. Theories as to their origin range from lunar erosion

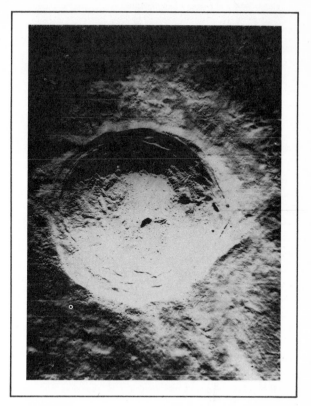

FIGURE 3.4 *This photograph was taken by the Lunar Orbiter V, in its mission to provide supplemental photographs of suggested Apollo sites. The area shown contains the crater Aristarchus. The excellent detail shows erosion of the crater walls and rilles in the floor of the crater. Aristarchus is 57 kilometers in diameter. This photograph was taken at a height of 121 kilometers. (NASA photo.)*

processes to collapsed lava tubes. The rays are probably ejected material from associated craters during their formation. Crater chains are, as the name suggests, linear crater groups, while wrinkle ridges may represent lava extrusions along fracture zones. The lunar surface is strewn with small rocky material, meteorite dust, mineral grains, and glass, all of which compose the lunar regolith, or soil. The material is up to 6 meters thick in places.

During the Apollo 17 investigation in December 1972, an intriguing "orange soil" was found by the two astronauts. The discovery prompted speculation that

FIGURE 3.5 *This is a portion of the first closeup photograph of the crater Copernicus, one of the most prominent features on the face of the moon, taken by Lunar Orbiter II. Looking due north from the crater's southern rim, detail of the central part of Copernicus can be seen. Mountains rising from the flat floor of the crater are 305 meters high, with slopes up to 30°. A ledge of bedrock is visible in the central part of the mountain chain on the floor of the crater. The 914-meter mountain on the horizon is the Gay Lussac Promontory in the Carpathian Mountains. Cliffs on the rim of the crater are 305 meters high and are undergoing continual downslope movement of material. From the horizon to the base of the photograph is about 80 kilometers. The horizontal distance across the part of the crater shown in this photograph is about 27 kilometers. Lunar Orbiter was 46 kilometers above the surface of the moon and about 250 kilometers due south of the center of Copernicus when the picture was taken. (NASA photo.)*

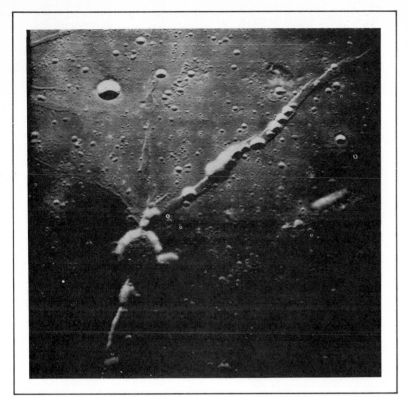

FIGURE 3.6 *Crater chains along the Hyginus Rille (Rima Hyginus). Hyginus is the large crater (11 kilometers in diameter) at the junction of the two segments of the rille. The disconnected crater chains are clear evidence of lunar vulcanism; thus, many similar and evenly spaced craters situated along an obviously fault-produced rille could hardly have been formed by impact. Note the small domes in the floor of the crater Hyginus. The dark material on the mare surface south and east of Hyginus may have erupted from that crater. Location: 8°N, 6°10'E. Framelet width: 3.3 kilometers. (NASA photo.)*

FIGURE 3.7 *Blacked-in area of Mare Crisium shown on the moon's front face is underlain with mass concentrations of high-density material. Mathematicians at the National Aeronautics and Space Administration's Jet Propulsion Laboratory in Pasadena, California, located the mascons by plotting spacecraft acceleration data from the moon-circling Lunar Orbiter V. The presence of mascons beneath the maria suggests a relationship between the two. (NASA photo.)*

27

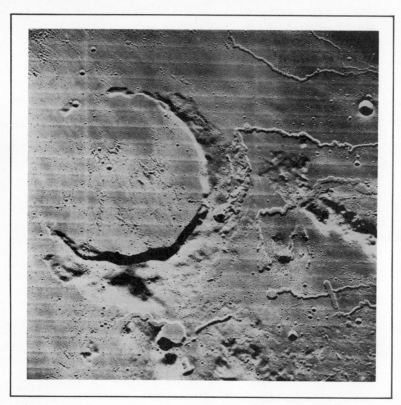

FIGURE 3.8 *On August 18, 1967, Lunar Orbiter V took this photograph of the moon's Harbinger Mountains, located at 43° west longitude and 27° north latitude, from an altitude of 138 kilometers. At the center of the photograph is Rima Prinz I (Prinz Valley I); at left is Rima Prinz II. Prinz II starts at a crater about 3.3 kilometers in diameter and runs north in a channel 66 kilometers long. The valley is 2,438 meters wide near the head and narrows to about 454 meters near the end. Lying inside the broad valley is a smaller crater about 1.6 kilometers in diameter, with its accompanying channel less than 305 meters wide. The valley drops about 1,212 meters from the flank of the Crater Prinz to the mare basin floor. This channel system seems to record an older broad-valley stage and a younger inner-valley stage of development. An interpretation is that these are records of volcanic eruptions and the flow of fluidal materials downstream, as in terrestrial streams in desert environments. Horseshoe-shaped Crater Prinz is shown at the bottom of the photo. (NASA photo.)*

the soil indicated vulcanism and oxidation by water on the lunar surface. Subsequent examination showed the orange soil to be formed of droplets, spheres, and sphere fragments, all composed of glass and having a brownish or burnt-orange color. The material did not represent a hydrous alteration or a volcanic action involving water. Lunar analysis teams have concluded that the orange soil is due to impact by a meteor that collided with the lunar surface 3.71 billion years ago. The explosion created a droplet spray, which upon cooling, formed solid glass beads from 0.002 to 0.05 centimeter in diameter. The orange color is due to high concentrations of titanium and iron.

The moon has a very weak magnetic field, and some lunar rocks have remanent magnetism. The mode of the magnetism on the moon is questionable and may be due to shock magnetization by impact rather than to an internal field. A gravity field does exist, although its strength is $\frac{1}{6}$ that of the earth. Studies of the gravitational field have revealed subsur-

face mass concentrations called mascons below the maria.

Geologic studies related to the Apollo lunar landings have revealed a great deal about the moon, including many similarities to the earth and certain marked differences. Due to the lack of an atmosphere on the moon, gradational processes, such as chemical weathering of rocks, precipitation of rain or snow, and erosion by streams, wind, waves, or glaciers, are absent. Of all the types of erosion prevalent on the earth, only gravity transfer seems applicable on the moon. Even evidences of lunar landslides on the steep mountain slopes are rare, possibly because of low lunar gravity. One area investigated during Apollo 17, however, was a probable landslide zone.

Meteorite impact has pitted the surface and splattered rock fragments in long streaks. Because weathering, as we know it, is nonexistent and erosion is exceedingly slow, this peculiarly cratered lunar landscape has changed little. It has preserved the largest impact scars for a few billion years and even the smaller craters for at least hundreds of millions of years.

Most of the lunar rocks are igneous in origin, with a composition like the common earth igneous rock basalt (Fig. 3.9). Lava flows from ancient volcanic eruptions are widespread. Other rocks are meteorite fragments.

Some common earth features are missing from the moon, including accumulations of layered materials deposited by wind or water; however, layered volcanic materials are present. On earth, parts of the surface have been deformed into mountain ranges. On the moon's surface, mountain chains do exist, but the formation process is not adequately known. The moon's interior does not seem nearly so restless as the earth's. Faults are less common there, but fractures are

present. Moonquakes do occur, and so some interior energy is present, unless the recorded moonquakes resulted from meteorite impact. Velocities from the seismic waves of these lunar quakes do not closely compare to the velocities recorded for various earth materials. The velocities do show a better correlation with other materials found (or eaten) on earth (Table 3.2).

The interpretation of seismic data as a result of instrumentation left during the Apollo programs indicates the following characteristics of the lunar crust: the crust has a layered structure to a depth of 65 kilometers. The uppermost layer (2 or 3 kilometers) indicates broken lunar rocks and soil. The second layer is composed of lunar basalt to the depth of 25 kilometers, either as a series of lava flows or as a thicker, more massive unit. The third layer, at a depth of 25 to 65 kilometers, apparently consists of igneous rock

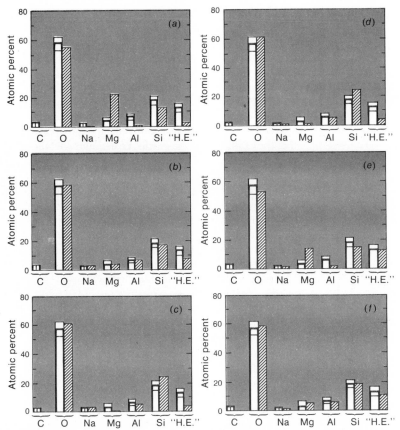

FIGURE 3.9 *Comparison of the observed chemical composition of the lunar sample (open bars) with the average composition of selected materials (hatched bars): (a) dunites; (b) basalts; (c) granites; (d) tektite (Indo-Malayan body); (e) low-iron chondritic meteorites, (f) basaltic achondrite meteorites (4). H. E., elements heavier than silicon. (From Turkevich, Franzgrote, and Patterson, 1967,* Science, *vol. 158, no. 3801, pp. 635-637, copyright 1967, American Association for the Advancement of Science.)*

types quite similar to those found on earth at comparable depths. At 65 kilometers, a distinct change in seismic characteristics is noted and seemingly indicates the base of the lunar crust and the beginning of the lunar mantle.

Metamorphic rocks are evidently absent on the moon, except those formed by impact shock from meteorites.

TABLE 3.2 **Comparison of Compressional Velocities of Lunar Rocks and Various Earth Materials**

Material	v_p (km/sec)
Lunar Rocks and Cheeses	
Sapsego (Swiss)	2.12
Lunar Rock 10017	1.84
Gjetost (Norway)	1.83
Provolone (Italy)	1.75
Romano (Italy)	1.75
Cheddar (Vermont)	1.72
Emmenthal (Swiss)	1.65
Muenster (Wisconsin)	1.57
Lunar Rock 10046	1.25
Sedimentary Rocks	
Dolomite	5.60
Dolomite	4.69
Limestone	5.06
Limestone	5.97
Graywacke	5.40
Graywacke	6.06
Sandstone	4.90
Metamorphic Rocks	
Schist	5.10
Slate	5.39
Gneiss	4.90
Marble	6.02
Quartzite	5.60
Igneous Rocks	
Granite	5.90
Diorite	5.78
Andesite	5.23
Gabbro	5.80
Gabbro	6.80
Diabase	6.33
Minerals	
Corundum	10.80
Garnet	8.53
Quartz	6.05
Hematite	7.90
Olivine	8.42

SOURCE: Schreiber, E., and Anderson, O. L., 1970, *Properties and composition of lunar materials: Earth analogies,* Science, vol. 168, pp. 1579-1580.

Absolute-age dating by radiometric means of the oldest moon materials gathered by astronauts reveals an age of 4.7 billion years, which exceeds by 0.7 billion years the most ancient rocks yet known from the earth's crust.

(d) Asteroids The pattern of planetary spacing has a gap between Mars and Jupiter that is occupied by the asteroids. The asteroids vary in size from small blocks to bodies as much as 800 kilometers in diameter and may be parts of an ancient planet that disintegrated, perhaps by collision of two small planets. The total number may exceed 50,000. The finer debris from such an asteroid-producing catastrophe is considered a possible source of many of the meteors encountered by the earth.

(e) Comets Perhaps the strangest bodies in the solar system are comets. They move chiefly in long elliptical orbits, ranging in distance from near the sun to beyond Pluto (Fig. 3.3). The total mass of a comet is very small. As it approaches the sun, some of the frozen mass vaporizes and streams out as a tail millions of kilometers long. Comets are not hot objects; they glow from the sun's reflection and from ions produced in them by the electromagnetic radiation from the sun.

(f) Meteorites As the earth moves around the sun and along with the sun through space, it frequently passes near small solid objects traveling in interstellar space. These may be sufficiently attracted by gravity that they fall toward the earth. An estimated 8 billion of these meteoroids enter the earth's atmosphere daily. Most meteoroids are very tiny, sand size or smaller, but on a clear night one can see the path of these bodies as they fall through the sky. The bodies ionize the atmosphere in their plunge toward the earth and as a result are often referred to as "shooting stars." The scientific name for the luminous stage is meteor. Meteors that strike the ground are called meteorites. Most of these escape man's attention, but probably hundreds of tons of meteorite dust are added to the earth daily.

The largest known meteorite is one found in South Africa weighing 59,000 kilograms, but the majority weigh only a few kilograms. Most of the meteorites found consist of an iron-nickel alloy. They appear as a metallic chunk, with edges rounded and pitted from searing heat of friction encountered during their descent through the air. Another type is a stony meteorite, which resembles certain rocks in appearance, composition, and susceptibility to weathering.

Many of them may escape notice because of these similarities to common rocks on the ground.

Solar System Motions   The sun, planets, and satellites show a remarkable similarity of motion. The earth rotates in a west-to-east direction, and so the stars and sun appear to move westward with the passage of the hours. All nine planets revolve west-to-east about the sun in nearly the same orbital plane, and the sun slowly rotates in the same west-to-east direction. All planets, except Neptune, rotate in the same direction as their revolution. (Neptune has its axis of rotation pointing almost toward the sun.) Furthermore, 25 of the 32 known satellites orbit in the same direction as their central planets rotate. The probability of obtaining such a similarity of motion by chance is extremely remote. Consequently, most astronomers think the accordance of solar, planetary, and satellite motions implies a common origin for all of them, probably from a single cloud of interstellar matter that coagulated to form the sun. The planets and satellites had enough orbital velocity to avoid being drawn into the sun by its gravitational pull.

Nearly all the sun's energy reaching the earth is electromagnetic radiation, ranging in frequency from x-rays to ultraviolet, visible light, radiant heat, and radio waves. The earth's atmosphere is opaque to most of these wavelengths, but radiant heat and visible light are able to penetrate it. Heat causes important climatic effects, powers the hydrologic cycle, plays a role in weathering and erosion, and makes the earth habitable for life. Except for this energy source, there would be no streams to erode mountains, nor any glaciers to carve the landscape. Without the winds resulting from the sun's heat, even the waves on the ocean would not shape the coastlines.

Continual explosive activity on the sun also causes expulsion of ions and protons into space. This flow of charged particles past the earth is called solar wind. While our atmosphere absorbs the particles, it is possible that on the moon, this solar wind may be an erosional agent. The motions of the earth in the solar system, combined with the effects of solar energy, allow for the development of seasons on the earth.

The axis of the earth's rotation is inclined $23\frac{1}{2}°$ from the perpendicular to the plane of its revolution (Fig. 3.10). Consequently, at different times during a 365-day revolution about the sun, the most direct solar rays will fall on different portions of the earth. Figure 3.10 illustrates the situation in the Northern Hemisphere in December. Note that solar radiation is spread over a much larger area in the Northern Hemisphere than is similar radiation in the Southern Hemisphere. The oblique radiation in the Northern Hemisphere means colder temperatures and, therefore, a winter season. During this time the North Pole is in

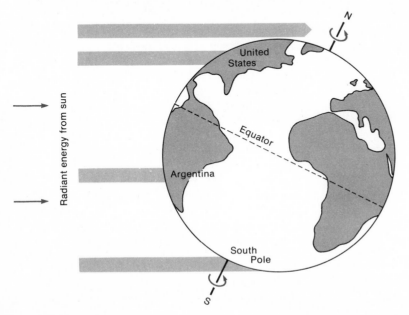

FIGURE 3.10   *The seasons, showing summer in the Southern Hemisphere and winter in the north.*

darkness throughout the 24-hour period of rotation, and the South Pole is in continuous daylight (midnight sun). The very oblique angle of incidence of the sun's rays at the South Pole causes only slight warming.

Now mentally shift the earth to the other side of its orbit 6 months later, in June (Fig. 3.10). Then the northern latitudes of the earth are in nearly direct sunlight, and the Southern Hemisphere is lighted obliquely. The resulting temperature changes produce summer in the Northern Hemisphere and winter in the Southern Hemisphere.

Halfway between these summer and winter periods, the sun's direct rays are on the equator. This is the time of the spring or fall equinox, with moderate temperatures in both the Northern and Southern Hemispheres and a day and night of equal length all over the world.

Another effect of solar system motions and gravitational attraction within the system is the tides, which are motions of the water within the earth's oceanic basins.

As shown in Figure 3.11, one side of the earth is closer to the moon than the other. Lunar gravitational attraction is 3 percent stronger on the side facing the moon than at the center of the earth. This fact, coupled with the effect of other motions of the earth-moon system, causes the oceans to exhibit two bulges several meters high, one pointing toward the moon and the other away from it. As the earth rotates and the moon revolves, these tidal bulges tend to follow the moon so that at a given place on the ocean, two high and two low tides pass every 24 hours and 50 minutes if no complicating factors intervene. However, complicating factors do appear and include the following:

The attraction of the sun and the position of the sun relative to the earth-moon system

The natural period of oscillation within an oceanic basin

The size, shape, and configuration of the oceanic basin

All the characteristics of the moon's position relative to the earth

All these and more must be taken into account to predict, or describe, the tides at any one place or time.

FIGURE 3.11   *Spring and neap tides. (a) Spring tides occur when the sun and moon are in line and produce exceptionally strong tidal pull. (b) Neap tides occur when the weak tides caused by the sun partially cancel the effect of lunar tides.*

Life on the Earth

The earth is probably the only planet in the solar system that has highly developed life as we know it. Certain conditions are essential for life. Carbon is probably the only element with suitable chemical properties that would make organic substances possible. Carbon dioxide and water vapor are necessary components of an atmosphere that will permit photosynthesis. The earth is precisely the right distance from the sun for water to occur in a liquid state. Finally, in order for highly developed life to evolve, life-supporting conditions must be maintained for a span of hundreds of millions of years or more. The steady energy output by the sun for eons was necessary for complex forms, such as man, to develop on the earth.

The addition of life has greatly modified the geology of the earth. Without photosynthesis our atmosphere would be deficient in oxygen, and animal life would be impossible. Weathering of rocks to form soil is partly dependent on oxygen in the air. Thus, before the land areas had an extensive plant cover, prior to 400 million years ago, erosion must have been much more rapid. Recently, man has destroyed much of this plant cover, and the erosion rate has increased greatly during historical time. Coal and petroleum could not have been formed without life.

The sun is only one of an estimated 1×10^{20} stars in the known universe, and conditions sufficient for intelligent life are likely on planets orbiting other stars besides the sun. Shapley (1962) estimates 100 million planetary systems suitable for organisms. Von Hoerner (1961) thinks that 1 in 3 million stars may be accompanied by intelligent civilizations. A search has already begun by radio telescopes and spacecraft to detect possible signals from other cultures. The Jupiter Probe, with its faceplate showing human physiology, is an attempt to communicate with other intelligent beings.

REFERENCES

American Association for the Advancement of Science, 1970, *The Moon issue*, Science, vol. 167, Washington, D. C.

Berlage, H. P., 1968, *The Origin of the Solar System*, Pergamon Press, New York.

Dietz, R. S., 1961, *Astroblemes*, Scientific American, vol. 205, no. 2, pp. 50-58.

Faul, H., 1966, *Ages of Rocks, Planets, and Stars*, McGraw-Hill Book Company, New York.

Freeburg, J. H., 1966, *Terrestrial Impact Structure—A Bibliography*, United States Geological Survey Bulletin 1220.

Gingerich, O., 1970, *Frontiers in Astronomy* (Readings from Scientific American), W. H. Freeman and Company, San Francisco.

Inglis, S. J., 1961, *Planets, Stars, and Galaxies*, John Wiley & Sons, Inc., New York.

James, J. N., 1963, *The voyage of Mariner II*, Scientific American, vol. 209, no. 1, pp. 70-84.

———, 1966, *The voyage of Mariner IV*, Scientific American, vol. 214, no. 3, pp. 42-52.

Leighton, R. B., et al., 1969, *Mariner 7 television pictures: First report*, Science, vol. 165, pp. 788-795.

Mason, B., and Nelson, W. G., 1970, *The Lunar Rocks*, John Wiley & Sons, Inc., New York.

Mayall, N. V., 1962, *The story of the Crab Nebula*, Science, vol. 137, no. 3542, pp. 91-102.

Rosen, S., Siegfried, R., and Dennison, J. M., 1965, *Concepts in Physical Science*, Harper & Row Publishers, Incorporated, New York.

Runcorn, S. K., 1966, *Corals as paleontological clocks*, Scientific American, vol. 215, no. 4, pp. 26-33.

Sander, W., 1965, *Satellites of the Solar System*, American Elsevier Publishing Company, Inc., New York.

Shapley, H., 1962, *Of Stars and Men*, 2d ed., Beacon Press, Boston.

Shoemaker, E. M., 1964, *The geology of the moon*, Scientific American, vol. 211, no. 6, pp. 38-47.

Sloan, A. K., 1966, *The scientific experiments of Mariner IV*, Scientific American, vol. 214, no. 5, pp. 62-72.

Strahler, A. N., 1971, *The Earth Sciences*, 2d ed., Harper & Row Publishers, Inc., New York.

Von Hoerner, S., 1961, *The search for signals from other civilizations*, Science, vol. 134, pp. 1839-1843.

Wells, J. W., 1963, *Coral growth and geochronometry*, Nature, vol, 197, no. 4871, pp. 948-950.

Whipple, F. L., 1968, *Earth, Moon, and Planets*, 3d ed., Harvard University Press, Cambridge, Mass.

Wyatt, S. P., 1964, *Principles of Astronomy*, Allyn and Bacon, Inc., Boston.

GEOLOGIC TIME

Time is the period during which processes operate, changes occur, or events take place in irreversible sequence. Its reckoning, therefore, is one of the fundamental measurements of science, along with mass and distance. From his earliest beginnings, man has been aware of days, moons, seasons, and years in their cycles, and has progressively invented improved devices for their measurement. These devices range from the crude counting procedures of hour glasses and various pendulum clocks to modern laboratory clocks based on rhythmic electrical and magnetic oscillations in cesium atoms. However, he has never been able to give a precise definition of time. The ancient Egyptians learned to measure the length of a year with great accuracy based on observation of the sun's circuit in the zodiacal constellations. In 1964, scientists adopted the vibration rate of cesium atoms as a standard for measuring time. According to international definition, the second is equal to 9,192,631,770 natural vibrations of cesium-133. Cesium clocks keep time with an error of less than 1 second per thousand years and were recently used aboard east and west round-the-world flights to test relativistic time in Einstein's theory of relativity.

AGES OF ROCKS

In geology, we are concerned both with the relative, or comparative, ages of rock as well as with their absolute ages in years. Relative age is comparatively easily determined by observation, whereas absolute-age determination requires lengthy, complex laboratory procedures. Only since about 1950 have reliable measurements in actual years been practical.

Relative Age

Observing the relationships between rock units in the field can generally be used to tell the relative ages of rocks. Given two different rock masses in any situation, A and B, the following relative ages are possible: (1) A is younger than B; (2) B is younger than A; or (3) A and B are the same age, or contemporaneous. The principal criteria used to determine relative ages are superposition, intertonguing, intrusions, deformation, metamorphism, and faunal succession.

Superposition   When different sedimentary rocks accumulate in layers in an environment on the surface

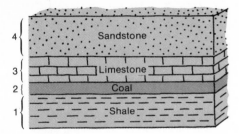

FIGURE 4.1 *Using the rule of superposition, it can be concluded from this cross section of rocks in the earth that the relative ages from oldest to youngest are shale, coal, limestone and sandstone.*

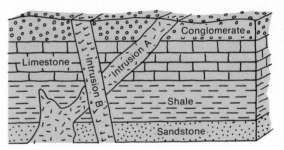

FIGURE 4.3 *Cross-cutting relations of intrusion establishes relative ages. The sequence of rock formation in this cross section is deposition of sandstone, shale, limestone, and finally, conglomerate sedimentary rocks, followed by intrusion A and then intrusion B of molten rock that cooled and solidified to form igneous rock masses.*

of the earth, each successive layer is laid on top of (superimposed on) the next older layer (Fig. 4.1). Unless the rock layers have undergone deformation after deposition, a simple relationship applies in that the older layers are on the bottom and the younger on top.

Intertonguing Sedimentary layers that interpenetrate are contemporaneous. The two contemporaneous rock types formed in related but different environments (Fig. 4.2).

Intrusions Igneous rocks injected initially as molten material in fractures cut across pre-existing rocks. The intrusive rock is always younger than the rock it penetrated (Fig. 4.3).

Deformation Rock units that are bent or broken must have originated prior to the time of deformation (Fig. 4.4).

Metamorphism Metamorphic rocks are rocks that have been changed from a pre-existing rock to a new rock in response to a changed environment. The metamorphism necessarily occurs after the formation of the original rock.

Faunal Succession Fossils (the preserved remains of life) are common in sedimentary rock layers all over the world. In a local area the fossils of any one layer are distinctive, and the sequence of fossils can be established by applying the previous relative-age rule of superposition to undeformed rocks of that area.

FIGURE 4.2 *The intertonguing shale and sandstone have the same age. Using the rules of superposition and intertonguing, the sequence of rock ages in this cross section is coal, contemporaneous shale and sandstone, and limestone.*

FIGURE 4.4 *The sedimentary layers in this cross section formed in horizontal layers in the sequence limestone, sandstone, coal, and finally, shale, and later they were deformed by folding and faulting. It is impossible to tell whether the faulting happened before the folding.*

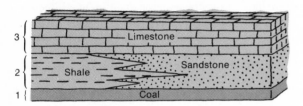

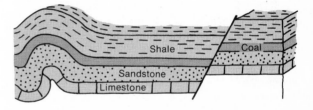

Then this fossil sequence can be used to tell the relative ages of other rocks. The method can be used even if the rocks have been severely deformed or if the two rock units are widely separated geographically. Furthermore, the succession, or relative age, of fossils is similar in different parts of the world. Rocks with similar fossils are interpreted as approximately contemporaneous. A given species, man for example, lived during only one time interval in the history of the earth; therefore, if a rock contains that particular species, the rock must have been formed during the time interval when the earth was inhabited by that species.

This fossil succession proceeds generally from simple to more complex forms of life with the passage of geologic time. This trend holds for plants, vertebrates, and invertebrates. The evidence of faunal succession guided Charles Darwin in formulating the theory of evolution in 1859. Evolution is a theory to explain the fact of succession of fossils, and the idea of evolution is not necessary for paleontologically dating the relative ages of rock layers. However, a working knowledge of evolution does permit paleontologists to establish more precise relative dates than was previously possible.

Absolute Age

In absolute-age dating, geologists seek to measure the actual ages of rocks in terms of years, rather than relative to other rocks. From earlier examples, we found we could determine that sedimentary rock layer A is on top of layer B; therefore, layer A *is younger than layer B*. We would now like to determine that layer A is 50 million years old and layer B is 60 million years old. There are several methods that have been and are being used to determine absolute age in geologic situations like the one cited.

Some of the early methods to determine absolute ages of rocks were based on theoretical measurements of the rates of processes. This approach was to observe changes taking place today and to assume that the rate observed today is the same as that in the past. Only crude estimates of the duration of geologic time are obtained in this way, but they do serve to indicate that the earth is many millions of years old.

Cooling Earth Astronomers in the nineteenth century generally considered the earth to be a solidified derivative of the sun. Later in that century, physi-

cists had learned enough about the way materials cool to estimate the time that would be required for a molten earth to cool to its present temperature at the ground surface. The most detailed calculation was made by Lord Kelvin, who concluded in 1899 that the earth's surface had been solid for 20 to 40 million years. This idea had two problems: First, astronomers began to suspect that our planet formed slowly by accumulation of cold rather than hot interstellar matter; secondly, other workers pointed out that Kelvin's figure would be only a minimum age because the heat from radioactivity decay in the earth's interior should slow down the cooling process, making our planet much older than he had calculated.

Salt in the Sea The rivers of the earth contain small percentages of dissolved salt (sodium, Na^+, and chlorine, Cl^-), and the total amount added annually to the sea can be estimated fairly accurately. The total sodium content of the sea (15×10^{15} kilograms) divided by the annual increment (15.1×10^7 kilograms per year) gives an approximate age of the ocean of 99 million years. We now know that even the 99 million years originally obtained by Joly in 1898 is too short. Sources of error include the presence of other salt sources and processes of removal. Salt has been removed from seawater by evaporation in isolated inlets of the ocean, as indicated by vast accumulations of rock salt in ancient sedimentary rocks. Also, some of the salt in the rivers is derived from the leaching of these ancient salt deposits and interstitial salty water from marine sediments now on land. Salt spray that enters the atmosphere from the ocean also is added to rivers by rainfall, thus recycling salt from the sea. Last, the continents stand unusually high above sea level now, and so the rate of erosion is probably accelerated. All these factors suggest that the amount of salt added to the sea is greater now than the average at other times in the geologic past.

Sediment Accumulation It is possible to estimate the rate of accumulation of different sediment types on the ocean floor and then to extrapolate this information to determine the time required to build up the total thickness of all ancient marine deposits. Working with the thickness of the most complete sedimentary records in different parts of the world, Schuchert in 1931 calculated that the Cambrian

Period, which is marked by the appearance of the first abundant invertebrate marine fossils, began about 540 million years ago. Perhaps he was just lucky in his assumption about the rates of sedimentation, but his estimate is remarkably close to the 600 million years now estimated for the beginning of the Cambrian Period.

Other methods that have been used to determine absolute ages are the direct measurements of annual accumulations. These include: annual tree rings in fossil wood, growth markings on fossil corals, and varves in glacial lake deposits, discussions of which follow.

Tree Rings Trees in a seasonal climate grow faster in spring and summer than in fall and winter, producing annual growth rings. Hesitations in growth may result from drought and secondary spurts may result from unseasonable rainfall, but despite these complications, one can date the approximate age of a log in years by counting its annual growth rings. This fixes the time span in years that the tree lived but does not pinpoint the specific year.

A succession of sometimes good, sometimes poor growing seasons, however, produces a distinctive pattern of sizes of growth rings that can be matched from log to log. Starting with living trees, tree-ring patterns have been matched in logs at ancient archeological sites whose rings overlap backward through time until a complete tree-ring calendar was constructed (Fig. 4.5). Other recent studies have analyzed the severe disturbances in the tree-ring sequences caused by surface faulting. The dating of the disturbance by, and in relation to, the nondisturbed annual rings served to time an episode of faulting. This has been quite successful in Alaska, where the faulting associated with an earthquake in 1958 had an effect on tree growth.

Growth Markings Using annual growth markings on fossil corals in their natural growth position in ancient rock layers, geologists can determine rates of sediment accumulation by measuring the number of years required for a growing coral to be buried by sediment. Fossil mollusks and marine calcareous algal colonies (stromatolites) also show annual and daily growth markings, just as living forms do. An interesting development concerning the number of days per year has resulted from a study of the growth markings on corals. Theoretical calculations of energy dissipa-

tion by the braking drag of tidal friction against shores and empirical observations on fossil corals indicate that the earth's rotation is slowing down. In the past there were more days per year than there are now (Runcorn, 1966).

Wells (1963) showed that modern corals display annual growth markings, as well as finer markings corresponding to daily growth in the calcium carbonate secretions of the organisms. Modern corals have an average of about 360 such daily growth markings per year; these faint markings are very difficult to count. In the Devonian period, some 400 million years

FIGURE 4.5 *Schematic drawings of wood cross sections showing how matchings can be made to extend a chronology from recent trees back into prehistoric time. (After Zeuner, 1958.)*

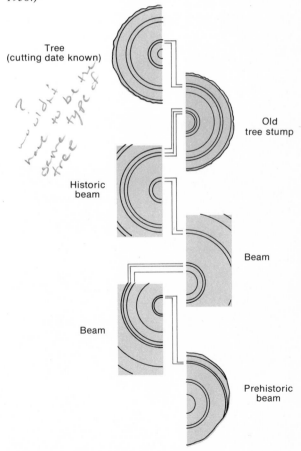

Tree (cutting date known)

Old tree stump

Historic beam

Beam

Beam

Prehistoric beam

, there were about 400 daily growth markings ...rth rotations) per year. In Carboniferous time, 300 million years ago, there were about 380 growth marks per year.

Varves Lakes in cold climates near the margins of glaciers annually deposit superimposed layers of sediment called varves. During the summer months, water from the melting glaciers carries abundant silt and mud into a lake, and a generally light-colored layer of sediment accumulates. In winter, there is no influx of meltwater. The tiny suspended particles slowly settle to the bottom along with dead algae from the summer's growth. It is then that a thin layer of finer mud, higher in organic carbon, is deposited. An annual couplet of a light and dark layer, called a varve, nor-

FIGURE 4.6 *Each light and dark lamination pair is a varve corresponding to a year's accumulation of sediment in glacial lakes in Canada. (Geological Survey of Canada.)*

mally represents 1 year. Can you think of a case where it might not? The time required to accumulate a thickness of such varved lake sediment can be measured by counting the varves (Fig. 4.6). Furthermore, the patterns of varves can be matched in different lake sites along a line of glacial retreat, much as in tree-ring chronology.

Radioactive Dating Many radioactive isotopes occur naturally, and the manner of their transformation or decay is well known. Radioactive transformation takes place at a characteristic and constant rate, regardless of temperature, pressure, or chemical environment.

The time required for a radioactive transformation of any element to go half way to completion is called the half-life of the reaction (Fig. 4.7). In other words, after one half-life of time, 50 percent of the original radioactive material will be transformed into a new isotope of some element: half of the remaining original radioactive substance will be transformed after the passage of a second half-life span of time, and so on. The absolute age in years is calculated from the formula:

$$\text{Age in years} = \text{number of half-lives} \times \text{years per half-life}$$

The age of a naturally radioactive substance can be measured by two methods: (1) the amount of original radioactivity remaining, and (2) the amount of decay product produced. Obviously, there cannot be subsequent addition or removal of the original or the decay product materials after the radioactive decay clock starts, if the date of the original radioactive material is to be accurate. The principal methods and uses of radioactivity dating are listed in Tables 4.1 and 4.2 and are explained below. Research into the possible use of other nuclide dating systems is being conducted. *(a) Tritium Dating* Neutron bombardments of the atmosphere produce rare or trace amounts of radioactive hydrogen with two neutrons, H^3, or tritium. Tritium combines chemically with oxygen, as do the other isotopes of hydrogen, to form water. The half-life of tritium is only 12.5 years, and so rain entering the gound surface or the ocean becomes less concentrated in H^3 with the passage of years. This loss of concentration gives a method to study the rate of movement of

TABLE 4.1   Major Methods In Geochronometry

1. Uranium-lead system (U-Pb)
   A. Decay series U^{238} to Pb^{206}:

      Half-life—4.5×10^9 years
      Effective range—10^6 to 4.6×10^9 years
      Minerals—zircon, pitchblende, uraninite

      U^{235} to Pb^{207}:

      Half-life—0.71×10^9 years
      Effective range—same as U^{238} to Pb^{206} series
      Minerals—same as U^{238} to Pb^{206} series
      A related series is the Th^{232} to Pb^{208} series with a half-life of 13.9×10^9 years

   B. Errors ± 1 to 2%
2. Potassium-argon system (K-Ar)
   A. Decay series K^{40} to Ar^{40}:

      Half-life—1.3×10^9 years
      Effective range—10^4 to 4.6×10^9 years
      Minerals—micas, hornblende, feldspars

   B. Errors ± 1 to 2%
   C. Limitations: Diffusive loss of argon at high temperatures; the date obtained may represent only a thermal event
3. Rubidium-strontium system (Rb-Sr)
   A. Decay series Rb^{87} to Sr^{87}:

      Half-life—47×10^9 years
      Effective range—10^6 to 4.6×10^9 years
      Minerals—micas and feldspars

   B. Errors 1 to 2%
   C. Limitations: Diffusion of strontium at elevated temperatures or by base exchange
4. Radiocarbon system (C-N)
   A. Decay series C^{14} to N^{14}:

      Half-life—$5,710 \pm 30$ years
      Effective range—100 to 50×10^3 years
      Materials—carbon substances or carbon-bearing substances

   B. Errors about 1%
   C. Assumptions: Rate of C^{14} formation in atmosphere is constant; mixing of C^{14} in environments is rapid; no formation or addition of C^{14} after death

TABLE 4.2   Results and Uses of Radioactive Systems

1. Extinct radioactive systems
   A. Accretion interval of meteorites $_{53}I^{129}/_{54}Xe^{129}$ (16 million years)
   B. Melting of accreted parent body $_{13}Al^{26}/_{12}Mg^{26}$ (0.74 million years)
2. Current use of the present systems
   A. Age of differentiation of meteorites (earth) Pb^{207}/Pb^{206} and Rb^{87}/Sr^{87}
   B. Possible rate of degassing of the earth's interior K^{40}/Ar^{40}
   C. Magmatic ages   All systems except C^{14}
   D. Metamorphic ages K^{40}/Ar^{40} and Rb^{87}/Sr^{87}
   E. Sedimentary ages (authigenic minerals) K^{40}/Ar^{40}
   F. Provenance and paleogeography of sedimentary rocks   All systems except C^{14}
   G. Pleistocene ages and correlation C^{14}, K^{40}/Ar^{40}, and daughters of the system U and Pb
   H. Groundwater flow H^3 and C^{14}
   I. Archeology C^{14} and K^{40}/Ar^{40}
   J. Various geochemical studies involving the mobility of elements   Use natural radioactivity as a tracer

water in the ground or to investigate the circulation of ocean water.

(b) Radiocarbon Dating   The most widely used technique for dating geologic and archeologic materials of the past 50 thousand years is the C^{14} method. This radioactive isotope of carbon originates indirectly from cosmic rays entering the earth's atmosphere from outer space. The rays collide with the upper atmosphere, producing neutrons. The neutrons react with atmospheric nitrogen to form radioactive carbon: N^{14} + neutron→C^{14} + proton. However, this carbon is unstable and decays with a half-life of 5,710 years to form nitrogen again: $C^{14} \rightarrow N^{14}$ + 1 electron. The C^{14} in the atmosphere is generated at a rather uniform rate, and the earth's cosmic ray bombardment is assumed to have remained at a steady intensity for hundreds of thousands of years; thus atmospheric carbon dioxide contains a fixed amount of C^{14} to ordinary nonradioactive C^{12}. Because all living plants and animals incorporate carbon into their bodies, their life processes maintain an equilibrium of their carbon isotopes with those of the atmosphere. Upon the death of the organism, however, the C^{14} clock starts. The concentration of C^{14} radioactivity is progressively decreased in accord with the decay curve of Fig. 4.7. Thus, carbon from a fossil log or a beam in an ancient dwelling 5,710 years old would contain only half as

TABLE 4.3   **Concordant Analyses of the Age of Uraninite, Black Hills, South Dakota**

Method of dating	Probable age (10^6 years)
U^{238}/Pb^{206}	1,580
U^{235}/Pb^{207}	1,600
Th^{232}/Pb^{208}	1,440
Pb^{206}/Pb^{207}	1,630

SOURCE: Wetherill, G. W., 1956, Transactions, American Geophysical Union, vol 37, pp. 320-326.

much C^{14} as carbon from modern trees. After the passage of about nine half-lives, the remaining concentration is very faint; consequently, the method can be used for only about the last 50,000 years.

(c) Uranium-Lead Dating U^{238} and U^{235} each decay spontaneously at different rates to specific isotopes of lead (Pb). The relative abundance of the several isotopes of U and Pb in a uranium-bearing mineral is determined by mass spectrometry. A less useful but related scheme is the decay of Th^{232} to its lead product Pb^{208}. Much of the lead in any specimen may be ordinary lead, Pb^{204}, of another origin, and so it, too, has to be considered in calculations. An advantage of the U/Pb method is that natural uranium consists chiefly of U^{238} with a small admixture of U^{235} and usually Th^{232}. Hence, a single mineral crystal of uranium ore contains the ingredients for making two or more independent age measurements on the same specimen. In addition, the ratios of the different isotopes of lead to each other provide internal checks on their consistency. If two or more results agree reasonably well, as shown in Table 4.3, then the analyst is confident that the system has not been chemically, thermally, or mechanically altered, and he considers the dates reliable. The possible errors in the half-lives used and the instrumental errors are thought to be less than 2 percent.

Unfortunately, uranium minerals have a limited distribution and are lacking in many rocks a geologist would like to date. They occur mainly in rocks formed by crystallization of molten material, either deep inside the earth or in volcanic layers. Where found, they serve best to date rocks 10 million to several billion years old.

(d) Potassium-Argon Dating   A small fraction of the

FIGURE 4.7   *Decay curve for radioactive substances*

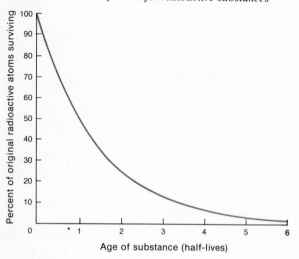

potassium found in minerals and rocks is radioactive K^{40}; the major part is the more common nonradioactive K^{39}. Potassium is the seventh most abundant element in the earth's crust (2.6 percent). It occurs mainly in potassium feldspar, micas, hornblende, and sedimentary minerals such as glauconite and sylvite (KCl). Radioactive K^{40} decays to argon by electron capture and to calcium by emission of an electron. The half-life is 1.29 billion (1.29×10^9) years. The resulting calcium is similar to nonradioactive calcium, and so it cannot be distinguished and used in an age determination. The argon gas, however, is a characteristic isotope, different from natural argon. This argon isotope becomes trapped in the parent crystal, and so its measured quantity can be compared with the remanant K^{40} present. Unfortunately, part of the accumulated argon, which is a gas, is apt to escape from the crystals so that the determined ages frequently turn out to be somewhat younger than the age of the material obtained by other methods. Argon losses are greatest when the rocks have been heated because of deep burial or igneous intrusions or have been recrystallized during metamorphism. Some minerals will lose argon easier than others; hence, analyses commonly are made on whole rocks instead of on separate minerals. Notwithstanding the slow rate of radioactive decay, which yields scant amounts of argon, the K/Ar method can be used on rocks less than 1 million years old, perhaps as little as 100,000 years.

(e) Rubidium-Strontium Dating Rb^{87} decays radioactively to Sr^{87}, with a half-life of 47 billion years. Small quantities of rubidium occur in mica and many other potassium-bearing minerals, and so the Rb/Sr ratio provides a check on the K/Ar method. If the two agree, argon leakage has been negligible; if they disagree, the K/Ar dates are suspect. The ages of mica crystals calculated by the two methods agree fairly well.

(f) Fission Tracks The energy of particles released by spontaneous fission (nuclear decay to smaller atoms) of U^{238} atoms is sufficient to make tiny markings or tracks in mineral crystals which contain trace amounts of uranium. The number of tracks is proportional to age and uranium concentration. The age is determined by comparing the number of natural fission tracks observed in the crystal with the number of fission tracks produced in the same crystal by artificial neutron bombardment. Artificial bombard-

ment is a known number of times more intense than natural neutron bombardment (Fleisher and Price, 1964). The procedure involves careful chemical etching and study under a high-powered microscope to identify the markings. The method is new, but it seems to yield results consistent with radioactive age determinations. In most minerals, the fission tracks are destroyed if the crystal has been heated to above 400°C after its first crystallization. The method has been used to date different geological materials ranging from 4,000 to 1 billion years old, with results as accurate as radioactive dating.

(g) Geomagnetism Dating Strange as it may seem, the position of the north magnetic pole does not remain fixed. Instead it migrates a few degrees over a period of thousands of years at a rate measured in meters per year, centering near the earth's pole of rotation (Fig. 14.8). Archeologists working in the American Southwest have correlated C^{14} dates with faint but different magnetic orientations preserved in fired pottery fragments and burnt clay found in ancient campfire sites (Weaver, 1967). This permits independent dating of pottery fragments and firepits found at other places where not enough carbon remains to obtain a C^{14} date.

Some iron-bearing minerals, notably magnetite (Fe_3O_4), have a slight magnetism oriented according to the earth's magnetic field at their time of formation. The relative positions of the magnetic poles and the continents have shifted thousands of kilometers during geologic time. Radioactive dating can be used on certain rocks to establish when the magnetic pole was at the particular position recorded by the magnetically oriented mineral grains. In theory at least, by using these rocks as standards, other iron-bearing rocks can be dated by measuring their magnetic orientation and comparing their orientation with that of rocks of known age. Future developments may see geomagnetism used more widely both in geologic dating of ancient rocks and in archeological applications.

DEVELOPMENT OF THE GEOLOGIC TIME SCALE

The rules for establishing relative ages of rocks were all recognized by 1800. Geologists in Western Europe were able, in the formative years of geology, to list the approximate sequence in which the rocks of the region were formed. They recognized a need for arbitrary age designations, and so they developed a geologic time scale. The scale arranged the divisions of earth history from oldest at the bottom to youngest on top, following the pattern of superposition. Table 4.4 is the geologic time scale as it was recognized in 1877. The largest time divisions, called eras, refer to the stages of life development: Paleozoic, ancient life; Mesozoic, medieval life; and Cainozoic, recent life. Eras were divided into periods, and periods into epochs. The time divisions are not based on equal amounts of actual time involved. The period names Laurentian, Cambrian, Silurian, Devonian, Permian, and Jurassic refer to localities where the rocks of these ages occur. The other period names allude to some characteristic of the rocks where originally recognized, such as Cretaceous from the Latin, *creta*, meaning chalk.

Note that there is no mention of dates in years, because in 1877 there was no way to determine absolute ages. These early-named geologic divisions are still the chief basis for the modern geologic time scale. The modern scale has refinements of relative ages, supplemented with absolute-age dates.

The absolute ages of rocks obtained from minerals or wood are most useful when they can be attached to the appropriate divisions of relative ages in the geologic time scale.

The matching of nearly all measured dates with the relative scale of geologic time often depends on the fossil assemblages in layers of sedimentary rocks. Unfortunately, occurrences of fossils in sedimentary

TABLE 4.4 Geologic Time Scale as Recognized in 1877

Era	Period	Epoch
	Post-Tertiary	Recent Post-Pliocene
Tertiary or Cainozoic	Pliocene Miocene Eocene	
Secondary or Mesozoic	Cretaceous Jurassic Triassic	
Primary or Paleozoic	Permian Carboniferous Devonian Silurian Cambrian Laurentian	

SOURCE: Lyell, Sir Charles, 1887, *Principles of Geology,* 11th, ed., D. Appleton and Company, New York

rocks whose age can also be measured radioactively are very rare. Most sediments either lack significant amounts of radioactive minerals (limestone, for example) or are a mixture of solid mineral grains derived by erosion of a variety of older rocks of differing ages. Individual mineral crystal ages differ widely from the age of the associated fossils that provide a tie to the relative geologic time scale.

A time-bracketed igneous intrusion can provide adequate dates for the relative time scale. Figure 4.8 provides an example of determining a date in years for the beginning of the Oligocene epoch. In the Caucasus region of Russia, a granitic igneous rock intrudes mid-Eocene fossiliferous sedimentary rocks. After a short period of erosion, the exposed igneous intrusion was buried beneath fossiliferous layers of early Oligocene sediments. From these field relations, the granite must be younger than mid-Eocene and older than Oligocene. The granite yields a K/Ar date of 38 million years and gives a fairly precise date for the end of the Eocene epoch (Kulp, 1961).

Modern Geologic Time Scale

The modern geologic time scale still divides relative time into eras, periods, epochs, and even smaller sub-

FIGURE 4.8 *Schematic cross section of an intrusion in Russia that yields a measured date for the approximate end of the Eocene Epoch. (After Kulp, 1961.)*

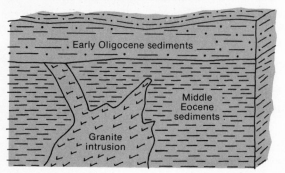

TABLE 4.5 Modern Geologic Time Scale Showing Divisions of Relative Time Compared with Three Estimates of Values for Measured Time

Millions of years before present	Era	Period	Epoch	Measured years (millions) before present			Events in earth history
				Kulp (1961)	Holmes (1959)	Harland, Smith, Wilcock (1964)	
0	CENOZOIC	Quaternary	Pleistocene	1	1	1	← Paleoindians migrate to America; Cascade Mountains
		Tertiary	Pliocene	13	11	11	
			Miocene	25	25	25	← Columbia Plateau volcanic eruptions
50			Oligocene	36	40	40	
			Eocene	58	60	60	
			Paleocene	63	70	70	← Folding of Rocky Mountains
100	MESOZOIC	Cretaceous					← Gulf of Mexico extends north to Illinois
150		Jurassic		135	135	135	
				181	180	180	
200		Triassic					
				230	225	225	← Folding of Appalachian Mountains
250	PALEOZOIC	Permian			270	270	← Marathon Mountains (Texas)
		Pennsylvanian		280			← Great coal swamps in eastern United States
300		Mississippian					
350		Devonian		345	350	350	← Acadian Mountains
							← First land vertebrates and forests
400		Silurian		405	400	400	← Great coral reefs in Michigan and Illinois
450		Ordovician		425	440	440	
500		Cambrian		500	500	500	← Seas inundated most of United States
550							
600	PRECAMBRIAN	Keweenawan		?	?	600	← Killarney Mountains
		Huronian					← Iron ore formed in Michigan and Minnesota
2,500		Timiskaming					Oldest rocks in Minnesota and Canadian Shield
		Keewatin					
3,400							

43

divisions. About 90 useful absolute dates scattered throughout geologic time have been determined from all over the world. These dates have been fitted in with the relative scale so that we have a fairly good representation of the distribution of time over the named divisions. These dates are thought to be accurate within 5 percent for K/Ar and within 2 percent for most other age determinations. Each specialist in geochronology evaluates the significance of these measured dates differently. There are several slightly varied modern estimates of the "best" time scale (Kulp, 1961; Harland et al., l964; Faul, 1966).

Table 4.5 is a summary of the geologic time scale, showing measured dates estimated by five specialists for the beginning of each period and for the epochs of the Cenozoic era.

The geologic time-scale nomenclature in Table 4.5 has evolved appreciably as compared to its form nearly a century before as shown in Table 4.4. The vast time span before the Cambrian period is sometimes called simply the Precambrian. Other workers divide it into the Archeozoic era, meaning "beginning," and the Proterozoic era, meaning "earlier life,"

or use only local names. The Ordovician period was introduced in 1879 to resolve a controversial overlap in ages of the Cambrian and Silurian as originally defined. The spelling of Cenozoic is a slight modification of the earlier Cainozoic, and several new epoch divisions have been added. Some modern geologists prefer to divide the Quaternary period into two epochs—the Pleistocene, as the time of the ice ages, and the Holocene, sometimes informally called Recent. The latter began about 7,000 years ago when rapid ice melting ceased at the end of the latest ice age.

The oldest reliable date for earth rocks from Greenland was 3.9 billion years. Dates of 4 billion years have been tentatively reported from Africa. The Apollo 15 expedition to the moon collected somewhat older rocks (4.6 to 4.7 billion years) than the most ancient rocks yet discovered on earth. They have been preserved because destructive geologic processes are absent or very slow on the moon. The oldest moon rocks will probably date almost from the time of origin of the solar system (5 billion years). On earth the original rock materials long since have been destroyed by erosion and metamorphism.

TABLE 4.6   A "Concept" of Geologic Time

January 1–March 31	The creation and early development, but *no sign* of life has appeared
April 1–May 15	The single-cell amoeba appear as first life form and evolve into multicelled sponges-corals
May 15–June 30	The development of the first vertebrates, fishes; land plants appear around June 30
July 1–August 31	First sea vertebrates move to a land environment; the reptiles develop and flourish
September 1–November 15	The dinosaurs begin a dominance in land, sea, and air
November 15–November 30	The dinosaurs decrease their dominance and the Rocky Mountains start to form
December 1–December 23	The Rocky Mountains have formed; birds and mammals become a dominant life form
December 23–December 25	The Colorado River has started to carve the Grand Canyon
December 26–December 30	The North American landscape starts to develop as we see it today
December 31 A.M.–1 P.M.	Man develops as life form
1–3 P.M.	The Ice Ages arrive
3–10 P.M.	The Ice Ages return and leave for four intervals, finally waning about 9:30 P.M.
10–11 P.M.	Man reappears in evidence
11:45 P.M.	Stone Age man arrives
11:46 P.M.	The Dawn of Civilization starts
11:58:38 P.M.	The Christian Era begins
11:59:40 P.M.	Columbus discovers America
11:59:52 P.M.	The Declaration of Independence is signed
11:59:59 P.M.	World War II

REFERENCES

Berry, W. B. N., 1968, *Growth of a Prehistoric Time Scale Based on Evolution,* W. H. Freeman and Company, San Francisco.

Boltwood, B. B., 1907, *On the ultimate disintegration products of the radioactive elements. Part II. The disintegration products of uranium,* American Journal of Science, ser. IV, vol. 23, pp. 77-89.

Clark, D. L., 1968, *Fossils, Paleontology and Evolution,* Wm. C. Brown Company Publishers, Dubuque, Iowa.

Cloud, P. (ed.), 1970, *Adventures in Earth History,* W. H. Freeman and Company, San Francisco.

Dott, R. H., Jr., and Batten, R. L., 1971, *Evolution of the Earth,* McGraw-Hill Book Company, New York.

Eicher, D. L., 1968, *Geologic Time,* Prentice-Hall Inc., Englewood Cliffs, N.J.

Engel, A. E. J., 1969, *Time and the earth,* American Scientist, vol. 57, no. 4, pp. 458-483.

Faul, H., 1966, *Ages of Rocks, Planets, and Stars,* McGraw-Hill Book Company, New York.

Fleischer, R. C., and Price, P. B., 1964, *Techniques for geological dating of minerals by chemical etching of fission fragment tracks,* Geochimica Cosmochimica Acta, vol. 28, pp. 1705-1714.

Hamilton, E. I., and Farquhar, R. M., (eds.), 1968, *Radiometric Dating for Geologists,* Interscience Publishers, a division of John Wiley & Sons, Inc., New York.

Harland, W. B., Smith, A. G., and Wilcox, B., 1964, *The Phanerozoic time-scale,* Quarterly Journal of the Geological Society of London, vol. 120, pp. 1-458.

Holmes, A., 1959, *A revised geological time scale,* Edinburgh Geological Society Transactions, vol. 17, pp. 183-216.

Hurley, P. M., 1959, *How Old Is the Earth?* Anchor Books, Doubleday & Company, Inc., Garden City, N. Y.

Jeltzky, J. A., 1956, *Paleontology, basis of practical geochronology,* American Association of Petroleum Geologists Bulletin, vol. 40, pp. 679-706.

Kitts, D., 1966, *Geologic time,* Journal of Geology, vol. 74, pp. 127-146.

Kulp, J. L., 1961, *Geologic time scale,* Science, vol. 133, pp. 1105-1114.

Libby, W. F., 1963, *Accuracy of radiocarbon dates,* Science, vol. 140, pp. 278-280.

Lyell, Sir Charles, 1887, *Principles of Geology,* 11th ed., D. Appleton and Company, New York.

Runcorn, S. K., 1966, *Corals as paleontological clocks,* Scientific American, vol. 215, no. 4, pp. 26-33.

Schuchert, C., 1931, *Geochronology of the age of the earth on the basis of sediments and life,* in *Physics of the Earth,* vol. 4, *The Age of the Earth,* pp. 10-64, National Research Council, Washington, D. C.

Tilton, G. R., and Hart, S. R., 1963, *Geochronology,* Science, vol. 140, pp. 357-366.

Weaver, K. F., 1967, *Magnetic clues help date the past,* National Geographic, vol. 131, pp. 676-701.

Wetherill, G. W., 1956, *Discordant uranium—lead ages: I,* Transactions, American Geophysical Union, vol. 37, pp. 320-326.

Zeuner, F. E., 1958, *Dating the Past,* 4th ed., Methuen & Co., Ltd., London.

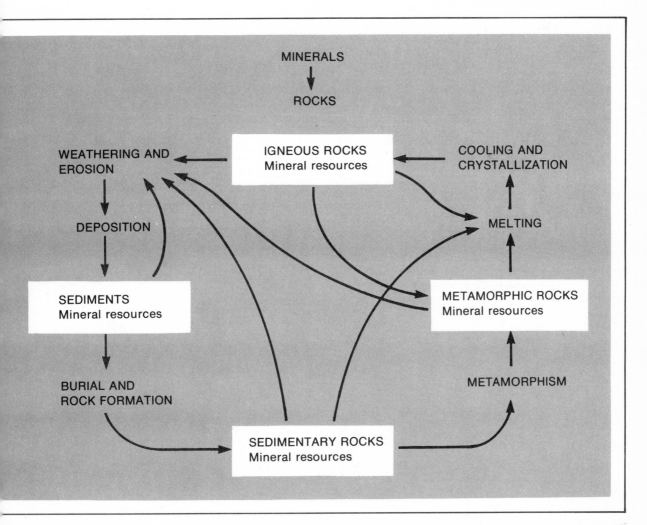

CHAPTER 5

MINERALS

In previous chapters, geology was defined as an earth science, and consideration was given to the general concepts of change, matter and energy, earth's place in the universe, and the geologic time scale.

We now turn to a study of earth materials: the materials of which the lithosphere is composed. The study will begin with minerals.

MINERALS

Minerals are the discrete crystalline particles out of which nearly all rocks are made. Several different minerals may occur together in many types of rock, or a single kind of mineral may make up the bulk of a rock as, for example, limestone or marble, which is composed of the mineral calcite.

In our study of earth materials, we are, in reality, considering a building-block process, from the smallest atoms and elements to compounds to specific molecules called minerals and, finally, to rocks. We have already considered atoms and elements, and so let us turn to the first major building block of the earth—a mineral.

A mineral, by definition, is a natural, inorganic, generally crystalline solid, composed of either a single element or a combination of elements in definite ratios, exhibiting characteristic physical properties. Substances of artificial or organic origin are excluded. This definition should be considered by an analysis of its component parts, considering first the compositional characteristics.

Composition of the Earth's Crust

Chemical analyses have established that more than 98 percent by weight of the earth's crust consists of only eight elements (Table 5.1). About 12 other elements make up most of the remaining 2 percent. Table 5.1 notes that the distribution of elements in the earth's crust by volume percent is also unique; five elements (O, Si, Ca, Na, and K) comprise approximately 99 percent of the earth's crust by volume. These two distribution characteristics of earth elements note that while the number of possible combinations for minerals is large, the number of important rock-forming minerals will be small. Several elements can occur alone, such as gold, copper, sulfur, and platinum, or the carbon of graphite and diamond, but most elements occur

TABLE 5.1   Chemical Abundances in the Cosmos and in the Earth's Crust

Element	Elemental abundance			Oxide abundance in earth's crust	
	Cosmos (wt. %)	Earth's crust (wt. %)	(vol. %)	Oxide	Wt. %
O	0.64	46.6	93.76	SiO_2	60
Si	0.05	27.7	0.86	Al_2O_3	16
Al	Insign.	8.1	0.47	FeO or Fe_2O_3	7
Fe	0.06	5.0	0.43	MgO	3.5
Mg	0.04	2.1	0.29	CaO	5.0
Ca	0.04	3.6	1.03	Na_2O	4.0
Na	1.05	2.8	1.32	K_2O	3.0
K	0.04	2.6	1.38	TiO_2	1.0
H	75.47	—	—		
He	23.10	—	—		

in chemical combinations. Since oxygen is present in the earth's crust in terms of 47 percent by weight and 94 percent by volume, some sort of chemical combination of an element with oxygen should be common. The distribution of oxide compositions, shown in Table 5.1, illustrates this. It can be seen that Si and Al, second and third in weight abundance, are the most important oxide groups; therefore, the silicate minerals, which are combinations of silicon and oxygen alone or with various other elements, such as sodium, potassium, aluminum, iron, and magnesium, are the major mineral group (Table 5.4). Other mineral compounds include the carbonates (CO_3^{--}), sulfides (S^{--}), sulfates (SO_4^{--}), chlorides (Cl^-), phosphates (PO_4^{--}), hydroxides (OH^-), nitrates (NO_3^-), and borates (BO_3^{3-}).

Chemical analyses reveal that every mineral has a chemical composition that can be represented by a chemical formula. For example, the mineral halite, which in processed form is common table salt, is represented by the formula NaCl. This formula is valid no matter what quantities of halite are considered; the chemical composition is always 1 ion of sodium to 1 ion of chlorine. The rule of definite proportions holds true for most minerals. There are exceptions where some substitution of elements is possible without creating a new mineral. This slightly modified chemical composition in minerals is called an isomorphic mixture and allows a greater diversity of minerals.

For example, some iron atoms may take the place of magnesium in the mineral olivine, which is an iron-magnesium silicate, and iron may substitute for part of the zinc in the mineral sphalerite, a zinc sulfide.

The most important isomorphic compounds are a silicate mineral group, the plagioclase feldspars. They form a continuous series of minerals from sodic albite ($NaAlSi_3O_8$) to high-calcium anorthite ($CaAl_2Si_2O_8$). The entire series of minerals is albite, oligoclase, andesine, labradorite, bytownite, and anorthite. From albite to anorthite, the sodium content decreases and calcium increases. The aluminum and silica content also varies. Albite contains about 69 percent silica, anorthite 43 percent, and the others intermediate quantities. This variation allows structural room and a charge imbalance for substitution.

Calcium, magnesium, ferrous iron, and manganese replace each other isomorphously in pyroxenes, amphiboles, and garnet, as do trivalent aluminum, ferric iron, and to some extent, chromium. The carbonate minerals, calcite ($CaCO_3$), dolomite [$CaMg(CO_3)_2$], magnesite ($MgCO_3$), and siderite ($FeCO_3$) all are structurally the same system but are chemically isomorphous only within certain limits. Thus calcite, for example, may contain a few magnesium or iron atoms, but it cannot grade all the way to dolomite, magnesite, or siderite because of space limitations in the structure for certain elements.

Chemical Bonding

The ions, atoms, or groups of atoms in minerals are held together by one of four types of bonding, or by a combination of these types: ionic, covalent, metallic, and van der Waals. Ionic and covalent bonding were explained in Chap. 2. Metallic bonding, occurring almost solely in metals, such as copper, results from roving electrons that statistically maintain an electrical balance. Van der Waals bonding is another weak electrical type formed between molecules, as between covalent sheets of carbon atoms in graphite.

The type of bonding affects not only the properties of cleavage and electrical conductivity but also hardness, heat conductivity, and refraction of light.

Physical Structure

To get a better understanding of this property, let us examine the mineral halite (Fig. 5.1). We see immediately that a typical specimen of halite has square edges and corners all around. In other words, it shows three rather distinct surfaces at right angles to each other.

FIGURE 5.1 *A cleavage block of crystalline halite (sodium chloride), showing its perfect cubic cleavage. (Ward's Natural Science Establishment, Inc.)*

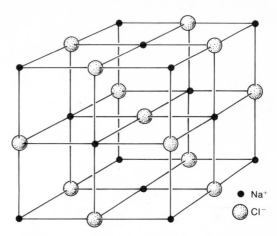

FIGURE 5.2 *Diagram showing the arrangement of ions of sodium and chlorine in a crystal of common salt (NaCl), the mineral halite. The ions alternate in three planes at right angles to each other.*

This specimen will be, in most cases, colorless or transparent, although white, pink, red, blue, green, or gray forms are found. Halite's luster is vitreous, or glassy; its hardness is 2 to 2.5; its specific gravity 2.1 to 2.3; its taste salty. Its chemical composition is sodium chloride, NaCl. It contains equal numbers of ions of sodium and of chlorine. How are these ions arranged?

The answer can be shown from x-ray studies. Von Laue (1879-1959) and Roentgen (1845-1923) reasoned that if x-rays were like light waves, except for their very short wavelengths, then x-rays should undergo diffraction (spreading), be thrown out of phase, and therefore show interference (either an increase or decrease in intensity) when they travel through a crystal. X-ray tests on sphalerite (ZnS), halite, and other minerals were very successful. The results illustrated that when the x-rays diffracted by a piece of halite are projected onto a photographic plate, they project into a symmetrical arrangement of spots known as a Laue diagram. The arrangement and spacing of this symmetrical arrangement from halite indicates an alignment of the ions in three rectangular planes. The ions of sodium and chlorine alternate in each of the three planes (Fig. 5.2), in what is termed a space lattice.

The Laue method, however, is not applicable to all minerals, and so other x-ray techniques have come

FIGURE 5.3 *Colorless hexagonal crystal of quartz with pyramidal end. (Ward's Natural Science Establishment, Inc.)*

tographic powder method and occur at angles characteristic of each mineral.

The atomic structures of hundreds of minerals are now known. With very few exceptions, there is a regular or systematic internal arrangement of atoms or associations of atoms in one of many possible mathematical groupings. This is termed the crystalline structure of a mineral and is its most characteristic feature. Other mineral traits include the shapes of crystals, the way minerals break, their hardness, their conductivity, and other physical properties.

The external appearance of this internal crystalline structure results in a crystal. Crystals are geometrical solids bounded by natural plane surfaces called crystal faces (Figs. 5.3 to 5.6). Their external forms reflect the internal crystalline structure. Studies of the general shapes created by the surfaces or

FIGURE 5.4 *Colorless hexagonal crystals of dog-tooth spar, a variety of calcite. (Ward's Natural Science Establishment, Inc.)*

into use. By the rotation method, a crystal is slowly turned on one of its structural axes, and the diffraction spots are formed in parallel rows on photographic film. Another technique is to use the powdered-mineral method, which is used to form a diffraction pattern of lines upon a strip of photographic film. The positions and intensities of the lines depend upon the space lattice of the substance tested.

Still another x-ray method uses a diffractometer, which records on paper the effects of diffraction when a powdered sample subjected to x-rays is slowly turned through known angles of arc. The internal reflections are picked up by a Geiger tube and put on a recording. Intensity peaks are analogous to the lines of the pho-

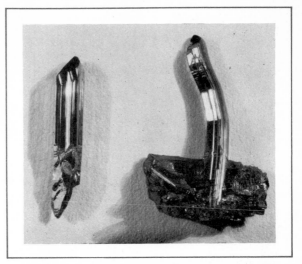

FIGURE 5.6   *Orthorhombic crystals of stibnite (Sb_2S_3), the principal ore of antimony. (Ward's Natural Science Establishment, Inc.)*

FIGURE 5.5   *Cubic crystals of galena (lead sulfide, an ore of lead). (Ward's Natural Science Establishment, Inc.)*

faces and their interfacial angles permit the classification of crystals into the six crystal systems. These are briefly outlined in Table 5.2 and Fig. 5.7. Some peculiarities in crystal form and development do occur. In some instances two or more crystals of the same mineral are intergrown in definite geometric positions; the result is a twinned crystal. Among the common minerals, orthoclase, plagioclase, calcite, fluorite, gypsum, and pyrite frequently exhibit twinning.

In plagioclase, thin leaflike plates are twinned alternately face-to-face and back-to-back, with the result that the narrow edges of the plates have alternating slants. The edges accordingly show striations, very fine parallel lines closely set, like ruled lines on a sheet of paper or the leaf edges of a book. Seen through a microscope, twinned plagioclase shows light and dark bands (Fig. 5.8).

TABLE 5.2   Crystal Systems

System	Characteristics	Representative minerals
Isometric (or cubic)	3 crystal axes of equal length at right angles to each other	Halite, galena, fluorite, pyrite, garnet, diamond
Tetragonal	3 perpendicular axes, only 2 of which are equal	Zircon, chalcopyrite
Hexagonal	3 equal axes in a common plane intersecting at 120°, and a 4th axis perpendicular to them	Quartz, calcite, tourmaline
Orthorhombic	3 axes of unequal length perpendicular to each other	Barite, sulfur, topaz
Monoclinic	3 unequal axes, 2 perpendicular and the 3rd inclined	Gypsum, orthoclase, augite, hornblende
Triclinic	3 unequal axes, all inclined obliquely to each other	Plagioclase feldspars

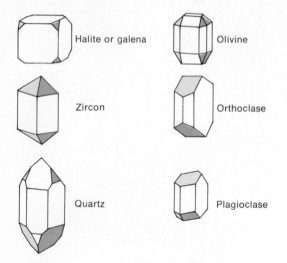

Halite or galena

Olivine

Zircon

Orthoclase

Quartz

Plagioclase

FIGURE 5.7 *Drawings of representative crystals in the isometric (cubic), tetragonal, hexagonal, orthorhombic, monoclinic, and triclinic crystal systems.*

Aggregates of crystals form characteristic structures in rocks. Numerous, variously oriented, small crystals give a rock a massive appearance. Other aggregates give rise to fibrous (like asbestos, Fig. 5.9), feathery, radiating, platy, columnar, concentric or concretionary (onionlike), dendritic (branched), and other types of structures.

Physical Properties of Minerals

The physical properties most useful in identifying minerals are hardness, specific gravity, cleavage, streak, color, and luster. Magnetism, tenacity, solubility in water, and other properties, such as reactions with acid, taste, feel, crystal form, and mode of occurrence are also useful in many instances. These properties, if known, enable one to identify many minerals upon sight or after a few simple chemical tests, which are sufficient for introductory mineral identification.

Identifications also can be made with a polarizing petrographic microscope. The procedure involves crushing the minerals to a powder or grinding them into very thin, translucent slices, mounting them on a glass microscope slide, and allowing polarized light to

FIGURE 5.8 *Multiple twinning in plagioclase (enlarged), as seen through a petrographic microscope. (Geological Survey of Canada.)*

FIGURE 5.9 *Asbestos, showing its fibrous structure.*

FIGURE 5.10 *Cleavage. Top row: orthoclase, two 90°
cleavages; gypsum, one perfect, one good, one fair cleavage.
Bottom row: halite, cubic cleavage; calcite, three slanted
(rhombohedral) cleavages.*

pass through them. In polarized light the vibration
paths lie in a common plane perpendicular to a ray.
The effect of polarized light on each transparent mineral is distinctive. For nontransparent, or opaque,
minerals, such as the metallic sulfide minerals, study
by a microscope using reflected light is necessary.

In difficult determinations of physical properties,
we can make use of (1) x-ray diffraction to determine
crystal structure, (2) spectrographic analyses or special probes to identify the component chemical elements, or (3) differential thermal analysis to measure
the temperature at which diagnostic changes take
place during heating.

FIGURE 5.11 *Outlines showing crystal forms and directions of cleavage (a) in pyroxene and (b) in hornblende.*

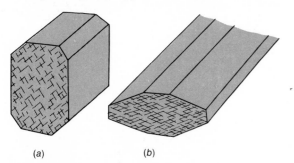

(a) (b)

Cleavage Cleavage is the ability to split readily
along certain planes and is generally recognized by the
flat surfaces resulting from breakage (Figs. 5.10 to 5.12).
Minerals may cleave in only one plane (like the micas),
in two planes (orthoclase, Fig. 5.10; pyroxene or
hornblende, Fig. 5.11), in three planes (calcite, Fig.
5.12), or not at all (quartz). The number of cleavages,
their angles, their ease, and their excellence (perfect,
good, fair, or poor) are to be noted. The lack of this
property in a mineral means the mineral will fracture.

Fracture Lacking cleavage, a mineral breaks, or
fractures, in an irregular nature. Many minerals fracture in a manner similar to broken glass. This type of
fracture is called conchoidal. Other minerals show
fibrous or splintery fracture.

Remembering the definition of a mineral, one can
determine the cause of cleavage and fracture: the arrangement of atoms in the crystalline structure and the

FIGURE 5.12 *A rhombohedral cleavage block of transparent calcite (Iceland spar), showing its property of doubly
refracting light. The images of the spot and of the two lines
that cross underneath the calcite are split in two by unequal
transmission of light within the calcite crystal. For this
reason calcite prisms are used in certain optical instruments.
Double refraction is used to identify many different minerals
under a petrographic microscope. (Ward's Natural Science
Establishment, Inc.)*

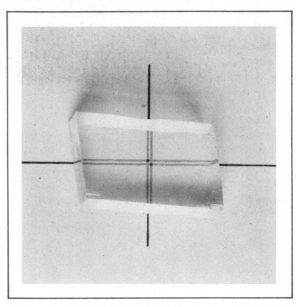

relative strength of the bonds between atoms or groups of atoms. It should also be noted that crystal form and twin crystals are also a reflection of the internal structure and, therefore, could be used as identifying physical properties.

Hardness The term hardness, in a sense, is a misnomer because hardness, as used to describe a mineral property, is actually the resistance of a mineral to scratching. Hardness is rated by its comparison with a standard scale. The scale, called Mohs' scale, includes 10 well-known minerals arranged in ascending order of hardness, as follows: (1) talc, (2) gypsum, (3) calcite, (4) fluorite, (5) apatite, (6) orthoclase, (7) quartz, (8) topaz, (9) sapphire (corundum), (10) diamond.

In testing hardness, one may use either known minerals of the Mohs' scale or other common items, such as a finger nail (hardness 2.25), a copper cent (3), a knife (about 5), or a piece of window glass (5.5). The relative hardness of an unknown specimen can be found by testing which item it can or cannot scratch, or which mineral will or will not scratch it. Care must be taken to distinguish a true scratch from a mark made by the softer mineral as it easily rubs off. This is another physical property whose characteristics are being determined by the internal atomic bonding and arrangement.

Streak Streak is the color of a mineral when powdered by grinding it or by rubbing it across a piece of unglazed porcelain. In many instances the streak is less variable and hence more useful in mineral identification than is the color of an intact mineral crystal. The streaks of several minerals are listed in Table 5.3. The streak of a mineral, or its lack of it, has a definite relationship to the mineral's hardness. Can you see this relationship?

Specific Gravity The specific gravity of a mineral is the ratio of its weight in air to the weight of an equal volume of water. This is computed most readily by weighing a small piece of mineral in air and then suspending it in water to find the weight of the water it displaces.

The specific gravity of many minerals ranges between 2.5 and 3.3, and so in ordinary practice only ex-

ceptionally light or extra-heavy minerals feel unusual when lifted by hand. Graphite has an unusually low specific gravity (2.2). Minerals with high specific gravity always have a high metal content. For example, pyrite, FeS_2, has a specific gravity of 4.9 to 5.2; galena, PbS, 7.3 to 7.6; and gold, Au, 16.0 to 19.0 (according to purity). One can see that this property is a function of the chemical composition of the mineral as well as the closeness or looseness of the internal structural arrangement of the atoms.

Other Properties There are other physical properties in minerals that can be used for identification purposes. The properties are grouped here, not because they are unimportant, but because they are more restricted to certain mineral groups or are more difficult to determine because of their variability or the equipment needed.

(a) Magnetism The only abundant, strongly magnetic mineral is magnetite. Magnetite is commonly found in "black sands" of river bars and ocean beaches, in small grains in crystalline rocks, or in larger masses in certain iron-mining districts. A number of other minerals are slightly magnetic.

(b) Tenacity The tenacity of a mineral is its cohesiveness. Tenacity is described by the terms brittle, (breaks with a snap), flexible (easily bent and stays bent), elastic (resilient, easily bent but springs back upon release), and malleable (can be shaped by hammering).

(c) Solubility A few minerals, like halite, are readily soluble in water; many others are only slightly soluble or not soluble at all. Acids dissolve a few minerals; dilute hydrochloric acid easily dissolves the mineral

TABLE 5.3   Streak Compared to Color

Mineral	Color	Streak
Quartz, calcite, halite	Colorless or white	White
Hematite	Reddish-brown, black, or steel-gray	Red (rust red)
Limonite	Yellowish-brown, dark brown, or black	Yellow-brown (rust brown)
Pyrite (fool's gold)	Pale brass-yellow	Greenish black
Sphalerite (zinc ore)	Yellow to brownis', black	Pale yellow

TABLE 5.4   The Silicate Minerals

Ferromagnesian (Fe,Mg) dark colored		Si:0 ratio		Nonferromagnesian light colored
Olivine $(Mg, Fe)_2SiO_4$		1:4 (Single tetrahedra)		
Pyroxene (augite) $Ca, Mg, Fe, Al Si_2O_6$		1:3 (Single-chain tetrahedra)		
Amphibole (hornblende) $Na, Ca, Mg, Fe, Al Si_8O_{22}$		4:11 (Double-chain tetrahedra)	3:8 (Infinite three-dimensional framework)	⎰ Plagioclase feldspar $(Ca, Na) Al Si_3 O_8$ ⎱ Orthoclase feldspar $K Al Si_3O_8$
Biotite mica $K,Mg,Fe,Al Si_4O_{10}$		2:5 (Sheet structure)	2:5	Muscovite mica $K Al Si_4O_{10}$
			1:2 (Infinite framework)	Quartz SiO_2

(vertical axis label between columns: —Increase in ratio of silica to oxygen—)

calcite ($CaCO_3$), with the production of bubbles of carbon dioxide (CO_2).

(d) Fusibility   The fusibility of a mineral is its relative ease of melting, as measured by the temperature of its melting point. The fusion, or melting, temperatures of minerals range from that of a match flame, for sulfur, to that of an electric arc, for asbestos. Many minerals dissociate, or decompose, upon heating and give off water, carbon dioxide, or other constituents at specific temperatures before their melting points are reached. More minerals melt at 1000 to 1500°C than at other temperatures.

(e) Color   The color of a mineral depends upon the wavelengths of light reflected from it after absorption of other wavelengths from the spectrum of white light. Many minerals are commonly white or colorless. Another large number are dark green or black. Some, such as quartz or calcite, may have almost any color—white, pink, red, blue, amethyst, brown, yellow, green, or gray, due to the presence of certain elements. Others show only one color, like the mineral sulfur, which is yellow, or the mineral malachite, green, or at most a few colors. These are generally in distinctive shades that are easily learned.

(f) Luster   The luster of a mineral is its visual characteristics as seen in reflected light. Luster ranges from dull or earthy (like clay), greasy, vitreous (glassy), silky, to adamantine (like diamond), or metallic. Other lusters are described as pearly, resinous, or submetallic.

(g) Special Sense Properties   Minerals may be salty, acid, alkaline, or bitter to the taste; harsh, greasy, or cold to the touch; or earthy in odor.

The Silicate Minerals

The most important rock-forming minerals in this group are the feldspars (orthoclase, microcline, and the plagioclase group), which make up about 60 percent of the earth's crust. Other abundant silicate minerals are quartz, the amphiboles (especially hornblende), the pyroxenes (especially augite), olivine, and the micas (muscovite and biotite). The building block for this mineral group is the silicon-oxygen tetrahedron, one silicon atom surrounded by four oxygen atoms. This unit, with a −4 ionic charge, can combine in five different ways (Table 5.4). Each combination, a different internal arrangement, will have different physical properties, such as with cleavage and crystal form. For example: olivine is composed of SiO_4^{4-} units bonded with iron or magnesium (Fe^{++} or Mg^{++}); the micas are composed of the SiO_4^{4-} tetrahedrons combined in sheets, or layers, bonded between layers by potassium (K^+), magnesium (Mg^{++}), or iron (Fe^{3+}). This between-layer bonding is weaker than the in-layer bonding and allows for a sheet-by-sheet, or layer, cleavage for the micas. Quartz, on the other hand, is composed of a framework of silicon and oxygen, with no excess charges; therefore, it has a relatively strong bonding in a three-dimensional framework and does not have cleavage, but fracture.

Chemically, as Table 5.4 illustrates, the silicates differ, but it should be noted that only the top eight elements by weight percent in the earth's crust are involved. Olivine, for example, utilizes magnesium and iron, while the pyroxenes and amphiboles utilize Mg^{++}, Fe^{++}, as well as Al^{3+}, Ca^{++}, and Na^+; the orthoclase feldspars utilize only K^+, and quartz utilizes solely its SiO_2 design.

ORIGIN OF MINERALS AND ROCKS

Most minerals crystallize from some sort of solution. They result from (1) cooling of magma (complex mutual solutions of rock materials in a molten state) or similar solutions of molten material called lava on the earth's surface; (2) subsurface solutions involving hot water or hot gases, including steam, as in many mineral veins and metallic ore deposits; or (3) hot vapors condensing to form minerals, such as sublimates of sulfur near volcanic vents. Other minerals form by (4) chemical reactions with previous minerals, as by hydrothermal alteration of feldspar to mica or by oxidation of iron-bearing minerals in the zone of chemical weathering at the earth's surface; (5) replacement or substitution of a mineral for an earlier one; (6) recrystallization of previous minerals as new compounds under changed temperature and pressure conditions; or (7) evaporation of water solutions.

The total number of known minerals is about 2,000, but less than 50 of these are abundant enough to make up any substantial part of the earth's crust. The silicate minerals containing Ca^{++}, Na^+, K^+, Al^{3+}, Mg^{++}, and Fe^{++} are by far the most important rock-forming minerals.

Most rocks are primarily aggregates of minerals. These aggregates, on the basis of their mode of origin,

are classified in three general groups: igneous, sedimentary, and metamorphic.

Igneous rocks are those formed by solidification of magma that originates within the earth. The principal mineral components of igneous rocks are the feldspars, quartz, pyroxenes, amphiboles, olivine, and micas. Granite and basalt are the two most common igneous rocks. Granite is a light-colored rock consisting of visible grains of quartz, feldspar, and minor amounts of mica or hornblende. Basalt is a fine-grained, usually dark-gray or black igneous rock composed primarily of calcium-rich plagioclase feldspar and pyroxene, with or without olivine. Gabbro, similar to basalt in chemical composition, is a dark-colored, coarse-grained rock derived from molten material that cooled and crystallized deep in the earth.

Most sedimentary rocks are derived from the weathered products of older rocks. One group is composed of clastic (broken), or physically derived, sediments. Such sediments are deposited as broken fragments of pre-existing rocks and minerals, in sizes ranging from gravel, sand, silt, to clay. Upon consolidation, these sediments become conglomerate, sandstone, siltstone, and shale. Another group of sedimentary rocks are the result of chemical precipitation. Limestone, a rock composed of the mineral calcite, is an example of a sedimentary rock formed this way. Some rocks, such as rock salt and gypsum, are evaporites from salt lakes or salt-water bodies isolated from the ocean. Peat and coal are sedimentary rocks of organic origin.

Metamorphic rocks are formed as a result of the transformation of pre-existing rocks. The transformation results from a change in either mineral composition or structure, or both, brought about by high pressure, high temperature, and chemically active fluids within the earth. For example, granite may change to gneiss, fine-grained limestone may change to coarse-grained marble, and shale to slate.

REFERENCES

Dana, J. D., 1971, *Dana's Manual of Mineralogy*, 16th ed., revised C. S. Hurlbut, Jr., John Wiley & Sons, Inc., New York.

Ernst, W. G., 1969, *Earth Materials*, Prentice-Hall Inc., Englewood Cliffs, N. J.

Mason, B., and Berry, L. G., 1968, *Elements of Mineralogy*, W. H. Freeman and Company, San Francisco.

CHAPTER 6

IGNEOUS ACTIVITY

Intermittently throughout geologic time, rocks within the earth have undergone melting. Sometimes the molten material makes its way through the covering rock and flows, or is thrown with violent force, onto the surface. Much of the molten material, however, remains deeply buried in the crust, where it slowly cools and crystallizes. At all times, when cooled and hardened, it forms igneous rock, so-called because it seemed to be born of fire. The term igneous activity comprises all movements of molten material, whether it occurs deep within the earth (intrusive) or on the earth's surface (extrusive vulcanism). This chapter will discuss the causes and types of igneous activity and the chapter to follow, the types of igneous rocks that are formed as a result.

CAUSE OF IGNEOUS ACTIVITY

The Heat Problem

The temperature of the earth increases from the surface downward at a rate of 1°C per 33 meters. This figure varies in many areas, but a figure of 20°C per kilometer is commonly used as the average heat increase, or thermal gradient.

Heat is carried to the surface of the earth largely by conduction through the rocks. This heat flow varies from place to place because various rocks differ in their content of heat-generating chemical elements, their conductivity, their water content, the ease of circulation of fluids through them, their nearness to hot masses underground, the movement of hot gases through them, and the recency of their mechanical deformation by earth movements. The average heat flow upward is about 1.4×10^{-6} calorie per square centimeter per second. This value is about the same beneath the oceans as beneath continents.

Although rocks are poor conductors of heat, the heat rising to the surface from the interior of the earth and radiated into outer space is about 7×10^{12} calories per second over the whole surface area. This is hundreds of times greater than that brought up by vulcanism, yet it is less than 1 percent of the energy supplied to the earth by the sun.

The melting point of basaltic lava at the earth's surface is about 1200°C. At depth it would be a little higher because of the pressure. What the temperature is at a depth of 48 kilometers or more, where basaltic magma originates, is not certain. Ordinarily it must be below the melting point of the materials because earthquakes indicate that the rocks at this depth are solid.

The principal source of the internal heat of the earth is thought to be radioactivity. The spontaneous disintegration of certain atoms, especially uranium, thorium, and an isotope of potassium, generates considerable heat energy. Radioactive decay results in the emission of particles and rays, the kinetic energy of which is converted to heat as these products are absorbed by the surrounding rocks. According to some investigators, 1 gram of U^{238} yields 0.71 calorie per year; U^{235}, 4.3 calories; Th^{232}, 0.20 calorie; and K^{40}, 0.21 calorie. Acidic igneous rocks contain enough radioactive material to yield about 8.10 calories per gram per million years, and basalt about 1.19 calories per gram per million years. The difference is thought to be indicative of the relative concentration of uranium, thorium, and potassium in granite (Table 13.3).

Other possible sources of heat are: (1) heat inherited from the time of the earth's origin as a planet, (2) heat by shrinkage from cooling or by recrystallization, (3) heat of compression and shear by mechanical deformation of the rocks at depth, and (4) exothermal chemical reactions. Each of these possible sources or theories for heat sources has strong opposition from some geologists. They envision an expanding rather than a shrinking earth and a progressively hotter rather than a cooling earth. Frictional heat produced by earth movements would be local, not worldwide. They theorize that exothermal chemical reactions, such as the reactions between volcanic gases or the release of the latent heat of crystallization, are superficial effects, and conclude that any heat liberated by crystallization of magma or lava is not new heat energy but a transmuted form of it already brought up from below.

Although the quantity of heat produced in a year by radioactivity seems minute, a layer of granite 19 kilometers thick around the earth could supply all of it. Since ocean basins have basaltic rather than granitic floors beneath their sediments, yet transmit as much heat as continents, radioactivity must be worldwide and not merely from concentrations in granite. The heat must be largely confined to an outer zone, perhaps tens of kilometers thick; otherwise, because of its poor heat conductivity, the earth below this zone would melt. In reality, earthquakes indicate that the earth is solid to a depth of 2,900 kilometers.

Characteristics of a Magma

According to recent studies on Hawaii, basaltic magma originates at a depth of 48 to 64 kilometers or more below the surface. Its source is within the zone termed the mantle that underlies the crust of the earth

FIGURE 6.1 *Hypothetical cross section through Mauna Loa and Kilauea, Hawaii, to show the zone of origin of magma in the earth's mantle and the shallow magma reservoirs that periodically swell and shrink as volcanic activity waxes and wanes. Surface uplift and outward tilting precede lava eruptions; subsidence and downbending follow them. (After MacDonald, 1961.)*

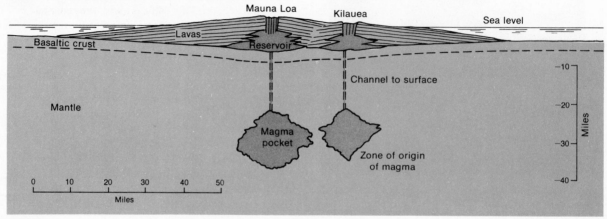

(Fig. 6.1). The composition of the upper part of the mantle is thought to be either peridotite or eclogite. Peridotite is a rock consisting almost entirely of ferromagnesian silicates, especially olivine and pyroxene. Eclogite is a rock composed essentially of pyroxene and garnet. Eclogite is equivalent to basalt in chemical composition, whereas peridotite has more iron, calcium, and magnesium. So if basaltic magma is made from peridotite, then preferential or selective fusion of only a certain fraction of the constituents of peridotite is needed.

In order to melt, rock must have room to expand, perhaps as much as one-tenth of its original volume in a solid state. This volumetric change is the reverse of that between ice and water. The expanded and, hence, lighter magma then is squeezed upward hydrostatically along fractures by the weight of the solid rocks around it until it solidifies by cooling under the surface, or it breaks out on the surface as a fissure type or central-vent type of volcanic eruption. Dissolved gases undoubtedly assist the upward progress of magma by decreasing its viscosity. Upon separation near the surface, during crystallization, the gases supply energy for nuées ardentes or other explosions.

EXTRUSIVE IGNEOUS ACTIVITY

Extrusive vulcanism is one of the most awe-inspiring acts of nature. It is geologically constructive, for it will build volcanic mountains, lava plains, and lava plateaus. Volcanic activity and eruptions may occur with a greater frequency than most people are aware of. Table 6.1 reports the listed volcanic activity during the period January 1972 to January 1973.

TABLE 6.1   Volcanic Activity January 1972 to January 1973*

Date	Location	Remarks	Date	Location	Remarks
January 1972	Soufriere, St. Vincent Island	Volcanic activity† initiated 31 October 1971; continued to 1 April 1972	June 1972	Alaid, Kuril Islands	Submarine eruption; continued to September 1972
	Nyiragono, Zaïre, Africa	Eruption; continued to April 1972	September 1972	Poás, Costa Rica Karthala, Comoro Island, Indian Ocean	Eruption Eruption; continued to 5 October 1972
February 1972	Villarrica, Chile	Eruption; initiated 29 December 1971		Sakurazima, Kyusyu, Japan	Eruption continued to 6 October 1972; renewed 16 November 1972
	Erta'ale, Ethiopia	Volcanic activity; continued to April 1972	October 1972	Mt. Merapi, Central Java	Eruption
	Kilauea, Hawaii	Eruption		Ritter Island, New Britain Islands	Submarine eruption
	Pacaya, Guatemala	Eruption		Pacaya, Guatemala	Eruption
March 1972	Puyehue, Chile	Eruption	November 1972	Acatenango, Guatemala	Eruption
April 1972	Mt. Smeru, East Java	Eruption	January 1973	Helgafjell, Heimaeny Island, Iceland	Eruption
	Anak Krakatoa Sunda Strait	Eruption; continued to September 1972			
	Kuril Islands, Pacific	Submarine eruption			
June 1972	Piton de la Fournaise, Réunion Island, Indian Ocean	Eruption			

*As reported on Smithsonian Institution Earth Science Event Cards.
†Volcanic activity indicates volcanic "restlessness" without eruption.

FIGURE 6.2 *Map of southern Italy, showing the sites of Mt. Vesuvius, Stromboli, Vulcano, and Mt. Etna.*

The word vulcanism comes from Vulcano, a small volcanic vent situated on one of the Lipari Islands off the coast of Italy (Fig. 6.2). This island was the site of the imaginary forge of Vulcan, the legendary god of fire and metalworking in the underworld. Better known is the neighboring active volcano Stromboli. Known as the "lighthouse of the Mediterranean," this 1,000-meter peak displays a cloud of steam by day and a glow reflecting hot lava in its summit crater by night.

Volcanic Products

Gases The materials ejected from volcanoes include all three physical states—gases, liquids, and solids. The principal gas, generally 75 to 90 percent of the total, is steam. Other gases include nitrogen, oxygen, hydrogen (which burns to form water), carbon dioxide, carbon monoxide, sulfur, hydrogen sulfide, chlorine, fluorine, boric acid, ammonia, methane, and argon, or their reaction products.

These gases have several important effects: (1) during crystallization of magma underground, they build up pressures to the breaking point and force eventual explosion, and as the gases emerge, they expel clots or fountains of fluid lava and various sizes of solid rock material (Fig. 6.3); (2) trapped within lavas, their expansion inflates the lava with bubble holes like a froth: coarsely vesicular lava puffed up in this way is called scoria, and that solidified as fine foam is pumice; (3) gases released below the surface may react with the rocks through which they pass and create new minerals; (4) at the surface the gases form sublimates of native sulfur, metallic chlorides, or sulfides of mercury, arsenic, antimony, iron, and other substances; and (5) hot gases may continue to escape in fumaroles, solfataras, and hot springs long after lava eruptions cease.

All solidified lavas contain hydrogen and oxygen

FIGURE 6.3 *The cloud of steam and volcanic ash being emitted from the crater of Mount Vesuvius during the 1906 eruption. The cloud at its widest point is 0.8 kilometer in diameter. (Photo by Perret, courtesy of C. A. Reeds.)*

FIGURE 6.4   *Lava stream moving rapidly downhill from a vent on the south-west rift of Mauna Loa, Hawaii, June 1950. The white-hot lava is very fluid at this stage. (MacDonald, U.S. Geological Survey.)*

that can be recovered as steam upon heating them in a laboratory. Obsidian, a quickly cooled, glassy rock, ordinarily contains some water. One variety of obsidian is perlite, which, upon heating, will puff up like popcorn, indicating that water is a constituent. Surface and near-surface water may be involved also, by participating in certain steam explosions, such as those that occurred at Krakatoa and Tarawera.

Lavas   Liquid lavas from fissure eruptions or from the Hawaiian volcanoes are basaltic in composition. They are comparatively low in silica (about 50 percent) and relatively high in iron, magnesium, and calcium; hence, they are said to be basic, or mafic. They have very low viscosity (Fig. 6.4). So long as their dissolved gases escape freely, their eruptions are relatively quiet, although at times their frothing leads to

lava fountains and the formation of vesicular spatter around the vents. Upon cooling, these basic lavas solidify to a glasslike texture or crystallize into masses of closely grown very fine-grained to microscopic crystals. They are composed of calcium-rich plagioclase feldspar, pyroxene (usually augite), and olivine. Dark-gray to black, fine-grained or partly glassy rocks of this composition are called basalt.

Some of the lavas extruded are andesites (intermediate rocks containing 60 to 65 percent silica) or rhyolites (silicic rocks containing 70 to 75 percent silica and very little calcium, iron, or magnesium). High-silica lavas are much more viscous than basaltic lavas. They tend to pile up on steep slopes, form plug domes, and cause explosions when the vents become clogged and the gases are concentrated underground.

Lava flows display a variety of characteristic

FIGURE 6.5 *Ropy pahoêhoê lava, Kilauea, Hawaii. This type of structure is formed when smooth sheets of lava stiffen as they cool and finally freeze solid. (Mendenhall, U.S. Geological Survey.)*

FIGURE 6.6 *Advancing front of a lava flow as it nears the coast on the west slope of Mauna Loa, June 1950. The white-hot, still-fluid lava within the flow pushes forward repeatedly and breaks through its solidifying crust. Blocks of broken crust (âã) topple and roll down the front. At this stage the lava advances only a few hundred meters per day. (MacDonald, U.S. Geological Survey.)*

FIGURE 6.7 *Devil's Tower National Monument in northern Wyoming. Large-scale columnar jointing formed by shrinkage during cooling is seen in this exposed volcanic neck (Johnson) or degraded laccolith (Darton). The feature is 264 meters high. Although single columns average 2 meters thick, some reach 5 meters. (National Park Service photo.)*

not more than a few percent of the total volume of material expelled.

In some places great masses of pumice, originally expelled from fissures in a series of glowing avalanches, were hot enough and thick enough to expel gases from the basal portion, flatten the particles, and agglutinate the mass into a lavalike rock. One such mass in the Taupo-Rotorua area, North Island, New Zealand, covers about 13,000 square kilometers and has an estimated volume of 830 cubic kilometers. Another in eastern California, called the Bishop Tuff, averages 167 meters thick over an area of 1,050 square kilometers. Many other tuff deposits have been identified in the western United States. They show columnar jointing and are easily mistaken for lava flows.

As soon as volcanic materials are exposed at the earth's surface, they are subject to the slow processes

FIGURE 6.8 *Pillow structure in lava of Precambrian age in Northwest Territories, Canada. This type of structure is formed when lava flows into a body of water and is suddenly quenched. This flow has been tilted and subsequently beveled by glacial action. The rounded tops of the pillows and the way they fit together indicate that the base of the lava flow is toward the right. The hammer handle is 38 centimeters long. (Geological Survey of Canada.)*

structures. Hawaiian and Icelandic lavas have either smooth ropy surfaces called pahoéhoé (Fig. 6.5), or rough surfaces of jumbled scoriaceous blocks, called áá (Fig. 6.6). Other structures include: columnar jointing, from shrinkage during cooling (Fig. 6.7); pillow structure, ellipsoidal masses as a result of being quenched in water (Fig. 6.8); and pressure ridges, uparched and fissured corrugations on lava surfaces. Tubular openings called lava caves are formed by the withdrawal of fluid lava from beneath a frozen crust.

The broken solids blasted out of volcanic vents are called pyroclasts (Fig. 6.9). These are mainly clots of lava that were torn loose by outrushing gases and expanded by their internal gases while in free flight. Their textures are generally very porous. Pieces of already solidified lava or any other rocks ripped loose along the underground conduit, such as pieces of limestone thrown out at Mt. Vesuvius, ordinarily make up

FIGURE 6.9 *Volcanic bombs. The characteristic spindle shape is caused by the ends being twisted during flight through the air while they are still soft; the spheroidal lumps are flattened upon landing.*

of weathering and erosion. Pyroclastic materials are especially vulnerable, because of their generally fine texture, large surface area, and high porosity. Lava flows, dikes, and volcanic necks (solidified conduits) are more resistant, but none is permanent.

Rainwash and stream erosion, with assistance from glaciers, groundwater, and wind, gradually gnaw away large parts of a volcanic cone (Fig. 6.10) and eventually reduce it to a group of low hills or even to a plain, surmounted perhaps by a stub of the former neck and wall-like remnants of associated dikes. Shiprock, New Mexico, is an example of a volcanic cone greatly wasted away (Fig. 6.11). Subsidence and explosion also contribute to the destruction of cones of active volcanoes. A dormant volcano may have a resurgence of activity that rebuilds it and offsets previous erosion. Cones built at sea are attacked by waves around their flanks. Small ones may be destroyed within a few months. Notwithstanding these destructive processes, great quantities of volcanic materials have persisted through long stretches of geologic time.

Let us now review the activity of several representative examples as case studies of extrusions, first the contrasting types of fissure eruptions and then of eruptions from central vents.

Fissure Flows

Plateau or Flood Basalts   Nearly 2.6×10^6 square kilometers of land are covered by prehistoric lava flows that issued in great volumes not from volcanoes but rather from fissures, such as the copious modern lava flows of Iceland. The Deccan lava flows of India cover an area of 5×10^5 square kilometers, with an average thickness of 667 meters and a maximum thickness of 3,000 meters. Erosional remnants around the edges indicate that their original area was about twice as large.

Similar flows in the Pacific Northwest (Fig. 6.12) extend over an area of 4×10^5 square kilometers, with an average thickness of perhaps 1,000 meters. Flood lavas as much as 5,000 meters thick also are found on Keweenaw Peninsula and Isle Royale of upper Michigan, in northern Wisconsin, and along the north shore of Lake Superior. Others occur in Argentina, Brazil, Ethiopia, Iceland, Siberia, northern Ireland, and Greenland.

FIGURE 6.10 *Iliamna Volcano, Alaska. Although this volcano still emits steam, its cone already has been deeply eroded. Troughs caused by glaciers can be seen in the foreground and at the right. (Spence Air Photos.)*

FIGURE 6.11 *Shiprock in northwestern New Mexico, the remains of a deeply eroded former volcano. The central solid core of the volcano is more resistant than the rock into which it was intruded, as are the vertical dikes (lava-filled cracks) that now form the wall-like ridges seen on the left and right. These resistant masses have been etched into relief by rainwash and stream erosion. Shiprock stands about 487 meters above the surrounding plain. The cone that formerly enclosed it is virtually gone. (Spence Air Photos.)*

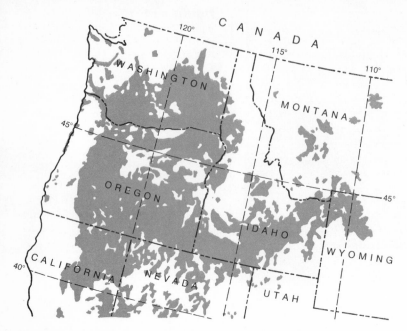

FIGURE 6.12 *Map of the lava fields of northwestern United States. The flows range in age from early to late Cenozoic. Nearly all are basalts or andesites. (After Geologic Map of the United States, U.S. Geological Survey.)*

The rocks from all these areas are characterized by relatively low silica (about 50 percent) and a high content of ferrous iron (8 to 12 percent), similar to the flows of the Hawaiian Islands. They evidently were formed from highly fluid lavas, as single flows can be traced for distances of many tens of kilometers. Individual flows are generally only 7 to 15 meters thick, and seldom more than 100 meters. Dozens of them accumulated, layer upon layer (Fig. 6.13). The geologists who study these lavas called them plateau basalts, not because of their altitude, which varies, but because of their unusual flatness. Some of them, however, have been warped or folded by later earth movements.

The rock that solidified in the fissures which served as feeders for the surface flows can be seen locally on the walls of canyons or where the edges of the flows have been eroded away. Within most of the fields of flood lavas, the fissure sources or dikes are either obscure or hidden.

(a) Laki, Iceland, 1783 A large-scale, well-documented fissure eruption occurred in 1783 at Laki fissure in southern Iceland. Lava welled up through a 32-kilometer rift, and in three main effusions, within 2 months, gushed forth about 12 cubic kilometers of basaltic lava. This rather free-running lava filled the 167-meter canyon of Skapta River nearby, and eventually inundated about 558 square kilometers of lowland. The flows spread out in lobate form 16 to 25 kilometers from the vent, and locally sent lobes still farther down the stream valleys. As the outflow of lava waned, gas pressure, probably induced by partial crystallization underground, was built up until gas-charged lava broke through along the fissure to form a row of about 20 small explosion cones of volcanic cinders atop the lava field. Basaltic ash, estimated at 3 cubic kilometers in volume, settled over the surrounding country, ruined pastures, killed thousands of livestock, and caused famine and the death of more than 10,000 persons.

A similar eruption took place at Eldja, Iceland, about A.D. 950, when about 8 cubic kilometers of lava were extruded from a fissure 29 kilometers long and covered an area of about 700 square kilometers.

(b) Tarawera, New Zealand, 1886 The Tarawera eruption is an example of an explosive eruption from a fissure. The eruption occurred in a flat area of previous silicic lavas broken by numerous fissures. In 1886, basaltic magma rose along a NE-SW trending fissure and broke through to the surface. Within 2 hours a dozen craters were active along its 30-kilometer

length. Where the fissures crossed three small lakes, explosions blasted away the soft lake sediments and previously steam cooked, hence weak, earlier lavas. A new basin was formed 1.6 to 3.2 kilometers wide and 5 kilometers long. The basin later was flooded by a single lake, Lake Rotomahana. One crater in this submerged basin was 3 kilometers long. Elsewhere the craters were about 270 meters wide, 120 meters deep, and 0.4 to 0.8 kilometer long, or in the case of Tarawera Chasm, 1.6 kilometers long. Red and black basaltic cinders, normal volcanic products, blown out of the craters fell over an area of about 10,000 square kilometers in a sheet as much as 75 meters thick near the fissure. The main explosions came within a span of 2 hours and ceased entirely within 24 hours. In the lake area, superheated underground water, suddenly changed into steam, contributed to these violent, though short-lived, blasts. Such explosions are termed phreatic.

Volcanoes

Volcanoes may be classified as effusive, explosive, or variable in behavior. They may take the exterior form of cones, cinder cones, or composite cones and be active, dormant, or extinct. The composition of the lava may be basaltic, andesitic, or rhyolitic.

Effusive eruptions like those on Hawaii build up gently sloping, basaltic lava cones (also called shield cones). Explosions, on the other hand, make cinder cones with slopes of about 30 to 40°, such as Wizard Island in Crater Lake (Fig. 1.2). Volcanoes that emit both lavas and cinders in more or less alternating layers on their slopes make composite stratovolcanoes of intermediate steepness, commonly 20 to 30°. Examples are Vesuvius, Etna, Mayon (Fig. 6.14), Rainier, and Shasta. A few volcanoes develop plug domes, as at Mt. Pelée, Martinique, and Lassen Peak, California.

The crater of a volcano is the bowl- or funnel-shaped vent through which material is ejected (Fig. 6.15). Its size ranges from a few meters to 0.8 kilometer. It may be widened near the top by explosions, by the sliding back of volcanic matter on the rim, by the lava column below the crater melting and dissolving the rock overhead, and by subsidence within the conduit.

The term caldera originated from the description of a huge pit in the Canary Islands, called La Caldera, which is more than 5 kilometers in diameter and is surrounded by cliffs nearly 1,000 meters high. The term is now applied to other large depressions in volcanic areas, some of which are immense. Aniakchak caldera, Alaska, for example, is about 10 kilometers in diameter. The one on Mauna Loa is 6 kilometers long and 3 kilometers wide. The caldera of Kilauea is 3.7 kilome-

FIGURE 6.13 *Exposure of lava flows in central Washington. The layered arrangement in the cliff results from a succession of flows, each 15 to 30 meters thick. Sloping piles of talus have accumulated at the base of the cliff. (State of Washington, Department of Natural Resources.)*

FIGURE 6.14 *Mayon Volcano in the Philippine Islands, 2,320 meters high. Its cone consists of both lava flows and pyroclastic fragments. (Photo by Perret, courtesy of C. A. Reeds.)*

ters long and 2.5 kilometers wide. Both these Hawaiian calderas were formed by collapse of the mountain summits following withdrawal of lava columns. The lava was diverted underground. The Katmai caldera in Alaska was the result of a collapse, in 1912, of the mountain top following internal solution, pumice explosion, and internal avalanching of the crater walls. The caldera of Crater Lake, Oregon, about 8 kilometers in diameter (Fig. 1.2), was formed by the collapse of former Mount Mazama after the expulsion of about 70 cubic kilometers of pumice.

Howel Williams (1941), a geologist who has studied volcanoes in great detail, maintains that with minor exceptions, volcanoes do not "blow their heads off," but rather they become disemboweled and their tops cave in. He bases his belief on the fact that ejecta surrounding most calderas are almost entirely pumice, not older rock fragments,

Of about 600 volcanoes known to have been active during historical time, most are concentrated around the borders of the Pacific Ocean (Fig. 6.16) in areas such as Indonesia, the Philippine Islands, Japan, the Kuril Islands, the Aleutians, Central America, and the Andes. Other concentrations of volcanoes occur in Italy, Greece, Asia Minor, eastern Africa, Arabia, and the West Indies. In these areas the volcanoes occur at sea as well as on land.

A new shield-type volcano, Surtsey, was born in November 1963, on the Mid-Atlantic Ridge southwest of Iceland, with spectacular explosions. By the reaction with seawater it produced showers of fragmental debris and dust-filled steam-born clouds that rose to 16,000 meters, nearly 5 kilometers into the stratosphere at that latitude. Extrusion of lava flows on top began 4 months later and ceased in May 1965. By that time Surtsey had grown to be about 190 meters high, 2 kilometers long, and nearly 2.8 square kilometers in area. A second island of fragmental material, 0.5 kilometer away, grew to 0.2 square kilometer in 4 months in 1965 and was destroyed in 5 months by wave erosion.

A volcanic eruption on January 23, 1973, from the dormant volcano Helgafjell, on the island of Heimaey, recalled to Icelanders the spectacle of the 1963 eruption that created Surtsey. Heimaey is located 20 kilometers south of Iceland and about 10 kilometers northeast of Surtsey. The volcano Helgafjell erupted in the early morning of January 23, after being preceded by small earthquake swarms on January 22 and 23. The largest quake registered 2.7 on the Richter scale of magnitudes. Lava and ash were emitted from the crater, and lava flowed from a 3-kilometer-long rift that opened across the 4-kilometer-wide island. Officials estimated that the entire island was covered by a fine-ash fall, and over one-tenth of the area was buried by lava. The flowing lava caused the abandonment of the town of Vestmannaeyjar and its 5,000 inhabitants. The lava flowing into the harbor area raised the water temperature to 44°C (111°F). On February 19, one side of the cone collapsed, allowing new flows of lava to spread out.

In addition to the volcanoes that protrude above sea level, the Pacific Basin has about 10,000 submarine volcanoes that are at least 1 kilometer high. Some are lined up as if along a fracture zone, some form groups, and others have a random distribution. Most of them never grew high enough to become islands; however, some may have been volcanic islands and are now submerged.

A few other case histories of volcanic eruptions are worth noting.

Mt. Vesuvius, Italy, A.D. 79

Mt. Vesuvius, perhaps the world's most famous volcano, has a long record of eruptions during historical time. This 1,400-meter mountain is situated beside the Bay of Naples about 11 kilometers from the city. It is the successor to a still earlier volcano, Mount Somma, a hollow-topped mountain that had lost its summit in prehistoric time. The top of Mount Somma is now only a ridge that partly encircles the hollow where Mt. Vesuvius was built up by later eruptions.

In A.D. 79, however, Vesuvius, which had been considered extinct, came to life again. It began to tremble with earthquakes and to spout volcanic ash from its summit. Finally, in one great catastrophic eruption, it sent out huge cauliflower-shaped clouds of hot gas and volcanic dust. A black cloud carrying a hail of rock particles descended upon Pompeii, a city of about 20,000 persons, 10 kilometers from the vent. At the time, Pompeii had just been rebuilt after damage from an earthquake in A.D. 63. The hot, dust-laden cloud of gas smothered or suffocated about 16,000 persons almost instantly. Pebbly pumice, hail-like particles called lapilli, and fine volcanic ash continued to settle until the city was buried to a depth of 6 meters. Rains on the ash-covered mountain slopes turned the newly fallen ash into torrential rivers of mud that buried the nearby town of Herculaneum.

In later centuries, Vesuvius has erupted violently about a dozen times. Between these times the people, heedless of danger, resettled on its slopes. One major eruption in 1631 produced lava flows, blasts of hot gases, and another destructive shower of volcanic ash. Some of this ash fell on Constantinople (now Istanbul), 1,300 kilometers away. Other major eruptions occurred at Vesuvius in 1872 and 1906.

Vesuvius has been intermittently active since 1906. In 1913, it built up a small cone of lava and cinders nestled within the previous hollow. In

FIGURE 6.15 *Amboy Crater, San Bernardino County, California, within the summit of a low cinder cone. (Spence Air Photos.)*

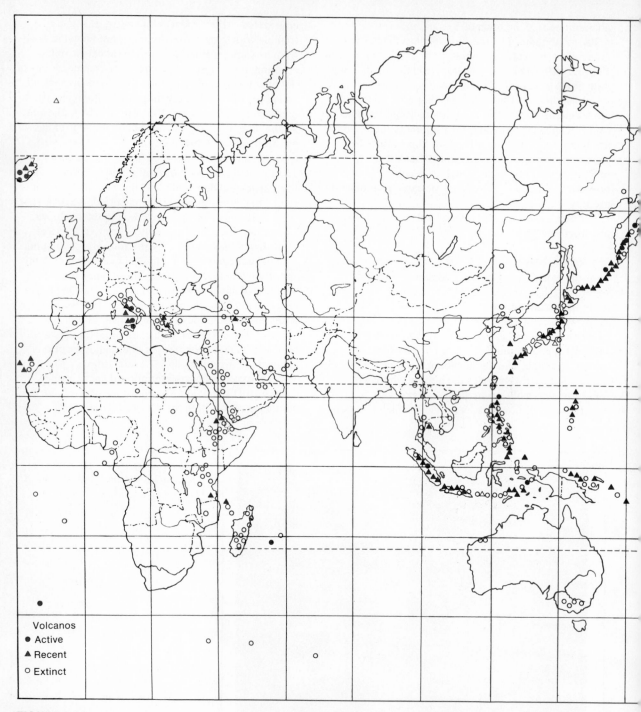

FIGURE 6.16 *The distribution of volcanoes over the earth. (From F. A. Vening Meinesz, 1964.)*

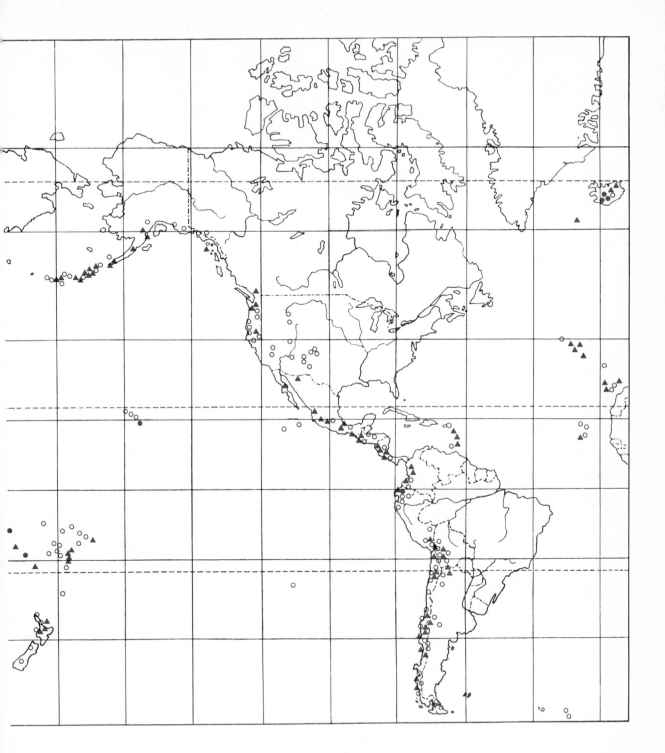

1927-1929, new lava overtopped the inner wall of Mount Somma and descended the outer slopes a distance of 1.6 to 3.2 kilometers. In 1944, lava rose through fissures on the crater floor, filled the bowl, spilled over the rim, and flowed down the mountain side over the village of San Sebastiano. Explosions beginning 2 days later spread dark ash over Naples, and mantled the ruins of Pompeii under a new layer nearly a half meter deep.

Krakatoa, Indonesia, 1883   The Krakatoa group of islands, situated off the west end of Java, Indonesia, was the scene of one of the world's greatest explosive eruptions. Earthquakes began in the area of Rakata, a volcanic island, in 1877 and gradually increased in frequency until May 1883, when Krakatoa began to throw out volcanic dust. The eruptions grew in violence, reaching a climax in August 1883.

A series of enormous explosions lasting 3 days blasted away part of the island and belched forth a huge volume of pumice. Clouds filled with stifling volcanic dust blacked out Jakarta, 160 kilometers away, in mid-day, and turned into mud-rain as condensation occurred. Fine volcanic dust was shot upward tens of kilometers into the stratosphere. It spread completely around the globe in about 2 weeks. Judged by the unsorted, unstratified layer of glassy pumice 67 meters thick spread over the summit of Rakata Island, the total quantity expelled was nearly 20 cubic kilometers. The bulk of the pumice fell to earth within a 16 kilometer radius. The sound of the explosions was heard in Australia, 5,000 kilometers away. Seismic sea waves dashed to heights of 40 meters against nearby islands.

As the underground reservoir of gas-charged lava was exhausted by these outbursts, the surface rapidly collapsed, leaving a basin about 6 kilometers wide, which in 1883 reached 230 meters, and in 1923, 300 meters below sea level. Engulfment in spasmodic stages and the resultant explosions probably accounted for the exceptional sea waves. Between 1927 and 1933, renewed activity produced a cinder cone, Anak Krakatoa, on the basin floor. On July 21, 1972, a ship's captain, whose vessel was in the Sunda Strait, observed an eruption of Anak Rakata. The eruptions lasted for periods of 20 to 30 seconds at intervals of 2 to 7 minutes. Hot ashes and lava were thrown about 250 meters into the air, causing the sky to take on a reddish tinge. This was just one of a series of eruptions that occurred during 1972 (Table 6.1).

Mt. Pelée, Martinique, 1902   Mt. Pelée, situated on Martinique, an island of the Lesser Antilles volcanic arc in the Caribbean Sea, had a disastrous eruption in 1902. Moderate eruptions had occurred previously in 1792 and 1851. In the spring of 1902, extremely viscous lava welled up in the crater and crusted over in the form of a massive dome that plugged the mouth of the conduit. Superheated steam, pent up beneath this seal, periodically broke out through cracks beneath the lid and sent puffs of hot dust-laden nuées ardentes. Nuées ardentes are density currents heavier than air (despite their heat). They moved rapidly down the mountain side in a valley to the west. On May 8, however, an exceptionally large blast projected a huge nuée ardente southward toward St. Pierre, a city of 28,000 persons. This hot dusty cloud, with a temperature estimated at 600°C, swept over the city with the violent fury of a firestorm, asphyxiated all the inhabitants (except one prisoner in a dungeon) within a few minutes, burned the city, and destroyed many ships lying at anchor offshore.

In October 1902, the plug was pushed up and broken into fragments, but a central spinelike column of rock slowly rose from the crater until, by the spring of 1903, it stood nearly 330 meters high. Its smooth sides, furrowed by vertical grooves, seemed to indicate plastic molding of nearly solid, pasty lava as the spine was extruded. Within a year it disintegrated into a pile of rubble.

Minor eruptions of Mt. Pelée recurred from 1929 to 1933. At times small plug domes and even short spines were formed, followed by occurrences of steam outbursts and nuées ardentes.

Parícutin, Mexico, 1943-1952   Parícutin Volcano, Mexico, had a spectacular 9-year history. Its history began one afternoon. It began as a hot spot in a cornfield, building up a 9-meter cinder cone by the next morning. By the end of the first week the cone had enlarged to 110 meters. It grew to 325 meters in the first year. Ultimately, the cone covered an elliptical base about 600 by 900 meters, reached a height of 420 meters, and had a crater 270 meters in diameter. Lava flows (Fig. 6.17) emerged from the base of the cone on several occasions and spread over an area of about 2.3 square kilometers. Dark-colored volcanic ash was distributed widely. The total volume of cinders, lava, and ash was about 0.8 cubic kilometer. In measurements made in 1945, steam was found to

FIGURE 6.17 *Advancing front of a lava flow at Parícutin Volcano, Mexico. The broken crust of the flow is pushed forward by the still-fluid lava in the interior. It topples and rolls down the front, as if shoved by a gigantic bulldozer, and is covered up as the front advances. Air-laid volcanic ash covers the foreground. (W. F. Foshag.)*

constitute about 1 percent of the total weight expelled. Geologists recorded its history virtually from birth to death.

Mauna Loa and Kilauea, Hawaii, 1919-   The spectacular eruptions of the Hawaiian volcanoes have provided new insights into volcanoes. For a century Kilauea was famous for its boiling "lake" of lava, Halemaumau (Fig. 6.18), nestled in a summit caldera 2 by 4 kilometers wide and hundreds of meters deep,

varying according to the level of the liquid lava within it. In 1924, the lava lake disappeared underground, and Kilauea had an unusual explosive eruption. The Hawaiian Volcano Observatory, established on the rim of Kilauea in 1912, has gathered voluminous data on Kilauea and its huge neighbor, Mauna Loa. Kilauea and Mauna Loa are the only two active centers among the five lava cones that make up the island of Hawaii.

Mauna Loa stands 4,500 meters above sea level and about 5,300 meters more above the sea floor

FIGURE 6.18 *Small circular lava lake, about 15 meters wide, at the foot of the wall of Halemaumau Crater, September 1920. The pool was kept liquid by superheated steam rising from fissures beneath it. (T. A. Jaggar, Hawaii Volcanic Observatory.)*

around its base, and so it is, in all, about 10,000 meters high. Its volume is more than 21×10^3 cubic kilometers, the largest bulk of any volcano. It has a broad dome-shaped summit and gently sloping sides, underlain by thin flows of basaltic lava. The lava ran freely on gentle slopes before solidifying. Much of the lava issued from fissures in rift zones, one on the southwest and another on the east-northeast.

The summit of Mauna Loa contains an elongate caldera, about 6 kilometers long and 1.6 to 3.2 kilometers wide, with walls locally 200 meters high. Several subparallel fissures traverse the caldera floor lengthwise, in which fresh lava rises at times to the caldera level, and occasionally overflows. Lava shoots up in fountains hundreds of meters high. More often the lava breaks out through rifts at lower levels and runs downhill in long slender streams. In 1919 and 1926, flows that emerged near the 2,500-meter level, reached the sea 2 weeks later, a distance of 20 kilometers.

Kilauea is a small shield-shaped lava cone, about 1,350 meters above sea level, low on the eastern slope of Mauna Loa, whose summit is 33 kilometers distant. Its caldera is a hollow about 2.4 by 4 kilometers in diameter and about 100 meters deep. The recurrently active lava pit of Halemaumau in the caldera floor is a little more than 0.8 kilometer in diameter (Fig. 6.18).

Kilauea shows a cyclic round of (1) slow rise of the underground lava column, (2) slight doming of the central part of the solid lava shield, (3) rapid sinking of the lava column by its escape into dikes (crack fillings) or flank eruptions, and (4) rapid settling of the dome. One such cycle occurred in 1924 and another in 1959–1960. New outbreaks on the rift zone of vents in 1969–1971 covered about 45 square kilometers and reached the sea repeatedly.

Fumaroles and Solfataras

Fumaroles are vents from which steam and other gases escape. They are common near both active and decadent volcanoes. Their temperatures range from the boiling point of water to as high as 650°C. Although most of the steam comes from heated groundwater of surface origin, the presence of other constituents indicates a magmatic source for some of the gases.

Solfataras are fumaroles that give off sulfur gases. Hydrogen sulfide, a common product, tends to oxidize on exposure to air to form water and native sulfur, and so rocks near solfataras often contain commercial quantities of sulfur. A well-known vent that yields sulfur compounds is La Solfatara, west of Naples; here, as in Sicily, Mexico, and Japan, volcanic emanations are worked commercially for sulfur.

Lakes of Volcanic Regions

Often as a result of volcanic activity, a closed depression on the land surface will form, within which ground or surface water may collect to form a lake.

Depressions in volcanic summits often hold lakes of water after the lavas have cooled and eruptions have ceased. Occasionally, the upper portion of a volcanic cone is destroyed by collapse and explosion, leaving a large basin called a caldera. Crater Lake (Fig. 1.2), about 2,000 meters above sea level, occupies the caldera of Mt. Mazama in southwestern Oregon. About 8 kilometers in diameter, it is nearly circular, without bays or promontories. The cliffs of dark volcanic rock encircling the lake rise precipitously to heights 200 to 700 meters above the lake. There are no streams entering the lake and no visible outlet. Precipitation in this region exceeds evaporation. The excess water must escape by seeping through the walls and floor of the caldera, possibly feeding several small streams in this region that rise from springs on the flanks of the mountain. The greatest depth measured 589 meters, making this the sixth deepest lake in the world, topped in the Western Hemisphere only by Great Slave Lake, Canada, which is about 26 meters deeper. Two small volcanic cones, one above water (Wizard Island) and one submerged, occur in the lake proper.

Lava flows may form dams and surface irregularities may form lake basins. Yellowstone Lake in Yellowstone National Park is an example of a lake formed by lava flows.

INTRUSIVE ACTIVITY

Field observations show that the great bulk of magma, nine-tenths or more, never reaches the surface as extrusive activity but solidifies underground instead. These original subsurface masses are intrusive in that they invade and cool within other rocks. Nearly all intrusions are revealed to us as a result of the removal by long-continued erosion of the rocks that formerly covered them. Intrusions may be emplaced at compara-

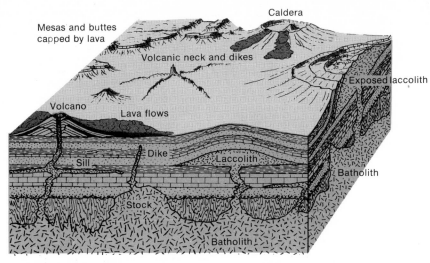

Mesas and buttes capped by lava

Caldera

Volcanic neck and dikes

Exposed laccolith

Volcano

Lava flows

Dike

Sill

Laccolith

Batholith

Stock

Batholith

FIGURE 6.19 *Diagram showing structural relations of various intrusive and extrusive igneous masses. (After F. P. Young.)*

tively shallow depths in the earth or as very deep plutonic masses.

Shallow Intrusions

Shallow intrusions adapt their shapes to the structures of the associated, or "country," rocks that they invade (Fig. 6.19). They have several forms, but the most common ones are dikes, sills, and volcanic necks. Many intrusions merely fill cracks discordant to the surrounding structures and become dikes (Fig. 6.20). Others squeeze their way concordantly between previous layers of rock and form sills. The magma that has solidified within the conduits of volcanoes, from the crater downward, makes a volcanic neck, or plug.

These intrusions vary considerably in size. Dikes range in width from 1 centimeter to tens (or rarely hundreds) of meters and in length to as much as tens of kilometers. Swarms of dikes follow parallel, intersecting, or radial fractures. Sills ordinarily are a few meters

FIGURE 6.20 *A light-colored dike with offshoots, intruded along rock fractures in Cornwall, England. The main part of the dike is 30 centimeters thick. (Geological Survey and Museum, London.)*

to over 100 meters thick. They extend through areas ranging from a small fraction of a square kilometer to hundreds of square kilometers. Necks are generally 0.1 to 0.5 kilometer in diameter or, exceptionally, 2 kilometers.

Batholiths

The deep-seated, or plutonic, masses are batholiths, large bodies of coarse-grained intrusive igneous rocks that occur in the cores of mountain ranges or in the erosional stumps of former mountain systems. By definition, their outcropping surface areas must exceed 100 square kilometers; similar, but smaller, masses are stocks.

Batholiths tend to be quite large. A series of them, for example, extends interruptedly over the full length of the Andes from Cape Horn to Panama, a distance of nearly 8,000 kilometers. Three units in this series, one in Peru and two in Chile, are continuously exposed for lengths of more than 1,300 kilometers each. The Coast Range batholith of British Columbia is about 2,000 kilometers long and 130 to 200 kilometers wide. The batholiths of central Idaho, the Sierra Nevada, and Baja, Mexico, are each about 700 kilometers long and 100 to 130 kilometers wide (Fig. 6.21).

Earthquake data indicate that batholithic mountains have roots extending several kilometers downward, and so their great depths and large surface areas, taken together, indicate enormous volumes. They were formerly thought to be bottomless, but geophysical evidence (mainly gravity and earthquake-transmission studies) indicate that they are rather shallow pods. Their tops are broad domes. Some exposed stocks may be protuberances on top of otherwise concealed batholiths. Offshoots in the form of dikes are common around the margins of batholiths.

Elongate batholiths in mountain cores tend to parallel the mountain structures, as in the British Columbia Coast Range or the Andes Mountains. In detail, however, many of them cut across folds and faults, and so they were emplaced somewhat later than most of the orogenic movements that made the folds and faults.

The positions of batholiths within areas of crustal deformation suggest that they have a genetic relation

FIGURE 6.21 *Batholithic intrusives of western North America and southern South America, intruded along the axes of ancient mountain ranges and subsequently unroofed by long-continued erosion. (After Geological Society of America Maps.)*

to accumulations of sediments and subsequent mountainmaking. Sediments and, in some instances, extrusive igneous rocks have accumulated all together in huge basins up to thicknesses of 13 kilometers. The strata in the basins eventually undergo horizontal compression and vertical uplift to form mountains into the cores of which batholiths are intruded. According to one widely held view, batholiths are formed by partial or complete melting of the silicic fractions of a thick sediment-filling basin at great depth. This interpretation of the Sierra Nevada is advocated by Bateman and Eaton (1967) and shared by Hamilton and Myers (1967).

An alternative view for the Sierra Nevada is proposed by Kistler, Everden, and Shaw (1971), who argue: (1) the site was an area of long-continued erosion, not sedimentation, (2) emplacement of granitic rocks took place in five major epochs at 30-million-year intervals, each epoch lasting 10 to 20 million years, (3) each epoch of intrusion corresponded with orogeny and regression of epicontinental seas, (4)

strontium isotope data show that the magma came not from crustal material but rather from deep levels in the earth, (5) the linear heat source also was deep-seated, as in mid-oceanic ridges, perhaps concentrated by solid earth tides, and (6) the intrusions were emplaced not at depth but at shallow levels, even 5 kilometers or less. With such a division of opinion, the problem remains for further study.

Batholiths consist almost entirely of granodiorite or granite, rocks which compare closely chemically to the average composition of sediments from which at least some granites have been transformed. Large batholiths are generally composite and, as a rule, are highest in silica and alkali (Na and K) feldspars, because of magmatic differentiation. Perhaps light material is also squeezed out from depths of tens of kilometers to add new material to certain batholiths.

The ages of batholiths range from early Precambrian to Miocene, though still younger ones, not yet old enough to have been uncovered by erosion, may exist.

REFERENCES

Barth, T. F. W., 1950, *Volcanic Geology, Hot Springs, and Geysers of Iceland,* Carnegie Institution of Washington Publication 587.

Bateman, P. C., and Eaton, J. P., 1967, *Sierra Nevada batholith,* Science, vol. 158, pp. 1407-1417.

Bullard, F.M., 1947, *The story of el Parícutin,* Science Monthly, vol. 65, pp. 357-371.

———, 1947, *Studies on Parícutin, Michoacan, Mexico,* Geological Society of American Bulletin, vol. 58, pp. 433-450.

———, 1962, *Volcanoes: in History, in Theory, in Eruption,* University of Texas Press, Austin.

Cotton, C. A., 1944, *Volcanoes as Landscape Forms,* Whitcombe and Tombs, Ltd., Wellington, New Zealand.

Darton, N. H., and O'Harra, C. C., 1907, *Description of the Devil's Tower Quadrangle, Wyoming,* United States Geological Survey Folio 150.

Dott, R. H., Jr., and Batten, R. L., 1971, *Evolution of the Earth,* McGraw-Hill Book Company, New York.

Eaton, J. P., and Murata, K. J., 1960, *How volcanoes grow,* Science, vol. 132, pp. 925-938.

Foshag, W. F., 1954, *The life and death of a volcano,* Geological Magazine, vol. 27, pp. 159-168.

———and González, R. Jenaro, 1956, *Birth and development of Parícutin Volcano, Mexico,* United States Geological Survey Bulletin 965-D, pp. 355-489.

Griggs, R. F., 1922, *The Valley of Ten Thousand Smokes,* National Geographic Society, Washington, D. C.

Hamilton, W., and Myers, W. B., 1967, *The Nature of Batholiths,* United States Geological Survey Professional Paper 554.

Herbert, D., and Bardossi, F., 1968, *Kilauea: Case History of a Volcano,* Harper & Row, Publishers, Incorporated, New York.

Johnson, D. W., 1907, *Volcanic necks of the Mount Taylor region, New Mexico,* Geological Society of America Bulletin, vol. 18, pp. 303-324.

Kistler, R. W., Everden, J. F., and Shaw, H. R., 1971, *Sierra Nevada plutonic cycle: Part I, Origin of composite granitic batholiths,* Geological Society of America Bulletin, vol. 82, pp. 853-868.

Leet, L. D., 1948, *Causes of Catastrophe*, McGraw-Hill Book Company, New York.

Maiuri, A., 1961, *The last moments of the Pompeiians*, National Geographic, vol. 120, pp. 651-669.

MacDonald, G.A., 1961, *Volcanology,* Science, vol. 133, pp. 673-679.

————, 1972, *Volcanoes*, Prentice-Hall, Inc., Englewood Cliffs, N. J.

———— and Abbott, A. T., 1970, *Volcanoes in the Sea—The Geology of Hawaii*, University of Hawaii Press, Honolulu.

McKee, B., 1972, *Cascadia: The Geologic Evolution of the Pacific Northwest*, McGraw-Hill Book Company, New York.

Menard, H. W., 1964, *Marine Geology of the Pacific*, McGraw-Hill Book Company, New York.

Muller, E. H., Juhle, W., and Coulter, H. W., 1954, *Current volcanic activity in Katmai National Monument*, Science, vol. 119, pp. 319-321.

Ollier, C., 1969, *Volcanoes,* The M. I. T. Press, Cambridge, Mass.

Perret, F. A., 1935, *The Eruption of Mt. Pele'e 1929–1932,* Carnegie Institution of Washington Publication 458.

Rittman, A., 1962, *Volcanoes and Their Activity*, 2d ed., John Wiley & Sons, Inc., New York.

Williams, H., 1941, *Calderas and their origin*, California University Department Geological Publications Science Bulletin, vol, 25, pp. 239-346.

————, 1941, *Crater Lake: The Story of Its Origin*, University of California Press, Berkeley.

Having already considered the forms magma or lava assume after cooling, we should consider in more detail the igneous rocks that comprise those forms.

Igneous rocks are classified by their texture (grain size or grain-to-grain relationship) and their mineral composition. When magma crystallizes, there is great diversity in the types of rocks created, because of the large number of variables involved. However, using the following points, it is possible to classify this potentially large number into four genetic types (Table 7.1):

1. Magmas can be mafic (high in Ca^{++}, Mg^{++}, and Fe^{++}), with associated amounts of Al^{3+}, silicon, and oxygen, or they can be siliceous, having K^+, Na^+, Al^{3+}, minor Fe^{++} and Mg^{++}, and higher amounts of silicon and oxygen. The terms basic and acidic, respectively, are used interchangeably with the terms mafic and siliceous.
2. The basic, or mafic, rocks are generally darker in color and have a higher specific gravity. Acidic rocks are lighter in color with a lower specific gravity.
3. The texture of all igneous rocks is dependent upon the rate of cooling: fine-grained, glassy, or pyroclastic rocks are indicative of fast cooling, while coarse-grained rocks are indicative of slow cooling.
4. Gradations between the extremes, or end members, presented in 1, 2, and 3 result in a great diversity of igneous rocks. In Table 7.1 a simplified classification of igneous rocks is presented. As you examine the table, notice what characteristics are most important in classifying igneous rocks. Classification of any set of things is difficult. Usually, the earliest schemes are modified, in some cases greatly, as additional information is obtained. This is true with the igneous rock classification scheme.

CAUSES OF DIVERSITY IN IGNEOUS ROCKS

The diversity in the chemical and mineralogical composition of igneous rocks can be explained by one of two theories: the product of original differences in

TABLE 7.1 Classification of Igneous Rocks

Geologic occurrence					Rock family	Chemical composition	
Extrusive			Intrusive				
Volcanic explosion debris	Lava flows and hot siliceous clouds	Lava flows	Small, shallow masses	Medium to large, deep-seated masses		Variation in element abundance	Average percent silica
	Very uncommon			Dunite Peridotite Pyroxenite	Ultrabasic (nearly all dark minerals)		41
Tuff (lithified ash) Volcanic breccia	Obsidian (glass) Pumice	Basalt	Intermediate textural and chemical-mineral varieties	Gabbro	Basic (dark minerals dominant)	Increase in Si, Al, Na, and K ↓ Decrease in Fe, Mg, and Ca ↓	48
		Andesite		Diorite	Intermediate (25-50% dark minerals)		57
				Granodiorite	Acidic (less than 25% dark minerals)		66
		Rhyolite		Granite	Acidic		72
Fragmental Pyroclastic	Glassy	Fine-grained texture	Porphyritic	Coarse-grained texture			

magmas; the subsequent changes in magmas by processes operating underground.

Primary Magma Differences

In support of original differences in magmas is the widespread occurrence of basaltic lava flows and great batholitic masses of granite or granodiorite. Many geologists interpret basalt to be the product of a primary magma derived by melting of the basaltic substratum or by selective fusion or partial melting of ultrabasic material at still greater depths. They interpret granite and granodiorite to be the products of magma derived by selective melting of the siliceous crust or, more specifically, fusion of siliceous sediments buried to depths of 13 or 16 kilometers.

Changes in Magma

The principal changes in composition that magmas undergo include reaction, differentiation, and mixing.

Reaction Series The reaction principle, as developed by Bowen (1928), involves reactions between crystals and fluid magma (Table 7.2). The ferromagnesian minerals constitute a discontinuous reaction series, beginning with olivine and ending with biotite. Olivine crystallizes early from a magnesium-iron-rich magma. As the remaining liquid magma cools, the liquid reacts with the olivine crystals to make pyroxenes, with the pyroxenes to make amphiboles, and with the amphiboles to make biotite. If the reactions are incomplete, remnants of olivine, pyroxene, or amphibole will be found in the cores of the later crystals which form. The plagioclase feldspars form a continuous reaction series from calcium-rich anorthite to sodium-rich albite. The change in this plagioclase series takes place gradually, rather than by discontinuous steps. Incomplete reactions form what are known as "zoned" plagioclase crystals. Such a crystal may have andesine on the inside surrounded by a rim of oligoclase. The discontinuous and continuous series are shown in Table 7.2.

The minerals opposite each other in the series tend to crystallize at nearly the same time; hence, the common association of pyroxene and labradorite in basalt, or of amphibole and andesine in andesite, and of biotite and oligoclase or albite. In 1962 T. Barth, a famous Norwegian igneous petrologist, added a third series—a continuous reaction series involving the potassium feldspars and the plagioclase feldspars.

When crystallization nears completion, quartz also may crystallize. The final residue of steam or hot-water solutions may continue to react with the minerals to form other minerals like chlorite, muscovite, serpentine, and zeolite, or these solutions may replace earlier rocks, or enter cracks to form pegmatite dikes, quartz veins, and ore deposits.

Differentiation Magmatic differentiation is the sum of the processes by which a single homogeneous magma separates into fractions to form different types of igneous rock. A discussion of one of these processes, crystallization, will serve to illustrate the main process of differentiation. Fractional crystallization is the separation of magma into two parts, crystals and liquid. Early crystallization of olivine, pyroxene, and labradorite from a basaltic magma, for example, leaves the liquid fraction poorer in magnesium and calcium but enriched in silica, iron, and sodium. This remaining liquid would have the approximate composition of an andesite. If it were separated from the crystal fraction by gravity or earth movements, it would behave as an andesitic magma. A separation of minerals lower in the reaction series would change its composition further, ending as rhyolite magma. A few intermediate to silicic lavas may have formed from an originally basaltic magma in this way. Basaltic magma, however, does not contain enough alkali, especially potassium, to form much rhyolite.

Mixing of Magmas The mixing of two magmas that are at different stages of crystallization results in an assemblage of minerals out of equilibrium. Pumice, found in the Valley of Ten Thousand Smokes, Alaska, is a mixture of white rhyolite pumice and dark andesitic scoria derived from two effervescent magmas expelled together.

In summary, whatever may be their origin, magmas are subject to reactions, to multiple processes of differentiation, and possibly to assimilation of foreign material or to mixing. They finally generate a wide variety of igneous rocks.

Textures

The texture of an igneous rock refers to the size and shape of the mineral grains or crystals, if any, and to the pattern of their arrangement. The main textures of igneous rocks are: coarse grained, fine grained, porphyritic, pegmatitic, pyroclastic, amygdaloidal, and glassy.

Coarse Grained Coarse grained igneous rocks result from slow cooling of magma at considerable depth, where the crystals have time to grow to such size that they are easily seen and identifiable in hand specimens without the aid of a microscope. A steam-rich magma or one high in fluids helps to make large crystals more readily than a relatively stiff dry one.

The most common coarse-grained rock is granite (Fig. 7.1). Granite normally is composed of potash feldspar (orthoclase or microcline), sodic

TABLE 7.2 **Binary Reaction Series**

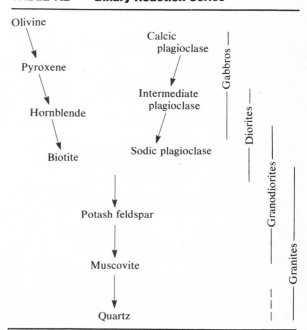

Modified from Bowen, N. L., 1956, *The Evolution of the Igneous Rocks,* Dover Publications, Inc., New York.

plagioclase, and quartz, together with minor quantities of hornblende or mica, or both. It is the coarse-grained equivalent of rhyolite. The average granite contains about 60 percent feldspar, 30 percent quartz, and 10 percent dark minerals. Many granites have a pinkish color from an abundance of pink or reddish feldspar. Others are salt-and-pepper gray from a combination of a white feldspar background with a sprinkling of dark biotite or hornblende. The quartz found in granite is usually colorless. Granite is widely used for building stone, monuments, decorative stone, and crushed rock.

In addition to granite, there are many other types of coarse-grained igneous rocks. A coarse-grained combination of intermediate plagioclase and hornblende is a diorite. Granodiorite consists of quartz, intermediate plagioclase, and a little orthoclase, together with hornblende, pyroxene, or mica as mafic constituents. It has a mineral and chemical composition between diorite and granite. Calcic plagioclase (usually labradorite), pyroxene (usually augite), and olivine, plus a few accessory minerals, like magnetite and apatite, make up a dark-colored rock called gabbro (Table 7.1). Other coarse-grained, dark-colored igneous rocks may be composed almost entirely of pyroxene, hornblende, or olivine. Rocks of this type are relatively rare, however.

Fine Grained Fine-grained igneous rocks are crystalline, but the minerals are too small in hand specimens to be recognized by the unaided eye. They form by cooling at moderately rapid rates, as in lava flows, thin sills, or long narrow dikes emplaced at shallow depths. The tops of many fine-grained lava flows are also scoriaceous and may be partly glassy as well.

The most abundant fine-grained igneous rock is basalt (Fig.7.2). Basalt, which is dark gray in color, is composed mainly of calcic plagioclase (labradorite), pyroxene, and sometimes olivine. Basalt is a resistant rock, and consequently, it is used extensively as crushed rock.

Other fine-grained rocks differ from basalt in mineral and chemical composition. One composed of about equal quantities of intermediate plagioclase (andesine) and hornblende is andesite (named for its occurrence in the Andes). A light-colored, fine-grained igneous rock with a high feldspar and quartz content is termed a rhyolite (Table 7.1).

Porphyritic A porphyritic texture consists of large crystals, called phenocrysts, set in a matrix of smaller crystals or of glass (Fig. 7.3). Porphyritic texture indicates two stages of cooling and crystallization: one in which fairly large crystals had begun to grow in the

FIGURE 7.1 *Granite, a coarsely crystalline, silicic (or acidic), igneous rock formed deep in the earth's crust, where extremely slow cooling favored crystal growth. The light-colored minerals are feldspar and quartz, the dark ones hornblende and biotite.*

FIGURE 7.2 *A field of basalt, a dark-colored, fine-grained, extrusive igneous rock formed by the solidification of basic lava. This rigid and blocky flow of lava issued from Belnap Crater (out of view on the right) on the crest of the Cascade Mountains of Oregon. In the center, this flow surrounds an earlier hill. Because this lava flow overlies glaciated surfaces, we know that it is of a geologically recent age and may be only a few thousand years old. (Oregon State Highway Department.)*

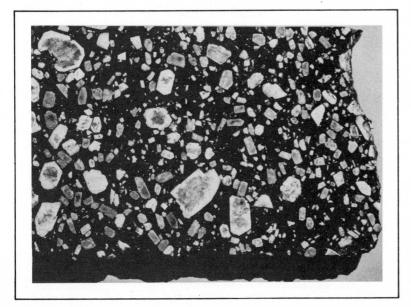

FIGURE 7.3 *Trachyte porphyry from Bannockburn Township, Ontario. Large crystals of feldspar are set in a very-fine-grained crystalline matrix. The two sizes of crystals were formed under different conditions and at different times. The phenocrysts (large crystals) were formed first in a deep-seated magma. Later, the matrix crystallized after this magma was extruded as a lava flow. (Ward's Natural Science Establishment, Inc.)*

FIGURE 7.4 *Scoria, a form of solidified frothy lava, showing numerous cavities formed by expansion of gas cavities in the lava before cooling. (Ward's Natural Science Establishment, Inc.)*

magma, followed by another during which only small crystals or none at all had time to form. Such a two-stage history results from movement of partly crystalline magma from a position of slow cooling at depth to one of rapid cooling in a near-surface or fully exposed position.

Pegmatitic A pegmatitic texture is formed by crystallization from fluids exceptionally rich in volatile materials. In such fluids, crystals can grow to lengths of a few centimeters, a few meters, or in certain instances tens of meters. The most common minerals found in pegmatites are orthoclase, quartz, and micas, but complex compounds of lithium, boron, fluorine, tantalum, columbium, zirconium, phosphorus, and thorium also occur as minerals in these unusual rocks. Pegmatites, which usually are in dikes, are commercial sources of feldspar, muscovite, gems (tourmaline, topaz, beryl, and zircon), and rare earth elements, as well as excellent mineral specimens. Rock or mineral names are commonly used as prefixes to describe pegmatites, (i.e., granite pegmatite or quartz-muscovite pegmatite). In some pegmatites, called graphic granite, the quartz and feldspar have a curious intergrowth resembling ancient cuneiform writing.

FIGURE 7.5 *Obsidian, a glassy, extrusive igneous rock. This type of natural glass occurs when lava has cooled too rapidly to allow crystals to form. (Ward's Natural Science Establishment, Inc.)*

Pyroclastic   Pyroclastic (fire-broken) textures are formed by explosive fragmentation. The shattered particles generally consist mostly of pumiceous or scoriaceous glass (Fig. 7.4). They range in size from huge blocks, bombs, cinders, lapilli, and volcanic sand, to volcanic ash or dust.

Pyroclastic pumice is used as an abrasive in scouring powders and as aggregate in light-weight concrete blocks. In Iceland, a large percentage of the road network is constructed from cinders.

Amygdaloidal   Amygdaloidal texture results where gas cavities are filled with mineral matter in scoriaceous lava. The minerals that most commonly occur as fillings in the cavities in the basalt are quartz, chalcedony, calcite, zeolites, and other minerals deposited from solutions. The solutions from which the amygdules are formed are generally from the lava itself.

Glassy   Glassy texture is the product of sudden chilling of lava, either in lava flows or in pieces exploded into the air, so that the atoms within the mass have no chance to organize themselves into crystalline minerals. Glass, having a random, noncrystalline atomic structure, is essentially a supercooled liquid. A common glassy igneous rock is obsidian (Fig. 7.5). When inflated naturally by internally expanding gases prior to solidification, volcanic glass becomes vesicular, either pumiceous (from numerous small cavities) or scoriaceous (from larger ones).

FIGURE 7.6   *Devil's Postpile near Mammoth Lakes, California. This igneous rock shows columnar jointing, the result of shrinkage during cooling. The individual columns are 18 meters long and up to 1 meter thick. (Frashers, Inc.)*

Structures

The structure of igneous rocks refers to their gross features, such as form, occurrence, columnar jointing, and flow banding, and to the fabric of their components. Columnar joints (Fig. 7.6) are present in many tabular flows, sills, and dikes, where shrinkage during cooling developed cracks normal to the cooling surfaces. In flat sills the joints are upright, palisadelike, and in vertical dikes they are horizontal. Lava flows commonly show banding or flow structure because they drag out gas vesicles, prismatic or needlelike crystals, and different types or colors of material into a streamline form as they move forward.

REFERENCES

Barth, T. F. W., 1962, *Theoretical Petrology*, 2d ed., John Wiley & Sons, Inc., New York.
Bowen, N. L., 1956, *The Evolution of the Igneous Rocks*, Dover Publications, Inc., New York.

Fenton, C. L., and Fenton, M. A., 1951, *Rocks and Their Stories*, Doubleday & Company, Inc., Garden City, N.Y.

Harker, Alfred, 1954, *Petrology for Students*, 8th ed., Cambridge University Press, London.

Pirrson, L. V., and Knopf, A., 1947, *Rocks and Rock Minerals*, 3d ed., John Wiley & Sons, Inc., New York.

Pough, F. H., 1955, *A Field Guide to Rocks and Minerals*, 2d ed., Houghton Mifflin Company, Boston.

Shand, S. J., 1951, *Eruptive Rocks*, 4th ed., T. Murby & Company, London.

Turner, F. J., and Verhoogen, J., 1960, *Igneous and Metamorphic Petrology*, 2d ed., McGraw-Hill Book Company, New York.

Wahlstrom, E. E., 1950, *Introduction to Theoretical Igneous Petrology*, John Wiley & Sons, Inc., New York.

Williams, H., Turner, F. J., and Gilbert, C. M., 1954, *Petrography*, W. H. Freeman and Company, San Francisco.

CHAPTER 8
METAMORPHISM AND METAMORPHIC ROCKS

Metamorphism may be defined as the physical and chemical alterations of solid rocks that transform their textures, structures, and mineral composition in response to changes in their environmental conditions within the earth. Changes resulting from surface weathering are excluded.

Each mineral and rock is the product of a particular environment, with a characteristic temperature, pressure, and fluid content. Many that are stable under one set of conditions become unstable under another set. The clay minerals of sedimentary rocks, formed by weathering at the earth's surface, are unstable when buried at great depth or subjected to compressive earth movements. Even the high-temperature minerals of igneous rocks respond to shearing stresses and to hot fluids. On the other hand, certain minerals, notably quartz, are stable under a wide variety of conditions.

Metamorphic rocks, then, are pre-existing rocks that have been changed in texture, structure, and/or mineral composition by physical and chemical processes (Table 8.1).

The common physical changes include: (1) crushing of grains, (2) recrystallization, (3) interlocking of grains, (4) increase in grain size, and (5) development of a parallel alignment of elongate or flat grains. The final texture is determined by the nature of the original material, the type of metamorphic process involved, and the intensity of the process.

Chemical changes involve the growth of a new mineral assemblage either by recrystallization of material already present or by the addition or removal of certain chemical compounds that are transported as ions in gases or liquids.

CAUSES OF METAMORPHISM

The immediate causes of metamorphism are pressure, heat, and circulating fluids. The pressure is either vertical rock load, caused by the weight of overlying rock, or the hydrostatic pressure of intergranular fluids. Both types of pressure increase with depth. Another type of pressure force is the dynamic, or unbalanced tangential, pressure associated with diastrophic earth movements. The heat may be earth heat, associated with the thermal gradient that is a heat increase of 20°C per kilometer of depth; the local heat of friction, as along the surface of a fault; the heat of magmatic intrusions; or the heat introduced by hot fluids. The ac-

TABLE 8.1   Classification of Metamorphic Rocks

		Rock name	Composition	Parent rock	Metamorphic process
Texture	Foliated	Slate	Abundance of dark, flaky and/or prismatic silicate minerals (micas, chlorite, talc, serpentine, hornblende); quartz	Shale; tuff	Increase regional ↓
		Phyllite		Shale; tuff	
		Schist (var. mica schist, chlorite schist, amphibole schist)		Shale; intermediate to mafic igneous rocks	
		Gneiss (var. garnet gneiss, granite gneiss)	Feldspar abundant; varying amounts of quartz and dark silicate minerals (such as amphiboles, pyroxenes, micas, and garnet)	Acidic to intermediate igneous rocks; arkose; graywacke; mica schist	Regional
	Nonfoliated	Metaquartzite	Quartz greatly predominant	Normal and quartzose sandstones	Regional or contact
		Marble	Calcite and/or dolomite, with or without Ca-Mg silicates	Limestone or dolomite, with or without impurities	Regional or contact
		Hornfels	Dark silicate minerals predominant	Shale; slate; intermediate to mafic extrusive rocks	Contact
		Anthracite coal	92–98% carbon	Peat, lignite, coal	Regional or contact

tive fluids include hot gases, especially steam and carbon dioxide, and hot solutions containing the ions of common elements such as sodium and calcium, as well as fluorine, boron, and sulfur compounds.

Various combinations of pressure, heat, and fluids allow a classification of specific metamorphic processes, or types.

Types of Metamorphism

Metamorphism may be geothermal, hydrothermal, contact, or dynamic (Table 8.2). A rock may be affected simultaneously or in succession and to varying degrees by more than one type of change. Its metamorphic history may be obscure. In some instances the alteration has been so profound that the identity of the original rock is in doubt.

Geothermal Alteration   General geothermal heating of rocks buried to great depths or adjacent to igneous intrusions and extrusions induces recrys-

tallization. At a depth of approximately 3,000 meters the temperature is about 100°C. New minerals grow at the expense of previous ones, such as chlorite from a ferruginous clay. Water is driven from the clay minerals of a shale; coal loses water and volatile gases. Some crystals become enlarged at the expense of small ones of the same composition, as in the development of a coarsely crystalline marble from a fine-grained limestone. The glass in igneous tuffaceous rocks crystallizes. The salts of evaporite deposits recombine into new and rare minerals, such as those found in the potash salt deposits at Stassfurt, Germany.

Hydrothermal Alteration   This type of metamorphism is produced by hot magmatic waters or by groundwaters that have been heated, mobilized, and chemically charged by igneous intrusions. It is illustrated by the conversion of hard, fresh feldspar into a soft chalky aggregate of kaolin clay or fine-grained mica, the change of hornblende to chlorite, or olivine to serpentine. The surface rocks near the hot springs and geysers in Yellowstone National Park have been bleached and softened by steam and hot water, and no doubt the action continues underground.

Hydrothermal metamorphism is often accompanied by the addition or removal of substances, or both. Replacement is common and is often accompanied by the deposition of metallic ores, as found in many of the mining districts of the West. A well-known example is found at Butte, Montana. The hot

waters that deposited the copper ore as veins and replacements within the granitic rock altered the granite so thoroughly that almost no unaltered rock remains in the vicinity of the ore.

Contact Metamorphism Contact metamorphism is the alteration that occurs around an igneous intrusion because of the resulting high temperature and the emanation of hot fluids. The metamorphism is greatest at the actual contact and gradually decreases

TABLE 8.2 Summary of Metamorphic Processes

Process	Dominant factors	Geologic provenance	Contributing processes	Typical minerals	Characteristic rocks
Geothermal	High or moderate temperature, low pressure	Adjacent to igneous intrusions and extrusions	Recrystallization	Andalusite Anorthite Olivine	Quartzite Marble
Hydrothermal	Moderate temperature, water or water vapor	Wherever heated water circulates	Hydration and replacement	Talc Serpentine Clay minerals Epidote	Soapstone Serpentine
Contact	1. High or moderate temperature, but with invading igneous fluids and gases	Pronounced only in the vicinity of large intrusions	Recrystallization and replacement	Metallic oxides and sulfides	Garnet rocks and ore deposits
	2. High-temperature mobile igneous fluids and gases	Regions invaded by batholithic intrusions	Recrystallization and replacement	Feldspars Pyroxenes Amphiboles Micas Garnet	Banded gneiss
Dynamic	1. Directed pressure, low temperature	Belts of folding, crush zones, thrust planes	Crushing and pulverizing	Muscovite Chlorite	Breccia Granulated rocks Some slate
	2. Strong directed pressure, high temperature	Tectonic belts, especially regions of geosynclinal sedimentation	Progressive recrystallization with rising temperature	Talc Chlorite Micas Amphiboles Kyanite Garnet	Phyllite Schist Gneiss Quartzite Marble
	3. High temperature, strong hydrostatic pressure	Lower levels of tectonic belts	Recrystallization	Feldspars Pyroxenes Garnet Olivine	Gneiss

outward, with decreasing temperature and fluid pressure. This varying degree of metamorphism can be seen in the sequence of new minerals developed in aluminum-rich rocks, near an igneous contact. At some distance away from the contact, spots of muscovite and chlorite appear, then biotite, and finally, nearest the contact, cordierite, a complex aluminum-magnesium-iron silicate can be seen. This sequence (muscovite, chlorite, biotite, and cordierite) represents zones of increasing intensity of metamorphism toward the contact.

The contact-metamorphic zone forms an aureole, or halo, a few meters to a few thousand meters wide around stocks, such as illustrated at Marysville, Montana (Figs. 8.1A and 8.1B), or a few hundred meters to a few kilometers surrounding batholiths. Around small dikes and thin sills, the invaded rocks are not changed extensively. Locally, however, beds of clay have been "fired," or burned into a hard, bricklike rock, by dikes that intersect them. In Colorado and Virginia, beds of coal have been converted into hard natural coke by sills injected along bedding planes near them.

In contact metamorphism, shales are baked to hornfels (a fine-grained, flinty rock of variable mineral

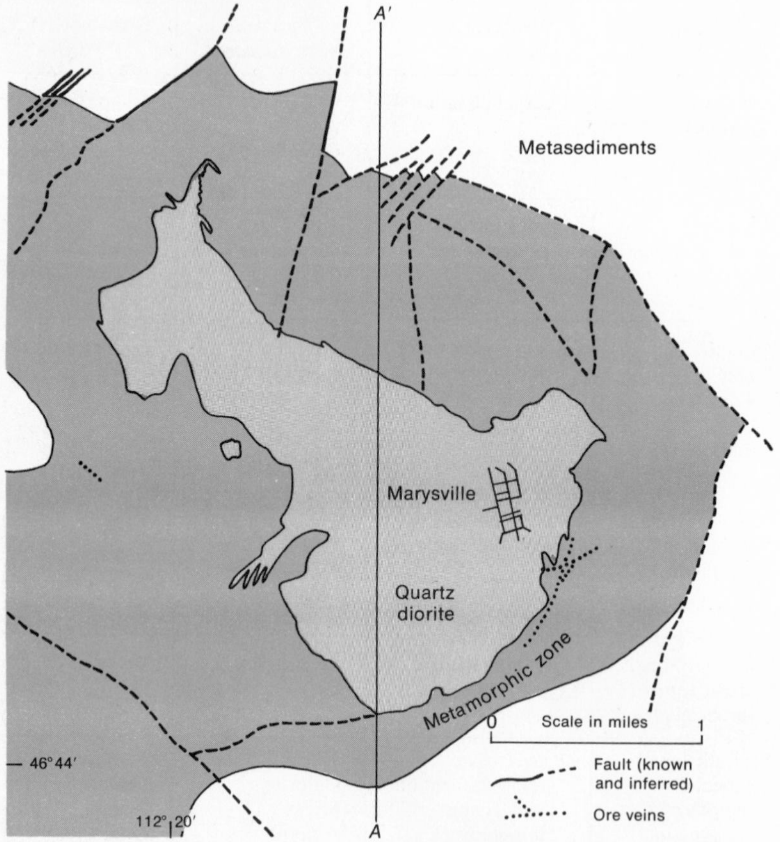

FIGURE 8.1A *Quartz diorite stock at Marysville, Montana, surrounded by a contact aureole developed at the contact with metasediments. Gold-bearing quartz veins occur near the contact. (After Barrell, U.S. Geological Survey.)*

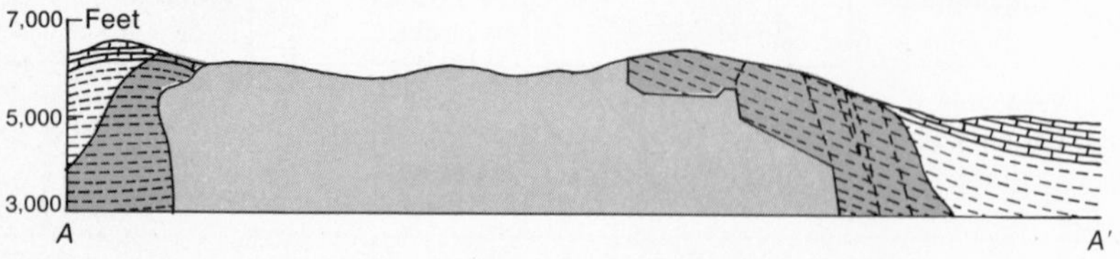

FIGURE 8.1B *Section along the line A-A' in Fig. 8.1A.*

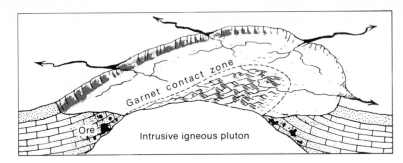

FIGURE 8.2   *Diagram showing a contact-metamorphic zone of garnet rock and ore in the garnet zone.*

composition) or to granulite (a rock composed mostly of even-sized, interlocking grains of quartz, feldspar, pyroxene or hornblende, and garnet). Limestones become coarse-grained marbles, impregnated with calcium silicates, and sandstones change to quartzite. Valuable deposits of iron, tin, tungsten, copper, and zinc have been formed as a result of contact metamorphism in limestones and calcareous shales (Fig. 8.2).

Dynamic Metamorphism   Dynamic metamorphism refers to changes induced by the orogenic movements caused by strong unbalanced or directed pressures. It is caused by faulting as well as by the intense folding of thick sections of sedimentary rocks that have been subjected to orogenic deformation. It often involves a larger area than the other types of metamorphism, and so it is sometimes called regional metamorphism.

Earth movements and accompanying dynamic metamorphism may take place as follows: differential pressures tend to shear rocks, crush them, and slice them along cleavage planes of their component minerals. These pressures also streamline the fragments into parallel bands by rotating and rearranging the pieces. The grains tend to flatten by plastic flow; then they heal together by crystal enlargement and by the growth of new crystals. New flaky minerals, such as chlorite, mica, and elongate minerals, like hornblende, grow lined up in definite planes controlled by the pressure system (Fig. 8.3). Equidimensional minerals, such as garnet and other late-forming minerals grown at the expense of previous minerals, may not show any preferred orientation.

Along deep-seated fault zones, the rocks near the faults are sometimes pulverized, the fragments are stretched out in the direction of movement, and finally they are welded together again in a strongly coherent

mass of microscopic mineral grains, forming a rock called mylonite.

TEXTURAL AND STRUCTURAL CHANGES

Moderate recrystallization to more intense alteration creates in pre-existing rocks a textural or structural feature called fissility or foliation (Fig. 8.4). This textural feature is the result of the parallel alignment of readily cleavable minerals, especially biotite, muscovite, hornblende, chlorite, or talc.

Granite and other coarse-grained feldspathic rocks can be changed to a vaguely foliated metamorphic rock called gneiss. A typical gneiss has light-colored streaks or lenses of feldspar and quartz alternating with streaks or bands of dark-colored biotite or

FIGURE 8.3   *Conversion of mudstone or shale to slate. (a) Mudstone or shale, greatly enlarged, consisting of fine quartz grains and smaller clay particles; (b) the same rock metamorphosed by pressure and recrystallization to form slate. The quartz grains are broken and flattened, and the fragments and minute lenses of quartz are oriented so that their long axes lie parallel to the direction of least pressure. The clay particles recrystallize to tiny mica flakes also arranged parallel to the direction of least pressure. This alignment of grains makes a slate readily cleavable.*

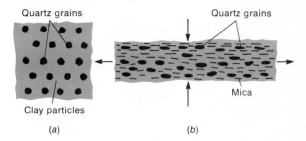

(a)   (b)

FIGURE 8.4 *Banded gneiss, near Embry Lake, Manitoba. The crinkled bands are vertical as a result of pressures applied horizontally at great depths within the earth. (Geological Survey of Canada, Ottawa, No. 86641.)*

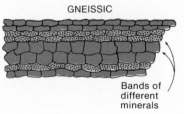

GNEISSIC

Bands of different minerals

FIGURE 8.5 *An outcrop of slate in Deep Creek Canyon, Montana, showing the platy character of the cleavage. (Walcott, U.S. Geological Survey.)*

hornblende (Fig. 8.4). The rock cleavage of gneiss is imperfect because platy minerals are not as abundant.

Slates, phyllites, schists, and gneisses are all said to be foliated (leaflike) because of the mineral alignment responsible for their ready splitting.

Foliated Metamorphic Rocks

The most abundant foliated metamorphic rocks are slate, phyllite, schist, and gneiss.

Slate Slate is a homogeneous, fine-grained rock that will split into thin sheets with relatively smooth surfaces (Fig. 8.5).

Because of its finer grain size, slate generally has a smoother cleavage surface than a schist. Some slates, however, have crumpled and wrinkled cleavage surfaces. The rock cleavage in slate usually intersects the bedding of the original shale at an angle. Traces of the bedding may appear as varicolored bands, often more or less plicated (folded), trending across the cleavage. The mineral components of slate are not distinguishable to the naked eye.

The colors range from gray through red, green, and purple, to black. The gray and black colors generally are due to carbonaceous material in the original rock, the carbon compounds having changed to graph-

FIGURE 8.6   *Complex folding and crumpling in quartzose schist on Boardman Hill, Clarendon, Vermont. The hammer handle is 38 centimeters long. (Dale, U.S. Geological Survey.)*

ite. The red shades are due to iron oxides, purple to manganese oxide, and green to ferrous iron silicates.

Phyllite   Phyllite is a foliated, fine-grained micaceous rock of nearly uniform composition. It is coarser and more lustrous than slate but too fine grained to be classified as a schist. The mica flakes in a phyllite are generally large enough to be barely discernible by the naked eye, but most of the other minerals are fine grained. A phyllite represents an intensity of metamorphism greater than that of a slate but less than that of schist. It may, however, grade laterally into either of these rocks.

Schist   Schist is a medium- to coarse-grained metamorphic rock that displays excellent cleavage along parallel planes of easily cleavable minerals. The grains are large enough to be readily seen and identified in hand specimens. The cleavage planes may range from smooth surfaces to bent and crumpled surfaces (Fig. 8.6).

Schists are classified according to the dominant minerals they contain, such as quartz-biotite schist, hornblende schist, muscovite schist, chlorite schist, or garnet schist (Fig. 8.7). They ordinarily contain little visible feldspar. Most schists are formed from shales, but some are formed by metamorphism of fine-grained

FIGURE 8.7   *Garnet metacrysts in schist, Stikeen River, Alaska. Garnetiferous schist is an intensely metamorphosed rock. (Courtesy of the American Museum of Natural History.)*

FIGURE 8.8 *Conglomerate schist with elongated pebbles, flattened by squeezing in the zone of rock flowage. This schist clearly was formed by dynamic metamorphism under conditions of high pressure and high temperature. (Geological Survey of Canada, Ottawa, No. 131803.)*

igneous rocks, such as volcanic tuff, rhyolite and basalt, or conglomerates (Fig. 8.8).

Gneiss Gneiss is a banded, coarse-textured, metamorphic rock with only a vague or rough foliation (Fig. 8.4). The alternating bands commonly represent different mineral composition. In most gneisses, feldspar is a prominent constituent; often the presence of large feldspar crystals serves to distinguish a gneiss from a schist. Metamorphic gneisses are derived mainly from coarse-grained feldspathic rocks (granite to gabbro), from conglomerates and feldspar-bearing sandstones, or from a higher grade metamorphism of the shale-phyllite-schist sequence.

Another form of gneiss occurs at the margins of batholiths, where granitic magma was injected in thin sheets along the cleavage surfaces of schists. The result is a hybrid rock, part schist and part granite (Fig. 8.9). The composite rock has a layered or gneissoid appearance. This type of rock is commonly referred to as an injection gneiss. The sill-like injections range in thickness from centimeters to several meters, and the rock between the injections is metamorphosed by high temperature, high fluid pressure, and reactions with hot fluids. Some injection gneisses are so intricately folded into unsystematic convolutions that geologists infer that both the granite and the host rocks were in a mobile state at the time the folding occurred.

In extreme instances the country rock grades from schist through an injection gneiss to rock with uniform texture like that of an igneous granite. Such a "granite," of metamorphic rather than magmatic origin, has been the subject of a controversy between geologists that is referred to as the "granite problem." The prevailing view is that many, if not most, granites with gradational contacts in the cores of highly folded mountain ranges have been transformed from previous rocks and are metamorphic, while other granites with sharp boundaries represent genuine magmatic intrusions. This interpretation eliminates the problem of making room for huge batholiths and is consistent with the field evidence for transformation in place.

FIGURE 8.9 *Migmatite, a mixture of light-colored quartz and feldspar with dark-colored schist. The quartz-feldspar aggregate forms discordant veins or dikes (at left) and concordant injections along the cleavage of the schist (center). The mass was heated sufficiently to be softened but not enough to be completely fluidized as a melt. Migmatite zones bordering batholiths may be hundreds of meters wide. (Geological Survey of Canada, Ottawa, No. 157702.)*

Certain gneisses appear to have been formed by movement during the consolidation of granitic magma, probably just after the magma had cooled to a pasty mass but had not yet become completely solid. In such granites the feldspar and mica or hornblende crystals are strung out in lines so that the structure is much like that of a granite gneiss formed by metamorphism. Such a structure is called flow banding, and the rock is a primary gneiss. A primary gneiss is not to be classified as a metamorphic rock.

Nonfoliated Metamorphic Rocks

Included in this category are massive or nonfoliated metamorphic rocks, which lack the parallel arrangement of cleavage minerals. Examples of these rocks are marble, quartzite, soapstone, serpentinite, and anthracite coal.

Marble   Marble is a crystalline calcareous rock formed by the metamorphism of sedimentary limestone or dolomite. The principal mineral is either calcite or dolomite. The texture of marble ranges from fine to relatively coarse, with grains less than a millimeter to several millimeters in diameter. It is more compact than limestone, having had its porosity reduced by pressure and recrystallization.

Marble composed almost entirely of $CaCO_3$ is white, but impurities that are commonly found in limestone give marble a great variety of colors. Pink, red, yellow, and brown varieties owe their color to varying proportions of hematite or limonite. Carbonaceous organic matter produces gray to black colors, and serpentine and chlorite produce green colorations.

Quartzite   Quartzite is a dense, highly siliceous rock formed from quartz sandstone. It was thoroughly cemented by intergrowths of quartz crystals and by secondary quartz brought into the rock in solution and deposited around the original sand grains (Fig. 8.10). A broken surface of the rock shows a glassy luster and a conchoidal fracture through the original grains. Quartzites, formed from sandstones with a high clay content, grade into quartz-mica schists.

Serpentinite   Serpentinite is a massive rock composed almost entirely of the mineral serpentine. The rock is commonly the result of hydrothermal alteration of olivine in former olivine-rich igneous rocks. Its color is variegated or spotted in various shades of green and brown. In some places serpentinite contains chromite ores, talc deposits, or veins of asbestos, a fibrous form of serpentine.

Another variety of serpentinite is verd antique, a decorative stone associated with marble and having an array of green and white colors.

Soapstone   Soapstone is a massive rock composed essentially of talc but commonly containing some mica, tremolite, chlorite, and quartz as accessory minerals. The rock is soft, has a bluish-gray or greenish-gray color, and feels greasy. It is a product of hydrothermal alteration of serpentine or other magnesian rocks, and consequently, contains hydrous silicates.

Anthracite Coal   Where bituminous coal beds have been intensely folded, as in eastern Pennsylvania, anthracite coal is present. Anthracite is a shiny black, dense coal that breaks with a conchoidal fracture. It commonly contains more than 90 percent car-

FIGURE 8.10   *Photomicrograph (×40) of a thin section of quartzite, showing deposition of secondary silica on quartz-sand grains. The rounded shape of the original sand grains is indicated by the concentric dusty outlines within the white grains. If the silica had been deposited by hot waters, the quartzite would be considered metamorphic.*

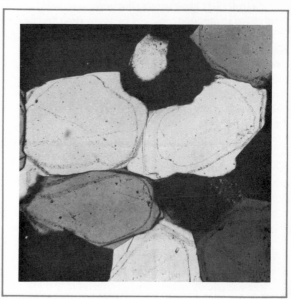

bon and about 8 percent incombustible ash. It contains little volatile gas-forming constituents.

Grain Sizes in Metamorphic Rocks

The size of grain in most foliated or nonfoliated metamorphic rocks varies from microscopic to a small fraction of a centimeter. Large well-formed crystals of garnet and other silicates that form later in metamorphism, occur commonly in schists and are analogous in appearance to the phenocrysts in igneous rocks. They are called metacrysts, or porphyroblasts. Some of them are up to a centimeter in diameter (Fig. 8.7).

Relics of previous structures found in the pre-existing rocks, such as bedding, concretions, fossils, and pebbles in sedimentary rocks, as well as phenocrysts in igneous rocks, sometimes survive (with distortion) the rigors of dynamic metamorphism. Phenocrysts and pebbles may be flattened and elongated in extreme instances (Fig. 8.8).

METAMORPHIC FACIES

A diversity of metamorphic rocks exists because of the variety of metamorphic processes and pre-existing rocks that can undergo metamorphism. However, geologists have quite successfully organized the diverse metamorphic rocks into distinct groups by the use of metamorphic facies.

Petrologists have found that certain mineral assemblages in metamorphic rocks occur time and again in rocks of different ages which have been subjected to similar metamorphic conditions. These assemblages, usually of two to six characteristic minerals, apparently represent metamorphic equilibrium under a particular set of environmental conditions. Each such cognate group of minerals constitutes a metamorphic facies.

An elaborate classification of 10 main facies and several subfacies has been worked out. Four of the facies represent increasing temperature effects in a contact-metamorphic zone, and six, the effects of increasing pressure and temperature in regional metamorphism. A full summary would involve the use of many unfamiliar names of minerals and rocks, but one can illustrate facies by the use of a few examples. Previous discussion has alluded to intensity zones in a contact-metamorphic zone, or halo.

Low-grade dynamic metamorphism of a shale produces an abundance of green chlorite and green epidote (a complex hydrous calcium-aluminum silicate) in association with quartz, albite plagioclase, and muscovite mica. This assemblage is called a greenschist facies. At slightly higher temperatures and pressures, some of the water of crystallization is eliminated. Biotite is formed from chlorite, and small crystals of the mineral garnet begin to appear. The result is a garnetiferous quartz-biotite schist, an abundant type of metamorphic rock.

At high temperatures and pressures, biotite changes to hornblende, garnet increases, and meta-

FIGURE 8.11   *Geologic map of the border of the Hudson Highlands near Poughkeepsie, New York, showing (by closer spacing of dashes) an increase in regional metamorphism from northwest to southeast. The boundaries of the biotite-, garnet-, and sillimanite-bearing rocks are indicated. The teeth on thrust faults are on the upper side. (After Balk, 1937, and Barth, 1938.)*

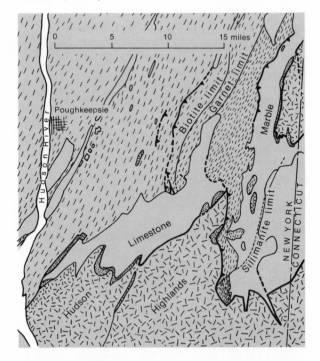

Gneiss   Quartzite   Limestone and marble   Slate and schist   Fault

crysts of a variety of metamorphic aluminum silicates develop. This aggregate forms a garnet-
amphibole facies that usually contains quartz, plagioclase, and muscovite. This sequence of chlorite, biotite, hornblende, garnet, and metamorphic aluminum
silicates forms an overlapping series of progressive
metamorphism, or a metamorphic facies.

A consistent series of metamorphic facies is
exhibited in Finland, Scotland, Canada, New England,
New York, Michigan, Idaho, and other places. On the
border of the Hudson Highlands in southeastern New
York (Fig. 8.11), muscovite- and chlorite-bearing
slates give way progressively to the southeast to facies
belts of phyllite and schist characterized successively
by biotite, garnet, and other metamorphic minerals.

Primary igneous granitic rocks are found even farther east in Connecticut. Pegmatite dikes and quartz
veins cut across all rock types, which also show along
their contacts alterations of chemical composition by
metasomatism. Hence the regional metamorphic zoning is attributed to decreasing distances inward toward
a magmatic source of heat and active fluids.

Metamorphic alterations in laboratory experiments suggest that the low-grade greenschist facies
forms at temperatures of 200 to 500°C, the garnet-
amphibole facies at 550 to 750°C, and a metamorphic
aluminum silicate phase near the upper limit at 750°C.

TABLE 8.3   Metamorphic Derivatives of Common Rocks

Original rock	Result of metamorphism
Sedimentary	
Shale	Slate, phyllite, schist
Shaly sandstone	Quartz-mica schist
Sandstone	Quartzite
Conglomerate	Conglomeratic quartzite, quartz schist, gneiss (if feldspathic)
Limestone	Marble
Shaly limestone	Slaty marble, calcareous schist
Coal	Anthracite
Igneous	
Granite, diorite, gabbro	Gneiss
Peridotite	Serpentinite, soapstone, talc schist, chlorite schist, hornblende schist
Basalt	Chlorite, hornblende, or biotite schist
Andesite	Hornblende schist
Volcanic tuff	Slate, phyllite, schist

The middle-grade biotite schist forms at intermediate
temperatures, perhaps 400 to 600°C(Fig. 8.12).

The final result of metamorphism depends upon
the chemical and mineralogical composition of the
original material and its texture, as well as on the type
and degree of metamorphism (Table 8.3).

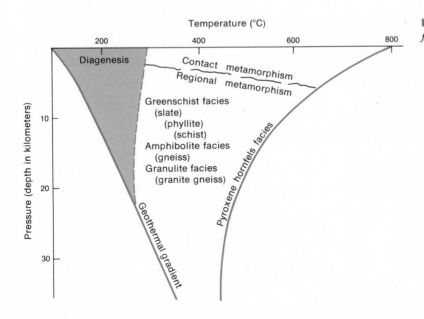

FIGURE 8.12 *Principal metamorphic facies.*

REFERENCES

Balk, R., 1937, *Structural Behavior of Igneous Rocks,* Geological Society of America Memoir 5.

Barrell, J., 1907, *Geology of the Marysville Mining District, Montana,* United States Geological Survey Professional Paper 57.

Barth, T. F. W., 1938, *Structural and petrologic studies in Duchess County, New York: Part II, Petrology and metamorphism of the Paleozoic rocks,* Geological Society of America Bulletin, vol. 47, pp. 775-850.

————, 1962, *Theoretical Petrology,* John Wiley & Sons, Inc., New York.

Ernst, W. G., 1969, *Earth Materials,* Prentice-Hall, Inc., Englewood Cliffs, N. J.

Fyfe, W. S., Turner, F. J., and Verhoogen, J., 1958, *Metamorphic Reactions and Metamorphic Facies,* Geological Society of America Memoir 73.

Harker, A., 1954, *Petrology for Students,* 8th ed., Cambridge University Press, London.

Mason, B., 1966, *Principles of Geochemistry,* 3rd ed., John Wiley & Sons, Inc., New York.

Ramberg, H., 1952, *The Origin of Metamorphic and Metasomatic Rocks,* The University of Chicago Press, Chicago.

Spock, L. E., 1953, *Guide to the Study of Rocks,* Harper & Brothers, New York.

Turner, F. J., 1948, *Evolution of the Metamorphic Rocks,* Geological Society of America Memoir 30.

———— and Verhoogen, J., 1960, *Igneous and Metamorphic Petrology,* 2d ed., McGraw-Hill Book Company, New York.

Winkler, H. G. F., 1965, *Petrogenesis of Metamorphic Rocks,* Springer-Verlag, New York.

CHAPTER 9

THE ATMOSPHERE AND THE HYDROLOGIC CYCLE

The study of geology thus far in the text has considered the solid earth, its materials or composition, and its processes and change. The discussion has emphasized internal features, internal materials, and internal processes. These do affect the earth's outer surface.

The following three chapters will consider the surface processes of weathering and erosion and the earth materials they change and form. The atmosphere initiates these processes.

Atmosphere is the envelope of air that surrounds the earth. Because it is readily compressed, half of it lies below about 6,000 meters (Fig. 9.1). Its mass is less than one-millionth that of the solid earth, but its influence is far reaching. It sustains the varied life of the earth and has other essential functions.

The atmosphere acts as a thermal shield that reflects or absorbs much of the radiation received from the sun. It distributes the heat received and tends to prevent the escape of heat, acting like an umbrella by day and a blanket at night. Its gaseous molecules and particles of suspended dust aid in the diffusion of sunlight. Furthermore, it protects the earth from excessive ultraviolet radiation and violent bombardment by meteorites.

The atmosphere itself is an important geologic agent. It is one of the chief sources of rock weathering. It reacts chemically with rocks to form new minerals. It serves as a medium for the transfer of water and solid particles. Some of the water transported is subsequently precipitated onto the earth as rain and snow. The rain and snow, in turn, form streams and glaciers that wear away the rocks and transport them seaward.

The wind is also an important agent of transportation of dust and sand. Wind is also the driving force in forming waves and shore currents, which also have significant geologic effects.

GENERAL FEATURES OF THE ATMOSPHERE

Composition

The atmosphere is a mixture of gases and suspended particles. Clean, dry air consists of about 78 percent nitrogen by volume, 21 percent oxygen, 0.94 percent argon, 0.03 percent carbon dioxide, and 0.003 percent of a mixture of neon, helium, krypton, xenon, nitrogen oxides, methane, and ozone. These proportions re-

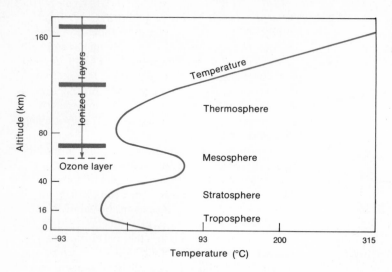

FIGURE 9.1 *Vertical section of the lower atmosphere showing its thermal layers.*

main constant to heights of tens of kilometers. Locally and temporarily, certain volatile organic substances, industrial fumes, carbon monoxide, sulfurous gases, and chlorine (from volcanoes) are also present in the atmosphere. Carbon dioxide, oxygen, and some nitrogen follow cycles that involve movement from the atmosphere through plants and animals and back again to the air, water, or rock. The carbon in coal, for example, was derived from the atmosphere via plants. Much of the mineral material in solution in water or deposited as limestone was previously dissolved with the aid of carbon dioxide.

Water vapor, though averaging 3 percent by volume, also is an important constituent of the atmosphere. It is equivalent to about 0.4 centimeter of liquid water if it were precipitated as rain. Most of it is below the 6,000-meter level. Its abundance in the atmosphere varies considerably. It may be nearly 0 percent to as much as 4 percent in some hot tropical areas.

Water vapor decreases the density of air, absorbs incoming solar radiation, and is the source of all precipitation. High humidity and high rainfall go together in tropical areas, as in the Amazon basin, whereas low humidity and extremely low precipitation characterize polar areas, such as Antarctica. High humidity is one of the factors that accounts for the lower daily and seasonal ranges of temperature in tropical areas as compared with extreme temperatures in deserts.

Suspended particles in the atmosphere consist of water droplets and ice crystals in clouds and salt crys-

tals from ocean spray, smoke, soot, pollen, spores, bacteria, volcanic dust, and meteoric dust. These particles, mostly microscopic or ultramicroscopic in size, are widespread in the atmosphere over both land and sea. All the particles, but especially the clouds, absorb, reflect, and scatter (diffuse) part of the sun's rays.

Many of the nonaqueous particles have an affinity for water, absorbing and retaining it. They function as nuclei, or centers, around which water vapor condenses to form particles of water or ice in clouds and rain or snow.

Atmospheric Stratification

The atmosphere is divided into natural layers (Fig. 9.1) on the basis of temperature. In the lower layer, the troposphere, the temperature decreases about 0.6°C for every 100-meter rise in vertical elevation in middle latitudes. This trend continues up to altitudes of 10 to 13 kilometers, where a zone of nearly constant temperature, −51 to −56°C, begins. This cold isothermal, or equal heat, zone is the base of the stratosphere. Above the stratosphere, which is 15 kilometers thick, the mesosphere and thermosphere occur. These zones of the outer atmosphere, while of interest to earth scientists, will not be considered in this book.

Troposphere The troposphere is the region of convectional circulation. The prefix tropo- means turning or overturning. It refers to movements of air caused by differences in temperature and corresponding differences in density between warm and cold air. Convection circulates air on a broad scale. Because temperature differences over land and water are widespread, so is convection. The land and sea breezes that occur around the earth are the result of these temperature differences.

The height of the troposphere decreases with latitude from about 18 kilometers at the equator to about 6 kilometers near the poles. It varies slightly with the seasons, generally being higher in summer than in winter.

Contrary to the usual steady decrease in temperature with increasing altitude in the troposphere, warm air sometimes overlies cool air. This occurrence is an inversion of temperature, one of the factors in the development of "smog" in industrial cities. Pronounced differences in temperature at different elevations of the ground surface cause corresponding differences in climate, vegetation, and habitability of places having the same latitude but different altitudes. In general, a change of altitude of 4.5 kilometers is about equal in its climatic effects to a change of latitude from equator to the poles, as shown for example by slowly descending snow lines and tree limits on high mountains as one goes northward.

Dust and water vapor are most abundant in the lower part of the atmosphere. The quantity of water vapor present is determined to a large extent by temperature. The lower the temperature, the less the amount of water vapor present. Consequently, no ordinary clouds exist above the troposphere, and most of those within it are low. Hence, precipitation on high mountains is greatest on intermediate slopes, not on top.

Stratosphere   The stratosphere, immediately above the troposphere, is a region of cold, clear, thin, dry air, with a nearly constant temperature near its base. Unlike the troposphere, the stratosphere is characterized by a continuous balance between absorbed and emitted radiation. Consequently, there is no convection within it.

Suspended matter is scant in the stratosphere. Nacreous (shell-like), or mother-of-pearl, clouds, presumably of ice crystals, are seen occasionally at 20- to 30-kilometer levels. Meteoric dust falls through the stratosphere continuously, and dust from volcanic and nuclear bomb explosions is blasted into it sporadically. Fine, slow-setting, volcanic-dust particles eventually

are distributed completely around the earth and give rise to colored sunrises and sunsets that persist for months after an eruption. Large quantities of volcanic dust in the atmosphere can affect the amount of solar radiation reaching the earth's surface. Geologists have proposed that suspended volcanic dust may be the cause of a worldwide cooling of climate and possibly the start of an ice age.

Atmospheric Heating

The atmosphere is heated almost entirely by the sun. Other sources of heat, including heat flow conducted outward from the interior of the earth, heat of hot springs, and eruptions of steam and other hot gases from inside the earth, total only a small fraction of 1 percent.

About two-fifths of the sun's radiation is light, and about half is in the range of heat rays. Of the total radiation received by the earth, about 34 percent is reflected back into space, 19 percent is absorbed by the air, especially by water vapor and clouds, and 47 percent reaches the surface (Fig. 9.2). Two-thirds of this radiation is effective in heating the atmosphere, the ground, and the surface of the sea (Fig. 9.2).

The heating of the atmosphere is largely indirect. Short waves of the sun's rays in the range from visible red light to ultraviolet are converted to longer heat waves by ground absorption, which then heat the atmosphere by earth radiation. The lower part of the atmosphere warms up more readily than the upper part because of this radiation from the earth's surface and because it has a greater density and water-vapor content.

The most effective absorbents of radiation in the atmosphere are water vapor, carbon dioxide, and ozone. Without them, atmospheric temperatures would be about 22°C cooler than now, unbearably hot by day and icy cold at night as on the moon. The retention of heat in the lower part of the atmosphere is called a "greenhouse effect." Some persons are concerned that the return of large quantities of carbon dioxide to the air from the burning of coal, petroleum products, and natural gas may lead to a warmer climate all over earth. This might cause increased melting of ice on Antarctica and Greenland, a rise in sea level, and flooding of coastal cities.

The effectiveness of solar radiation in heating the atmosphere varies with length of the day, angle of the

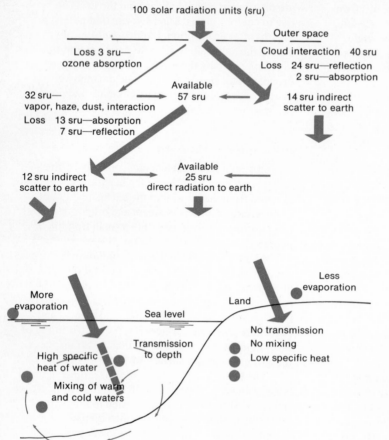

100 solar radiation units (sru)

Outer space

Loss 3 sru—
ozone absorption

Cloud interaction   40 sru
Loss   24 sru—reflection
        2 sru—absorption

Available
57 sru

14 sru indirect
scatter to earth

32 sru—
vapor, haze, dust, interaction
Loss   13 sru—absorption
        7 sru—reflection

12 sru indirect
scatter to earth

Available
25 sru
direct radiation to earth

More
evaporation

Less
evaporation

Sea level

Land

Transmission
to depth

No transmission
No mixing
Low specific heat

High specific
heat of water

Mixing of warm
and cold waters

FIGURE 9.2 *Upper: Distribution of solar radiation in the atmosphere and on the earth's surface. Lower: Differences in heating and cooling of land versus water. (From Strahler, 1969,* Physical Geography, *3d ed. Copyright © 1951, 1960, 1969, John Wiley & Sons, Inc.)*

sun's rays (as determined by latitude and seasons), cloudiness of the sky, humidity of the air, and elevation of the land. Tropical areas with direct sun throughout the year receive much more than average heat; polar areas receive only about one-fourth as much heat as do the tropics. The seasonal effect on the length of the day and on the height of the sun is greatest in middle to high latitudes. Mountains tops with thinner than average air above them are readily heated by day and just as easily cooled at night. A cloud cover both shuts out radiation from above and retains heat from below.

Barren rock surfaces absorb and later radiate heat more rapidly than areas covered with soil and vegetation, or with snow and ice. The land also warms up and cools faster than the sea, because rocks absorb and radiate heat more readily than water. Their specific heat is about one-fifth that of water. They also reflect less of

the radiant energy than water does, are less deeply penetrated by solar radiation, and are less affected by cooling due to evaporation. Finally, they are not subject to mixing, as water is (Fig. 9.2). Air temperatures over both land and sea are modified, however, by ocean currents and prevailing winds.

The quantity of heat received in a year at latitude 40°N is estimated to be equivalent to more than 5 million kilowatt-hours per acre. One begins to appreciate this tremendous quantity of energy when one witnesses the fast melting of snow on a warm day, the rapidity of evaporation after a rain, the power of the wind, or the fury of a storm.

Heat in the atmosphere is distributed by convection, conduction, and radiation. Thermal convection is especially strong in the tropics. It operates on a large scale globally and on a small scale locally. Winds, as well as rising and descending air currents, are also a part of the earth's convectional system.

Conduction, the transfer of heat by direct contact with warm or cold surfaces, is relatively unimportant because air is a poor conductor.

Radiation into outer space maintains a heat balance or equilibrium with the energy received from the sun. At latitudes north of 38°N, the outward radiation exceeds the absorption of heat. To balance this loss, a continuous poleward flow of heat moves in the upper atmosphere across the 40° latitude circle.

The heat balance at a particular place shifts with the rising and setting of the sun, the seasons, atmospheric conditions (cloudiness, humidity, and dust content), surface conditions (land or water, forests or grasslands, snow cover or barren ground), and the elevation of the land surface. Radiation is greatest at night, in winter, and through thin, clear, dry air at high elevations. In the long run, for the entire earth, the loss by radiation equals the gain, and a state of dynamic equilibrium prevails. Otherwise the earth would be warming or cooling.

Atmospheric Pressure

The atmospheric pressure at a particular locality is the weight of the column of air above it. At sea level the atmosphere, on the average, weighs about 1.04 kilogram per square centimeter (14.7 pounds per square inch), or enough to raise a column of mercury in a barometer to a height of 760 millimeters (29.92 inches). The mean annual pressures at sea level are shown in Fig. 9.3.

Atmospheric pressure decreases rapidly with increasing elevation. The pressure at a height of 5.6 kilometers is half that at sea level. Pressures fluctuate at a fixed elevation in response to changes in the temperature and water content of the air. Heat causes air to expand and become lighter, and water vapor, being rela-

FIGURE 9.3 *Mean annual atmospheric pressure at sea level. The subtropical highs and the Aleutian and Icelandic lows are especially noteworthy. (After Koeppe and DeLong, 1958.)*

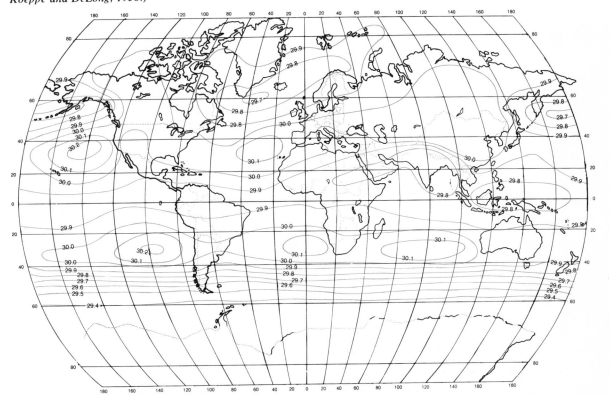

tively light in comparison with air as a whole, also reduces the pressure.

Air Movements

Air moves in response to pressure differences in the atmosphere. These differences are caused mainly by the unequally distributed energy from the sun. A difference of 2.6 millimeters (0.1 inch) barometric pressure is enough to initiate movement. The principal movements are winds and air currents.

Wind is air moving from an area of high pressure to one of lower pressure in a mainly horizontal direction under the pull of gravity. Its speed is proportional to the steepness of the pressure gradient, i.e., the rate of horizontal change of pressure at a given elevation. As in the flow of water, the steeper the gradient, the greater the velocity. Winds are modified in strength and direction by friction, turbulence, eddies, the earth's rotation, and centrifugal force.

Air currents are mainly vertical movements of air. Examples are the convectional rise of warm moist air and the subsidence of cold dry air. Winds converging

upon an area of low pressure from all sides eventually combine to form a rising current. Rising air is cooled by expansion, and subsiding air is heated by compression.

These air movements distribute heat and moisture over the earth. Water vapor in the air has a special role in distributing heat. The quantity of heat needed to evaporate water varies inversely with the temperature. It requires about 540 calories per gram of water at 100°C to nearly 600 calories per gram at 0°C. Hence water vapor is a carrier of substantial heat in a latent (hidden) state. When moist air later cools and the water vapor condenses, this latent heat of evaporation is released. Accordingly, movements of moist air assist the transfer of heat from place to place, as well as supply moisture for rainfall.

Prevailing Winds   The atmosphere circulates in a regular pattern (Fig. 9.4). Because of pronounced temperature and pressure differences, a complex convectional system operates between the equator and the poles. It includes (1) the doldrums belt at the equator,

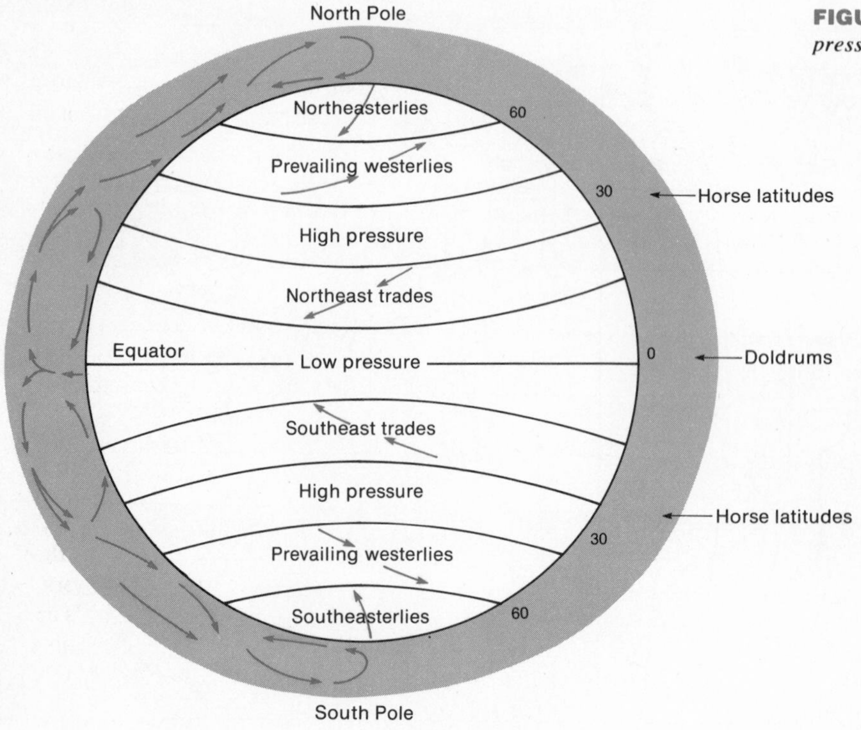

FIGURE 9.4 *Prevailing winds and pressure systems on the earth's surface.*

a low-pressure belt caused by extreme heating and characterized by calm or light winds and by rising moist air and frequent thunderstorms; (2) northeast trade winds in the Northern Hemisphere and southeast trade winds in the Southern Hemisphere, which are almost continuous; (3) subtropical high-pressure belts of dry, descending air near latitudes 30°N and S (the horse latitudes); (4) prevailing westerly winds in middle latitudes; (5) polar easterlies, cold winds spiraling out from polar high-pressure areas; and (6) vagrant belts of cyclonic storms along the edges of surging masses of this cold polar air (polar fronts). These prevailing winds and pressure belts are modified by differences between land and sea, relief of the land, seasonal changes, and local factors. The equatorial doldrums, trade winds, and subtropical high-pressure belts migrate north and south a few degrees of latitude, in keeping with the apparent seasonal movement of the sun.

The prevailing winds have profound effects on climate. They account for the heavy equatorial rains on the windward side of mountains and deserts on the lee side. Winds bring changeable weather in middle latitudes. They also cause ocean currents—a westerly drift along the equator and large gyres in the North and South Atlantic and North and South Pacific.

Secondary Winds   In addition to the large-scale air movements, there are many smaller ones. These include land and sea breezes, mountain and valley breezes, monsoons, hurricanes and typhoons, thunderstorms, tornadoes, and chinook winds. These various storms and surface winds sometimes cause extraordinary tide levels and directly or indirectly participate in the movement of heat, moisture, waves, shore currents, and sedimentary particles.

ATMOSPHERIC MOISTURE

Condensation and Sublimation

Condensation is the process of changing a gas to a liquid. Sublimation is the transfer of a gas directly to a solid state (or vice versa) without an intervening liquid state. Condensation and sublimation of water vapor are caused by cooling the air to the point of saturation. The temperature at which saturation occurs is called the dew point. If the dew point is above freezing, condensation forms clouds, fog, or dew; if below freezing, then sublimation forms ice-crystal clouds, or frost. The droplets of water in fogs and clouds and the ice crystals in clouds are so tiny that they easily remain suspended in air, especially in rising air. As previously noted, condensation is facilitated by water-attractive dust or salt particles in the air. This is the basis for cloud seeding in attempts to induce increased rainfall or to eliminate fog over airports.

Cooling to cause condensation or sublimation takes place by radiation, contact with cold surfaces, mixing with colder air, expansion in moving to places of lower pressure, as in a rising convectional current, or by lifting air over a mountain range. Rains from thunderstorms that accompany volcanic eruptions and large forest fires are special effects of condensation of moisture in swiftly rising air.

Precipitation

Droplets of water and crystals of ice in some clouds grow until they fall. Their growth is facilitated by collisions in turbulent air or by falling through clouds. Some drops remain suspended by updrafts and some evaporate on the way down. The others fall to the ground as precipitation.

The principal forms of precipitation are rain and snow. Hail, sleet, glaze (ice storm), dew, and frost are significant locally.

The total annual precipitation at a location is measured as the depth of a liquid layer in inches or metric units. The average rainfall on land is about 66 centimeters (26 inches) a year. It is higher at sea, but records there are inadequate. The average for the earth is estimated to be 99 centimeters.

The amount of precipitation is distributed very unequally over the earth. In Assam, India, it is locally more than 12 meters a year, whereas in the Sahara and other deserts it is less than 26 centimeters. In some desert areas, several years may elapse between showers, but these may be torrential. In the United States, precipitation ranges from about 100 centimeters a year on the Atlantic coastal plain to 76 centimeters in the northern part of the central interior, 50 centimeters or less on the Great Plains, less than 26 centimeters in parts of the Great Basin, and more than 250 centimeters a year on the Olympic Mountains of western Washington.

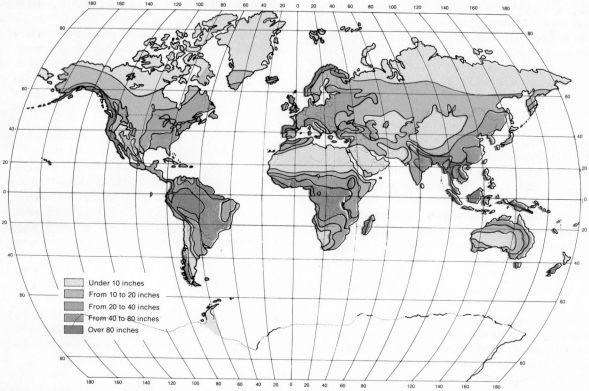

FIGURE 9.5 *Mean annual precipitation of the world. (After Koeppe and DeLong, 1958.)*

The worldwide differences of precipitation on land (Fig. 9.5) are influenced by latitude, nearness to the sea, topography (especially the presence of mountain ranges), prevailing winds, seasons, and frequency of storms, hurricanes, and typhoons. Rainfall is heaviest in the calm tropical belt, where moist air rises abundantly and daily showers are the rule, and on the windward sides of continents and mountain ranges. By contrast, many of the great deserts of the earth lie along the subtropical high-pressure belts, in continental interiors, or in the lee of mountains.

HYDROLOGIC CYCLE

As Solomon observed, streams flow to sea, yet the sea is not full, though the streams continue to flow. The explanation of this anomaly escaped the brilliant students of natural philosophy from Aristotle (384–322 B.C.) to Kepler (1571–1630). Studies by Perrault, Marioti, and Halley in the seventeenth century developed the concept of the hydrologic cycle. The term hydrologic cycle describes the transfer of moisture from the sea to the atmosphere, thence to the land, and back to the sea. The simplest form of the hydrologic cycle occurs at sea. It includes evaporation from the oceans, transportation and condensation in the atmosphere, and precipitation directly upon the sea surface. Although this is estimated to be nearly four-fifths of the total water in the hydrologic cycle, it is of little use to man; but the small fraction of the precipitation which falls on land is vital.

The continental hydrologic cycle (Fig. 9.6) involves evaporation from the sea or other water areas, followed by distribution as vapor or clouds over the land by winds. Precipitation follows, leading directly to runoff or to infiltration and emergence as seepage or springs, with eventual runoff to the sea. However, several subsidiary cycles on land serve as short circuits, since only about one-fifth of the precipitation on land returns to the sea as runoff.

The principal source of moisture in the air is evaporation from the sea, lakes, ponds, streams, soils, or other moist surfaces. The sea supplies nearly 85 percent of the total. This moisture is widely distributed by winds and is ultimately precipitated. Of the water that falls on the land as rain, snow, or other forms of precipitation, a part infiltrates the soil, loose rock, or solid rock to become groundwater. Some precipitation evaporates, even in mid-air; some is caught on the surface in snowfields, glaciers, ponds, lakes, or on vegetation. Some immediately runs off in streams toward low ground, a lake, or the sea.

Much of the precipitation on land infiltrates the ground. Often a rain may barely wet the top few inches of soil and produce no immediate runoff. Only when more water falls than can infiltrate does immediate runoff occur.

Of the fraction that enters the ground, much is returned to the air by evaporation from soil or by transpiration from plants. Later, some of the water that enters the earth emerges as springs or seepage. Upon emergence, it may evaporate, or it may join and maintain streams between rains.

Lakes and the Hydrologic Cycle

Once a lake is formed, its existence depends on a continuing source of water—surface inflow to the lake, precipitation directly upon the lake, or inflow of groundwater—to balance outflow or evaporation from the lake's surface. A lake above the water table depends mainly on rainfall or on the water from melting snow for its water supply. Its basin must therefore be of sufficient size and depth to conserve water from one precipitation period to another. Ponds and small lakes

FIGURE 9.6 *The hydrologic cycle. The heat of the sun evaporates moisture from the sea or from lakes, ponds, reservoirs, streams, soils, vegetation, and falling rain. Moisture is also supplied by transpiration from plants and by sublimation from snow and ice. Precipitation results when relatively light, warm, moist, maritime air is blown landward and upward and is cooled over the sloping front of a cooler and heavier continental air mass, in the convectional system of a thunderstorm, or in ascent over mountains. The precipitated moisture may be stored as snow fields, glaciers, ponds, lakes, soil water, or water infiltrated into the ground. Some water runs off immediately after precipitation. Groundwater returns to the surface to be evaporated or to join the runoff in streams that flow to the ocean. The runoff thus completes a cycle.*

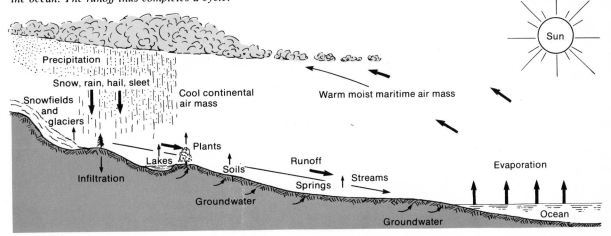

may be associated with perched water tables above the regional water table.

The disappearance or drastic decrease in size of several "permanent" lakes on Long Island, New York, during 1966 illustrates the close relationship between lake and groundwater levels. The lakes, representing windows into the shallow groundwater system, decreased in area and depth with the declining groundwater level (Fig. 9.7). This decline of the groundwater level and lake elevation, and the 1966 record- or near-record-low stream flows resulted from 6 years of below-normal precipitation in the northeastern United States.

Regulatory Effect of a Lake on River Discharge

The level of a lake regulates the discharge of its outlet stream. Flood waters of inflowing streams spread out over a lake, raising the water level slowly. Thus the discharge at the outlet rises slowly, reducing the possibility of flood damage along the outlet valley. During dry seasons, the level of the lake will drop slowly, and if it does not go below the sill level of the outlet, it helps in maintaining a permanent stream during dry seasons. The regulatory effect of

FIGURE 9.7   *Drop in level of Long Island lakes with lowering of the water table. (U.S Geological Survey.)*

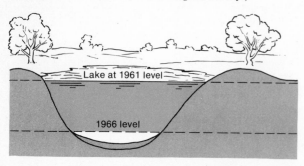

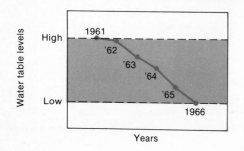

lakes on the flow of rivers has been a major reason for building man-made lakes or reservoirs where no natural ponding occurs. In some cases, large flood-control reservoirs can store entire floods.

Characteristics of Lake Water

Lake waters vary in their dissolved salt content from the fresh waters of a mountain lake fed by glaciers to the salty water of Great Salt Lake or the Dead Sea. The chemical composition of lake water depends largely on: the composition of the rocks that the water which feeds the lake has moved through or over, the extent to which the water in the lake has been concentrated by evaporation, and the presence of decaying organic material, either in the lake or in the area drained by streams feeding the lake.

Lakes that have outlet streams are generally fresh because the outflow prevents any significant concentration of salts by evaporation. On the other hand, in undrained basins, evaporation increases the concentration of dissolved materials in the lake until the point of saturation is reached. At this point, deposition of the chemicals occurs on the lake floor. The conditions favoring the formation of a saline lake are present where streams discharge into an undrained basin in an arid climate. Because such a lake is undrained, its area varies with relative balance between precipitation and evaporation. Some lakes exist for only a few days to a few weeks after storms. These intermittent lakes are called playa lakes. "Lake" Eyre of Australia, referred to by some as little more than a mirage, exemplifies the variability in area of playa lakes. In 1950 it covered an area of 8,031 square kilometers, and by 1952 it had completely evaporated. Conditions favoring saline lakes are prevalent over a large area in the southwestern states, the largest being Great Salt Lake in Utah.

Great Salt Lake is a shrunken remnant of Lake Bonneville, a vast fresh-water lake that once covered an area of 51,000 square kilometers, mainly to the west and south of the present lake shores. Lake Bonneville and many similar lakes in the Basin and Range province were formed during the glacial epoch when precipitation was somewhat higher and evaporation less: hence, the water level was higher. Lake Bonneville overflowed to the north through the Snake and Columbia Rivers into the Pacific Ocean. It left a record of its boundaries in the form of hundreds of kilometers of beaches, terraces, and wave-cut cliffs along its mountainous shores. At one time, Lake Bonneville had a maximum depth of over 300 meters. The site of the present Mormon Temple in Salt Lake City was under as much as 280 meters of water. With the waning of the Pleistocene epoch, approximately 10,000 years ago, the volume of water entering the lake by both rainfall and streams decreased, and the lake level fell below its outlet. Since that time, the dissolved solids in the original lake and subsequent inflow have been concentrated. Now the salinity is more than 200 parts per thousand, or more than 5 times the normal 35 parts per thousand salinity of seawater.

Saline lakes containing large amounts of sodium and potassium carbonates are commonly called alkaline lakes.

Man and the Hydrologic Cycle

Man can and does modify the runoff. He does so by land-management practices, by building dams across stream channels, and by diverting or even reversing stream courses. The type, extent, and pattern of vegetation and land use over a drainage basin may be adapted, for example, to produce more infiltration, slower melting of ice and snow, delayed runoff, and greater total runoff. The self-storing aspect of snow on high mountains, accumulating in winter and melting in late spring and early summer during the growing season, makes it an especially useful form of precipitation. To take greater advantage of this kind of storage, man is experimenting with methods for increasing the snow pack. In the State of Washington, experiments are being conducted to determine how to influence glacier melting rates in summer to yield more runoff.

Among the efforts to bring runoff to the parched areas of the earth is a proposed North American Water and Power Alliance. This is a multibillion dollar project to collect waters from the mountains of western Canada and distribute it to water-deficient areas of Canada, western United States, and Mexico.

Man can also modify other parts of the hydrologic cycle. Water can be removed from wells or injected into them for storage. Leaky surface-storage reservoirs may increase infiltration. A few such water reservoirs have been built deliberately, but a number of others were constructed unintentionally. Weather modifications may deliver more precipitation when and where it is needed and provide a measure of flood protection. Atomic energy may also be used to provide an artificial water cycle by turning salt water into fresh, allowing the use of seawater and large quantities of saline groundwater.

CLIMATES

The atmosphere is the medium of weather and climate. Weather can be determined as the daily set of conditions that includes temperature, air pressure, wind, humidity, cloudiness, and precipitation. Climate is the composite of weather and its variations over long periods of time. It is described in terms of mean annual temperature, range of temperature extremes, humidity, cloudiness, amount and seasonal distribution of precipitation, storms, and winds.

The varied climates of the earth (Table 9.1) illustrate the effects of several climate controls: (1) latitude, (2) the permanent high- and low-pressure areas, (3) prevailing winds, (4) interacting air masses and storms, (5) distribution of land and water, (6) nearness to the sea, (7) mountain barriers, (8) ocean currents, and (9) altitude. The differences in wind, rainfall, temperature, snow and ice, and other particulars of climate affect geologic processes active in different climatic areas. Because soils, vegetation, and water runoff by streams also are determined largely by climate, many geologic processes, especially the weathering of minerals and rocks, and the erosion of land by wind, water, and ice, differ greatly from one climate to another.

Just as they do today, the prevailing winds had geological consequences in the past. They modified the circulation of marine waters, especially in ancient seas. Prevailing westerly winds similarly drifted wind-blown sand toward the east to form extensive sandstone deposits in Utah.

The permanent subtropical high-pressure belts, where the descending air is warmed and hence disposed to drying, are the sites of seven great deserts: the Sahara of Africa, Thar of western India, Kalihari of southwest Africa, Arabian, Victorian of Australia, Atacama of South America, and the Sonoran of southwestern United States and northwestern Mexico. The Kalihari and Atacama are also favored by cool ocean currents offshore.

Extensive desert deposits of salt and gypsum, now found at middle latitudes in New York, Michigan, Kansas, and even Saskatchewan, suggest that these

TABLE 9.1   Classification of Climates

1. Tropical climates
   A. Tropical rain forest, in the equatorial low-pressure belt—the Congo and Amazon Basins
   B. Tropical savanna (with wet and dry seasons), in trade-wind latitudes that shift with the sun—Burma, and the Veldt and Sudan of Africa
2. Dry climates—potential evaporation exceeds precipitation
   A. Low-latitude desert and steppe, in the subtropical high-pressure belts—the Sahara, Arabia, and Australia
   B. Middle-latitude desert and steppe, in continental interiors or behind mountains—Iran, Mongolia, and the Great Basin of southwestern United States
3. Warm-temperate (mesothermal) climates—moderately warm
   A. Mediterranean (dry-summer subtropical)—Italy, Spain, and California
   B. Humid subtropical—southern China and southeastern United States
   C. Marine west coast, with long rainy seasons—Western Europe, and southeastern Alaska
4. Snow (microthermal) climates—moderately cool
   A. Humid continental, in the stormy middle latitudes—north-central United States, central Europe, and northern China
   B. Subarctic, at high latitudes—northern Alaska, northern Canada, and Siberia
5. Ice climates
   A. Tundra—the borders of the Arctic Ocean
   B. Icecap—Antarctica and Greenland
6. Highland climates
   Local climates of mountains and plateaus, varying according to altitude and latitude—the Andes, Himalayas, Rockies, Tibet, Kenya, and Mexico

high-pressure belts have shifted in latitude during the geologic past, either by wandering of the poles of rotation or by continental drift, or both.

Through its varied effects on soils, vegetation, water resources, agriculture, industry, housing, and human health and comfort, climate to a large degree determines man's use of the land in different parts of the world.

REFERENCES

Bates, D. R., 1964, *The Planet Earth*, 2d ed., Pergamon Press, New York.

Chorley, R. J. (ed.), 1969, *Water, Earth, and Man—A Synthesis of Hydrology and Socio-Economic Geography*, Methuen & Co., Ltd., London.

Critchfield, H. J., 1966, *General Climatology*, 2d ed., Prentice-Hall, Inc., Englewood Cliffs, N.J.

Hambidge, G. (ed.), 1941, *Climate and Man*, United States Department of Agriculture, Washington, D.C.

Koeppe, C. E., and DeLong, G. C., 1958, *Weather and Climate*, McGraw-Hill Book Company, New York.

Monkhouse, F. J., 1962, *Principles of Physical Geography*, 5th ed., Philosophical Library, Inc., New York.

Newell, R. E., 1964, *The circulation of the upper atmosphere*, Scientific American, vol. 210, no. 3, pp. 62-74.

Riehl, H., 1972, *Introduction to the Atmosphere*, 2d ed., McGraw-Hill Book Company, New York.

Strahler, A. N., 1971, *The Earth Sciences*, 2d ed., Harper & Row, Incorporated, New York.

Sutton, O. G., 1961, *The Challenge of the Atmosphere*, Harper & Brothers, New York.

Trewartha, G. T., 1968, *An Introduction to Climate*, 4th ed., McGraw-Hill Book Company, New York.

CHAPTER 10
ROCK WEATHERING
AND SOILS

Rocks at the earth's surface are modified by physical and chemical processes. Igneous rocks crystallize at high temperatures, many of them deep in the earth. As geologic time passes, events may bring these rocks to the surface, where the environment is markedly different. Natural equilibrium is disturbed, and adjustment occurs. The most obvious consequence is the formation of soil. Equally important, rock materials become fragmented, facilitating erosion of the material to a new location. Sedimentary rocks show a similar response when exposed at the ground surface.

WEATHERING

The earth's crust consists of a variety of rock types. Rock making up the firm outer part of the lithosphere is bedrock. Between the bedrock and its enveloping hydrosphere or atmosphere is a zone of interaction known as the regolith. The regolith is composed of soil and loosened fragments of bedrock (Fig. 10.1).

The regolith is formed by the different conditions present at the earth's surface, where the atmospheric moisture and organic activity produce great modifications of the bedrock itself. Most of these chemical and

FIGURE 10.1 *Cross section of the upper part of the earth's crust, showing bedrock and regolith divisions.*

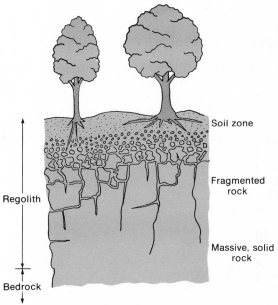

physical changes result from the atmospheric conditions associated with the weather. The processes of regolith formation, therefore, are collectively known as weathering.

Weathering generally takes place too slowly to be observed directly. Fortunately, early civilizations built structures from various rock materials so that historic and archeologic information can supply dates for evaluating the weathering rates of different rock types over long periods. Who has not curiously observed tombstone inscriptions and seen the gradual obliteration of the inscription over time as the gravestones weather?

Weathering affects organic materials even more readily than most rocks. Consequently, the remains of plants and animals are not readily preserved in the rock record, Even steel and concrete yield to weathering over time. Man has been successful in developing ways to build and preserve his handicraft faster than nature can destroy it. This very fact, however, creates great concern to environmentalists.

WEATHERING PROCESSES

Weathering changes in rocks are due partly to physical disintegration and partly to chemical decomposition. Physical breakdown of bedrock involves the fragmentation of minerals and rocks without changing their chemical composition. The varied processes that accomplish this physical breakdown result in disintegration of the rock. The processes of decomposition alter the chemical composition of the rocks to produce new minerals or to dissolve certain elements and ions. During decomposition, the minerals are at least partially broken down into component chemicals. The products of both disintegration and decomposition are more susceptible to removal by erosion than is bedrock itself. For that reason, weathering is generally the forerunner of erosion.

Disintegration and decomposition will be discussed separately; however, physical and chemical weathering usually overlap and take place simultaneously. In fact, one process enhances the effectiveness of the other.

Physical Weathering

Disintegration of rocks by physical weathering produces rock fragments and loose aggregates of mineral

FIGURE 10.2 *A live oak growing in a crack and splitting a mass of limestone near Kerrville, Texas. (Elting H. Comstock.)*

grains. This loose material facilitates plant growth because it permits easier root penetration and increases the moisture-holding capacity of the regolith. There are five physical weathering processes.

Organic Activity   Organisms cause rock disintegration in a variety of ways. Plant roots may penetrate pre-existing rock fractures, and root growth exerts enough force to enlarge such fractures (Fig. 10.2). Not only tree roots but the roots of grass and small plants that penetrate the upper regolith break up rock.

Burrowing animals cause a great deal of disintegration. Groundhogs, moles, chipmunks, and prairie dogs disintegrate rock materials as they excavate passageways. Earthworms are an even greater catalyst as they eat their way through the soil. Soil is swallowed and passed through the intestinal tract. The organic material can be removed for food. An acre of fertile ground may be populated by a million earthworms, and in a single year they can consume up to 50 metric tons of soil while feeding.

Animals or man, moving or using the ground surface, also promote physical weathering. The continual trampling of the ground by grazing animals pulverizes loose rock and soil, and so it is easily removed by gravity transfer, streams, or wind. Man's intensified cultivation in the past thousand years has resulted in a marked increase in the rate of physical weathering. About 6 percent of the world land surface is broken up annually by plowing, affecting the top few centimeters of the regolith. With the advent of large earth-moving equipment, man is now scraping down to bedrock. In many places this redistribution of broken rock and soil has caused serious problems. The artificial earth-fill material does not provide as firm a foundation as does natural regolith grading down into bedrock.

Temperature Effects   Alternate heating and cooling have limited effects. Nearly all solids expand upon heating and contract upon cooling. Each substance has a characteristic volume change when heated, known as its coefficient of expansion. Rocks consist of several mineral types, and so one would expect each rock type to expand and contract by different amounts as surface temperatures rise by day and fall at night. The long-term effect of such heating and cooling should cause fatigue and ultimate fracturing, a clear example of physical weathering. Laboratory tests, however, seem to show that the process is insignificant. Dry granite was subjected to alternate heating and cooling through a temperature range of 110°C for 89,400 cycles (simulating 224 years in nature), and no fracturing occurred. Apparently, temperature change alone will not normally cause disintegration.

A different story is told by desert travelers and mountain climbers. These people frequently report hearing loud snapping noises at sunset. The noise may be caused by rocks fracturing as the air temperature drops rapidly in a short time. Temperature fatigue may be a minor factor of rock weathering in these unusual environments where temperature variations are extreme and abrupt.

On the other hand, some geologists contend that the major cause of boulder fragmentation in the semiarid forests of western America is differential expansion of the rocks caused by the heat of forest fires. The shattering of rock by internal steam under such conditions can hurl debris tens of meters. One occasionally observes this phenomenon by a very hot camp fire.

In summary, simple temperature effects are rela-

tively minor causes of physical weathering. However, if the temperature drops below freezing and considerable moisture is present in the rock, expansion due to ice formation may produce disintegration.

Ice Expansion   Ice expansion breaks rocks effectively. Precipitation enters voids in the regolith and penetrates cracks in rocks to fill the spaces between mineral grains. During cold weather this water freezes from the surface downward. When water changes to ice, its volume expands 9 percent and the enclosing ground or rock is forced apart. Ice crystals can exert a pressure up to 2,000 kilograms per square centimeter, far exceeding the tensile strength of 246 kilograms per square centimeter for the strongest rock.

Consider briefly how ice may enlarge a previous fracture in a rock. If the crack is filled with rain water and the temperature drops below freezing, the water in the crack will first freeze at the surface. This seals the pressure, and any additional freezing will force open the crack.

A phenomenon known as frost heaving results from expansion of freezing ground moisture that is unevenly distributed. The top layers of the ground are buckled upward by ice to produce an irregular surface. Boulders and pebbles also tend to move upward through a soil. This presented a problem to the early New England farmers, who had to remove the larger stones from the ground every spring. Frost heaving is most effective in middle latitudes where the temperatures fluctuate about the freezing point more than a hundred times a year. Its effects are noticeable in the broken highway pavements developed each winter in northern states. In warmer climates, such as the southern United States, the temperature is not sufficiently cold, nor is freezing frequent enough, to produce much physical weathering through ice expansion. In polar regions or on very high mountains, temperatures remain so cold that repeated freezing and thawing do not occur.

Crystal Growth   The formation of new minerals as a result of weathering can exert disrupting forces. As chemical weathering proceeds, certain minerals are altered to new minerals that occupy different volumes than the original. Such examples include the alteration of various silicate minerals, such as feldspars, into clay minerals, the alteration of minerals that contain iron (biotite, pyroxene, and pyrite) to limonite or hematite, and the volume expansion when anhydrite changes to gypsum. The weathering and the physical-volume changes to new minerals proceed at different rates for various minerals so that they exert differential forces on each other. Finally, the grains loosen and cause the rock to crumble.

Unloading   Expansion from reduced pressure also breaks rocks. Bedrock materials often have minute fractures in their structure due to internal expansion. As erosion removes much of the upper regolith, this reduces the confining pressure on the bedrock underneath. The release of pressure permits slight expansion of the rock and the fractures. The fractures and joints are widened and accentuated by weathering processes (Fig. 10.3). The large slabs of once deep-seated rock at Half Dome, Yosemite National Park, and Stone Mountain, Georgia, are in part due to expansion because of reduced pressure. Subsequent weathering and erosion have intensified and further developed these features.

FIGURE 10.3   *Venus Needle (on left), more than 60 meters high, has been detached from the cliff (right) by crumbling of the intervening rock along vertical joints. Physical weathering was the primary cause. (New Mexico State Tourist Bureau.)*

Chemical Weathering

In the weathering zone a variety of chemical reactions occur because of differing chemical environments at the earth's surface compared to environments deeper in the earth's crust. Table 10.1 summarizes these contrasts. The first two changes are related to the heat within the earth and great pressure from the weight of the overlying rocks. The last four contrasts result from a relative abundance of chemical reactants at the surface that are scarce or absent in the earth—water, carbon dioxide, gaseous oxygen, and organisms.

There are many reactions of chemical weathering, but they can be grouped into five types: oxidation, hydration, hydrolysis, carbonation, and solution.

Oxidation   Iron and certain other metals oxidize in the weathering zone. Oxygen, one of the most reactive elements, is also the most abundant element in the earth's crust (Table 5.1) and second in the atmosphere. Deeper, within the earth's crust, all oxygen is chemically combined with the various elements that are present—especially iron, aluminum, potassium, sodium, and magnesium. Iron is the least reactive of these common elements. Oxygen combines preferentially with the other elements listed whenever there is not enough oxygen to combine with all the iron. The ionic forms of iron may be either ferrous or ferric iron. Iron tends to occur in the ferric state if oxygen is abundant.

Olivine is an isomorphous iron-magnesium silicate, and one variety is essentially a pure iron silicate. If such an olivine crystal, found, for example, in gabbro or peridotite, becomes exposed to oxygen ions in the weathering zone, the following chemical change can occur:

$$2Fe_2SiO_4 + O_2 \rightarrow 2Fe_2O_3 + 2SiO_2$$

olivine   oxygen   hematite   silica

Any magnesium present in the olivine crystals will weather by dissolving in carbonated water, leaving an iron oxide residue.

Ferrous iron with available oxygen oxidizes to ferric iron. Ferric ions are smaller in size than the original ferrous ions, and so they create a strain or a looseness in the crystal structure that allows the silica of iron silicates to be more readily available to water. Hematite from an olivine-bearing rock produces a reddish soil. Limonite is another iron oxide mineral. It produces a yellow or brown color in soil. Chemically, hematite is the same as red rust, and limonite is the

TABLE 10.1   Principal Contrasts between Weathering Environment and Conditions Deep in the Earth, with Changes that Accompany Chemical Weathering

At the surface of the earth	Deep in the crust of the earth	Effect in weathering
Low temperature	High temperature	Many minerals formed deep in the earth are unstable, and new minerals tend to form that are stable at cooler temperatures
Low pressure	High pressure	New minerals can form with lower density; porous soil develops from solid rock
Water abundant	Water scarce	Minerals form with water in their crystal structure, notably clays in soil and sediments; some minerals become dissolved
Abundant free oxygen (not chemically combined)		Oxidation in weathering, especially producing hematite and limonite with red and yellow colors in soil
Carbon dioxide widespread	Carbon dioxide generally absent	Minerals containing carbonate ion (CO_3^{--}) such as calcite and dolomite
Organisms present	Absence of life	Plant growth uses mineral nutrients and dies, to form humus in soil or accumulate as peat; digestive processes of tiny soil animals chemically modify minerals; organic acids are produced at the surface of the earth

same as a yellow-colored rust. Limonite forms by oxidation of metallic iron in the presence of water. Ferrous silicate minerals other than olivine that also weather to iron oxide (hematite or limonite) include amphibole, pyroxene, biotite, and chlorite.

Sulfides, such as pyrite, oxidize to form metallic oxide minerals and release sulfur dioxide to the air or water. Sulfide minerals are abundant in many metallic ores, and pyrite is frequently associated with coal. The weathering of pyrite (FeS_2) usually involves reaction with water along with oxidation:

$$2FeS_2 + \frac{15}{2}\,O_2 + 4H_2O \rightarrow 4SO_4^{--} +$$
$$\underset{\text{pyrite}}{} \quad \underset{\text{oxygen}}{} \quad \underset{\text{water}}{} \quad \underset{\text{sulfate}}{}$$

$$\underset{\text{limonite}}{Fe_2O_3 \cdot nH_2O} + \underset{\text{hydrogen}}{8H^+}$$

Limonite forms by oxidation of the iron and chemical incorporation of water into its amorphous structure, Quite often, oxidation of pyrite, which releases heat, proceeds rapidly enough to start fires by spontaneous combustion of mine waste dumps.

If the sulfate combines with H^+ ions in water, it forms sulfuric acid:

$$\underset{\text{sulfate}}{SO_4^{--}} + \underset{\text{hydrogen}}{2H^\pm} \rightarrow \underset{\substack{\text{sulfuric acid} \\ \text{(in water solution)}}}{H_2SO_4}$$

This acid, in turn, can dissolve the limonite so that waste waters from coal mines are in part a solution of iron in sulfuric acid. This solution devastates fish and other life in streams and makes the water unfit for human consumption or industrial use. As the solution is aerated and diluted by cleaner stream waters, limonite is precipitated as a yellowish-brown stain on the channel floors of polluted streams.

In many mining areas of the world this weathering phenomenon creates a serious environmental problem; however, stream pollution by such mine wastes can be stopped. A knowledge of chemical weathering suggests an answer. Cut off the supply of oxygen and the flow of water from abandoned mines. The extra expense of sealing old mine entrances and covering mine debris with earth to inhibit oxidation leads to cleaner streams and less pollution. Legislation now attempts to curb this type of pollution in many coal-mining areas.

Hydration and Dehydration   The formation of a new mineral with water in its structure is called hydration. The reverse reaction is dehydration. Water is so abundant in the weathering zone that it commonly reacts with anhydrous minerals. Consider the hydration and dehydration reaction with gypsum and anhydrite:

$$CaSO_4 \cdot 2H_2O \underset{dehydration}{\overset{hydration}{\rightleftharpoons}} CaSO_4 + 2H_2O$$

gypsum   anhydrite   water

The reaction shown is reversible, and experimental results indicate that both hydration and dehydration take place readily in nature. Gypsum and anhydrite exist indefinitely as museum specimens, indicating that the reactions must be extremely slow.

Hydrolysis   Even though the weathering of pyrite or gypsum is of interest, these minerals are scarce compared to the feldspars that weather by hydrolysis. Feldspars constitute about half the minerals in igneous rocks. Their chemical weathering produces clay minerals and solutions of silica:

$$4KAlSi_3O_8 + 22H_2O \rightarrow Al_4Si_4O_{10}(OH)_8 +$$

potash feldspar   water   kaolinite clay

$$4K^+ + 8H_4SiO_4 + 4OH^-$$

potassium ions   silicic acid   hydroxyls

$$4NaAlSi_3O_8 + 22H_2O \rightarrow Al_4Si_4O_{10}(OH)_8 +$$

sodium feldspar   water   kaolinite clay

$$4Na^+ + 8H_4SiO_4 + 4OH^-$$

sodium ions   silicic acid   hydroxyls

Several aluminum-bearing silicate minerals besides the feldspars also weather to clay minerals, including biotite, muscovite, amphibole, and pyroxene.

In the very severe chemical weathering common to the tropics, additional breakdown by removal of silica and the leaching of kaolinite clay can occur:

$$Al_2Si_2O_5(OH)_4 + nH_2O \rightarrow Al_2O_3 \cdot nH_2O$$

kaolinite   water   bauxite

$$+ 2SiO_2$$

silica
(dissolved in water)

Bauxite has a formula analogous to limonite, but without the iron, and serves as the principal ore of aluminum.

Carbonation   Certain elements, especially Ca, Mg, Na, and K, readily form carbonates or bicarbonates. Carbon dioxide in the air dissolves in rain water to form weak carbonic acid:

$$CO_2 + H_2O \rightleftharpoons H_2CO_3$$

carbon   water   carbonic
dioxide   acid

This acid ionizes mainly to H^+ and HCO_3^- ($HCO_3^- \rightleftharpoons H^+ + CO_3^{--}$) and reacts with the metallic ions in several minerals to form new carbonate compounds. As an example, consider the weathering of calcite, the principal mineral in the sedimentary rock limestone. In the process shown, crystalline calcite disappears:

$$CaCO_3 + H_2CO_3 \rightarrow Ca^{++} + 2HCO_3^-$$

calcite   carbonic   calcium   bicarbonate
             acid   ions   (in solution)

The abundant calcium, magnesium, sodium, and potassium bicarbonate solutions formed by carbonation of various minerals enter streams and are carried toward the sea. Chemical weathering of calcium silicates ultimately provides the Ca^{++} and CO_3^{--} for the large calcium carbonate accumulation on the sea floor. This calcium carbonate is in the form of shells and tests of organisms. Limestone weathers by reacting with carbonic acid, and the resulting calcium carbonate is then easily removed in solution. Calcite is readily dissolved by weathering in a humid climate. In semiarid and arid climates as in Wyoming, however, limestone is quite resistant to weathering and frequently forms bold cliffs. Water is too scarce there to carry on the solution process effectively.

Solution   Solution of certain constituents as a part of weathering of some minerals has already been described. Several other minerals are directly soluble in water or will react chemically with natural acids and then dissolve in water. The most obvious example is the mineral halite (NaCl), which sometimes occurs in sedimentary layers as rock salt. Halite weathers by dissolving in water. Consequently, halite crystals are not found at the ground surface except in extremely arid areas.

Natural acids may react with minerals that do not dissolve directly in water. We have already mentioned limonite, which is normally insoluble but which dissolves in sulfuric acid waters.

Natural acids produced by weathering may augment other weathering processes. Sulfuric acid liberated by the weathering of pyrite and other sulfides can react with minerals like calcite and feldspar. Various organic acids released by decay processes yield another unlimited source of weak acids. Regardless of the source of the acid, or its chemical composition, acid can react with various minerals, and certain of the products can, in turn, dissolve in water.

As this survey of chemical weathering has shown, the processes are very complex. Chemical reactants vary in different rock types and weathering environments, and various reactions also occur simultaneously.

DISINTEGRATION AUGMENTS DECOMPOSITION

Chemical weathering proceeds more rapidly as mineral grains are physically fragmented. Consider the illustration in Fig. 10.4. For simplicity, assume that it is a cubic crystal of halite 2.54 centimeters (1 inch) across. Chemical weathering of halite occurs by solution. The crystal dissolves on its surface where contact is made with water, rather than inside the mineral grain. Consequently, the rate of weathering is proportional to the surface area. If the cube has an initial volume of 16.4 cubic centimeters (1 cubic inch), its surface area is 38.7 square centimeters (6 square inches). Suppose that the crystal is cleaved into eight cubes, each 1.27 centimeters ($\frac{1}{2}$ inch) across. The total surface area is increased to 77.4 square centimeters (12 square inches), while the total volume remains

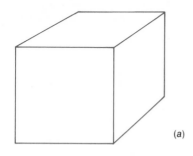

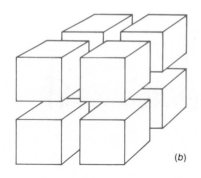

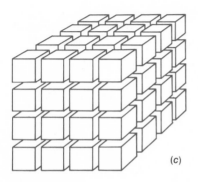

FIGURE 10.4 *Sketches showing how disintegration of a 2.54-centimeter (1-inch) cube increases the rate of weathering as the surface area enlarges.*

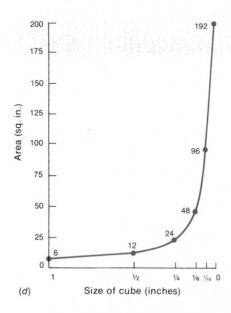

(a)

(b)

(c)

(d)

**TABLE 10.2   Goldich Mineral Stability Series (---)
           Compared with Bowen's Reaction Series (——)**

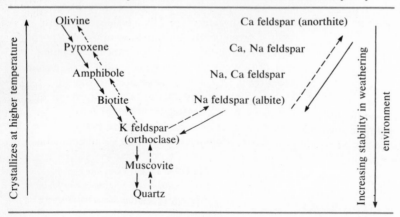

16.9 cubic centimeters (1 cubic inch). Further disintegration to 0.64-centimeter ($\frac{1}{4}$-inch) cubes yields a surface area of 154.8 square centimeters (24 square inches), with a constant total volume. Because chemical weathering occurs at the grain surface, solution proceeds 4 times faster in 0.64-centimeter ($\frac{1}{4}$-inch) than in 2.54-centimeter (l-inch) cubes. If the crystal is crushed to fine dust, solution takes place 10,000 times faster than in the original 1-inch cube.

Weathering by chemical reaction rather than by simple solution, such as transformation of feldspar to a clay mineral, is enhanced by disintegration. Nutrients are more readily obtained by plants in finely pulverized mineral matter where ions are more abundant. Soil nutrients are more easily obtained from pulverized rock than from hard bedrock.

Weathering of Representative Rocks

In 1938, Goldich pointed out a relationship between the sequence of crystallization in Bowen's reaction series (Chap. 7) and the rate of chemical weathering of these same minerals. He found that the minerals which crystallized first (at high temperature) are those which are least stable in the weathering environment. They weather first and more rapidly than the last minerals to crystallize from a magma (Table 10.2). This is a classic example of change and adjustment to equilibrium.

Disintegration and decomposition can be summarized by reviewing the weathering of three rock types in a humid climate: granite, basalt, and quartz sandstone.

Granite   If a granite mass, exposed at the earth's surface, has undergone physical weathering to such a degree that the granite has been broken into smaller fragments, decomposition may then increase rapidly. Granite is composed chiefly of K, Na, and Ca feldspars, quartz, biotite, amphibole, and muscovite. The feldspars will gradually be altered to clay minerals, plus solutions containing K^+, Na^+, Ca^{++} ions, and soluble silica. As the feldspars are weathered, the rock changes color and becomes less resistant. Quartz, which crystallizes last in the Bowen reaction series, is stable at and near the ground surface and may accumulate as a sand-sized residue. Biotite weathers to kaolinite, limonite, potassium and magnesium bicarbonate, and silica solutions. Muscovite weathers more slowly, but it too eventually alters to kaolinite. The final result of the weathering of granite is a yellowish-colored sandy clay soil.

Basalt   This rock is composed chiefly of plagioclase feldspar and pyroxene, with lesser amounts of olivine and magnetite. It crystallizes at much higher temperatures than granite, and it is normally affected by chemical weathering at a faster rate. The rate of chemical weathering is greatly increased as disintegration of the solid rock reaches an advanced stage. Upon weathering, the feldspars and pyroxenes in basalt produce clay minerals. The weathered material is usually colored by iron oxides resulting from the breakdown of olivine, pyroxene, and magnetite. Quartz is absent, and so the final decomposition product is clay minerals. In a tropical climate the clay itself may further decompose, leaving a mixture of limonite and bauxite in a soil type called laterite. Occasionally, laterite derived from basalt or peridotite is rich enough in iron to be used as a low-grade iron ore.

Quartz Sandstone   A quartz sandstone can be derived by the accumulation, deposition, and subsequent cementation of sand-sized grains of quartz, other minerals, and rock fragments. In a quartz sandstone the dominant mineral is quartz. Silica and calcium carbonate are common cements.

In this case, the mineral grains in the sandstone have already been through at least one cycle of weathering and erosion. The chief constituent is quartz, and so chemical weathering is not appreciable. Instead, the sandstone slowly disintegrates by physical weathering and produces a light-colored sandy soil deficient in clays.

Other Rock Types   Rock weathering is quite varied, depending chiefly on mineral composition but also on climate. Disintegration is more important in a dry climate than in a humid one. Figure 10.5 summarizes the rate of weathering of several rocks in such a climate. The examples compare the three major rock groups, igneous, sedimentary, and metamorphic.

SPECIAL EFFECTS OF WEATHERING

Spheroidal Weathering

Massive rocks without closely spaced fractures often weather into rounded forms. This phenomenon is called spheroidal weathering (Fig. 10.6).

Consider a cubic block of granite exposed at the ground surface (Fig. 10.4). The cubic block could represent a granite block several meters across. Recall that the principal weathering of granite involves the formation of clay minerals from feldspar and mica. These chemical reactions are most effective at the surface of the granite, rather than within it. Thus weathering takes place along each face of the cubic block. However, the edges, and the corners, weather faster than the faces of the cube, and eventually a residual spheroidal mass of granite is formed. Further chemical weathering proceeds uniformly inward from all directions, intensifying the spheroidal shape.

The reactions of water, oxygen, and carbon dioxide with chemically unstable rocks, such as granite, often proceed unevenly on the rock surface so that concentric rock layers crack loose. This process of concentric spalling off of rock layers as a consequence

FIGURE 10.5   *Rocks differ in resistance to weathering in a humid temperate climate. The arrows indicate the range of behavior of each rock type.*

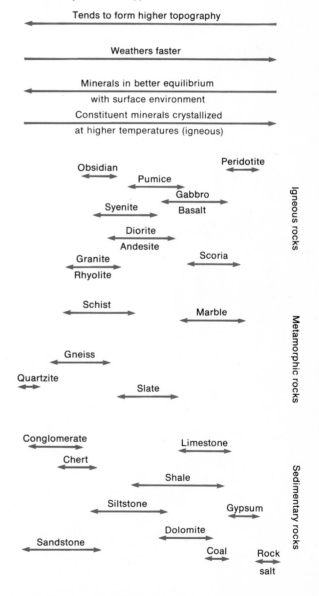

FIGURE 10.6   *Spheroidal weathering converting jointed blocks of rock into rounded boulders by exfoliation, along American River near Riverton, California. (Eliot Blackwelder.)*

FIGURE 10.7   *Tower of Babel, rock formation at the foot of Lake Moraine, in the Valley of Ten Peaks, near Lake Louise, Banff National Park, Alberta, Canada. Intense weathering has attacked the cliffs, and gravity transfer has carried the materials to lower levels near the lake. (Canadian Government Travel Bureau.)*

FIGURE 10.8 *Montezuma Castle, a national monument near Camp Verde, Arizona. Differential weathering etched the pitted recess in this cliff, which the cliff dwellers further modified into a home. (The Atchison, Topeka and Santa Fe Railway Company.)*

of chemical weathering is called exfoliation. Exfoliation generally results in and promotes spheroidal shapes.

Differential Weathering

Uneven weathering is common because many rock masses are not homogeneous, and weathering processes are not uniform over the entire rock. This results in irregular patterns and shapes of the rock masses through a process called differential weathering. The layers of various sedimentary rocks, such as sandstone and shale, often weather differentially, leaving the more resistant rocks standing in relief (Figs. 10.3, 10.7, 10.8, 10.9, and 10.10). On a large scale, tilted layers of rocks like sandstone may form ridges or other areas of high relief, while less resistant shales form valleys.

On a smaller scale, differences in rock composition and water circulation may cause a rock exposure to be pitted (Fig. 10.8). Occasionally, differential weathering produces shapes likened to sculptures of animals (Fig. 10.9). Other odd-shaped, weather-resistant pinnacles are called hoodoos (Fig. 10.10).

Often a rock fracture may become enlarged by weathering; these "chimneys" are often utilized by al-

FIGURE 10.9 *Camel Rock, near Santa Fe, New Mexico. This unusual shape originated by differential weathering. (New Mexico Tourist Bureau.)*

FIGURE 10.10 *Hoodoos developed by differential weathering in Yoho National Park, British Columbia. Large rock fragments in a breccia protected the underlying material from weathering and erosion. (Canadian Government Travel Bureau.)*

pine climbers. Sometimes an extremely pitted rock surface due to differential weathering forms isolated features such as small "window rocks" or large "natural bridges." Natural bridges originate not only by differential weathering but by many other erosional processes.

While weathering (and erosion) work to create particular differential weathering features, the processes do not stop once a feature is formed. Chimney Rock, a unique pinnacle formed by differential weathering, was once a landmark in northeastern Oklahoma. Early in 1973 (March 15), Chimney Rock, which was a guiding point for wagon trains in the early 1800s, collapsed. Rains in the area had weakened the base of the structure, and high winds subsequently blew it over.

Caliche

In humid climates, many of the products of chemical weathering are removed by the plentiful water. In drier regions, these products may accumulate in the soil horizons, forming a whitish crust called caliche. In semi-arid and arid terrains, calcite is the most abundant mineral in caliche. In extremely arid areas, caliche may also contain sodium carbonate, silica, halite, or nitrate salts.

SOILS

The word soil has various meanings. To the engineering geologist it implies the loosely coherent to unconsolidated material overlying firm bedrock; hence, it is synonymous with regolith.

To other geologists, soil is a loose accumulation of mineral grains and rock fragments that forms as the bedrock interacts with the atmosphere and water. The soil scientist emphasizes the interrelation of life with soil, recognizing the living and dead organisms in it and aware of the varied plants and animals that are nourished from the soil. This chapter will present the agricultural perspective.

The capability of providing a rooting place for plants distinguishes soil from nonsoil. As defined by soil scientists, soils are natural materials on the earth's surface, containing living matter and supporting or capable of supporting plant growth.

Soil Components

Soil scientists recognize four essential soil components. First of these is mineral materials. However, soil is not a homogeneous mixture of loose mineral particles, because the minerals usually are segregated into vertically arranged zones.

The second essential component is organic matter. Decaying organisms provide valuable nutrition for other plants and clearly modify other chemical and physical properties of the soil. For instance, earthworms markedly increase the availability of plant nutrients by chemically and physically modifying the mineral grains. Bacterial content influences decay processes, and some bacteria convert nitrogen from the atmosphere to forms that are soluble as plant nutrients. Larger plants of the legume family, such as peas, beans, clover, and locust trees, also add nitrogen to the soil in a symbiotic association with nitrate-producing bacteria.

Water is the third ingredient, and one obviously necessary for agriculture. Compounds dissolved in water serve as the food supply for plants.

Lastly, a productive soil must have adequate pore space for air and water movement between the grains. As a result of decay processes, soil air generally contains more carbon dioxide and less free oxygen than the atmosphere. Soil must be well ventilated or oxidation will be curtailed, including chemical weathering of mineral grains and the decay of organic debris. Deficiency of oxygen causes the iron to be in the more soluble ferrous state and favors the accumulation of organic matter. In extreme cases it will allow the formation of a peat soil.

Organic activity, disintegration, and decomposition interact so closely with erosion that thousands of varieties of soil have been identified in the United States alone. The condition of the soil in any place at any time is the result of interacting creative and destructive processes. Soil characteristics alter when the

soil-forming processes change or vary in their intensity. Table 10.3 summarizes the major soil-modification factors.

Soil Profile

The arrangement of soil components in a recognizable vertical sequence is a soil profile. This sort of zonation within the soil varies from place to place, depending on climate, the parent rock material, relief of the terrain (especially slope angle), the biologic content of the soil, and time.

Partially weathered bedrock becomes soil when it crumbles into loosened material capable of supporting plant life. This soil changes as environmental factors shift. Gradually the soil increases in fertility; then it declines. Surprisingly, parent-bedrock type influences soil type only to a minor extent. Climate, slope, and biologic factors are dominant; thus similar soils can originate from various bedrock types in similar environments.

Figure 10.11 illustrates a typical mature soil profile. The following horizons are generally present in a well-developed soil:

A_0. The topmost soil layer generally is dominated by a dark organic accumulation of fresh or partly decomposed organic matter (humus) overlying the mineral grains of the soil.

A_1. (zone of leaching) Organic material mixed with mineral grains. Organic debris in percolating water that contains dissolved oxygen

TABLE 10.3   Factors that Control Thickness and Character of Soil

Favors soil accumulation	Favors soil destruction
1. Disintegration of bedrock into regolith	1. Erosion of topsoil by water, wind, ice, or gravity transfer. Becomes more efficient as slope increases
2. Decomposition of parent minerals into clays and oxides	2. Leaching of soluble products, removing them to streams
3. Accumulation of organic decay products, generally producing dark color and increasing the looseness of the soil	3. Burning of organic material. Destroys water-holding ability, depletes organic content of soil, and obliterates the plant cover that shields the soil from erosion
4. Protective cover of plants shields the soil from rain and wind erosion	4. Harvest of crops depletes certain chemicals. Normally these are returned to the soil by plant decay or animal manure, but removal of crops to urban areas depletes the plant nutrients almost as effectively as leaching by rainwater
5. Long time favors development of thick soil	
6. Addition of fertilizers adds chemicals helpful for plant growth	

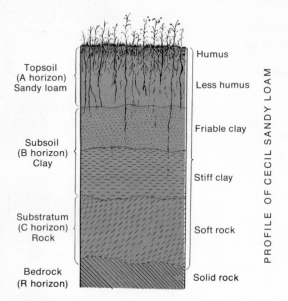

Topsoil (A horizon) Sandy loam

Subsoil (B horizon) Clay

Substratum (C horizon) Rock

Bedrock (R horizon)

Humus

Less humus

Friable clay

Stiff clay

Soft rock

Solid rock

PROFILE OF CECIL SANDY LOAM

FIGURE 10.11 *Profile typical of a mature, well-zoned soil in the Corn Belt of North America. Labeled zones are explained in text.*

(Fig. 10.12). Weathering converts silicate minerals to clay minerals and iron oxides, which are then physically carried downward by water. If quartz is present, it will concentrate in the A horizon.

B. (zone of accumulation) Aluminum (in clays) and iron (limonite or hematite) are deposited by downward percolating water and fill the spaces around partially weathered mineral grains. This zone may be red or yellow in color from abundant iron oxide and may form a "hardpan" that is resistant to plowing and root penetration.

C. Loosened rock and mineral grains, broken by physical weathering or fragmented by erosion, but slightly modified chemically.

R. Underlying bedrock (consolidated). This is the unweathered rock material.

The A_0 layer thickens if the rate of plant growth exceeds decay (Fig. 10.12). It is the soil layer most susceptible to erosion, and so slow erosion is essential for development of a thick A_0 horizon (Fig. 10.13).

This horizon can also be destroyed by forest or grass fires or by mixing with other materials by plowing.

Soil profiles are produced over hundreds of thousands of years, depending on the original material and the climate. As rocks with a diverse mineralogy weather, soil looseness and availability of plant nutrients increase until the well-zoned fertile soil is said to be mature. Additional weathering removes nutrient chemicals from the soil layers within reach of plants, and so organic productivity declines and the soil reaches an old-age stage. Old soils develop only where weathering is very severe and the average rate of erosion is slow. Animals suffer malnutrition in regions with old soils. This is a serious hindrance to development of adequate food supplies in low-latitude countries of Southeast Asia, India, central Africa, and equatorial South America, as well as southeastern United States. Many severely leached soils can be restored to productivity by appropriate addition of commercial fertilizers.

Figure 10.14 illustrates soils that can develop in a region underlain by granite. (Recall that bedrock type is relatively unimportant, and fairly similar soils may form from different rock types.) The aging of the soils is illustrated from left to right in columns I to IV.

If erosion is very active, loosened rock fragments are removed from the bedrock, and so soil zones do not form (column I). Instead, there is an outcrop of bedrock exposed at the ground surface. As time passes, and if erosion is not too active, the bedrock becomes covered with broken rock and mineral fragments, forming a C horizon. The loose material can be penetrated by plant roots, and chemical breakdown has not advanced far enough to release many soil nutrients. Such an infertile soil with slight development of a profile is said to be very youthful (column II).

In column III, an optimum soil zonation has developed. The upper layers are leached and a loose residue remains, and so roots penetrate effectively. The tiny clay and iron oxide particles washed downward from the A horizon have accumulated in the B horizon. Mineral nutrients are readily available to plants in the partially weathered grains of the B horizon, and the less weathered material in the C horizon provides good root support and water storage plus nutrients. Plant growth is favored, and so a top layer of organic debris develops an A_0 horizon. This well-zoned mature soil is at its optimum fertility.

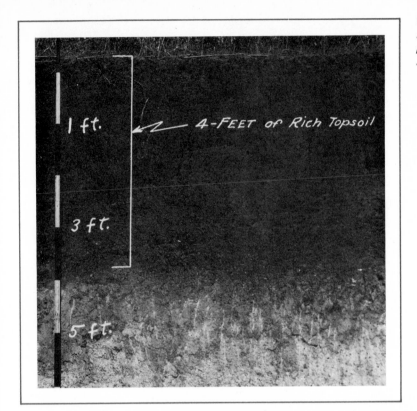

FIGURE 10.12 *Soil profile with thick humus layer in A horizon on flat topography in eastern Texas. (Soil Conservation Service photo.)*

FIGURE 10.13 *Soil profile on gently sloping surface near area shown in Fig. 10.12. Soil erosion has removed humus layer of the A horizon. (Soil Conservation Service photo.)*

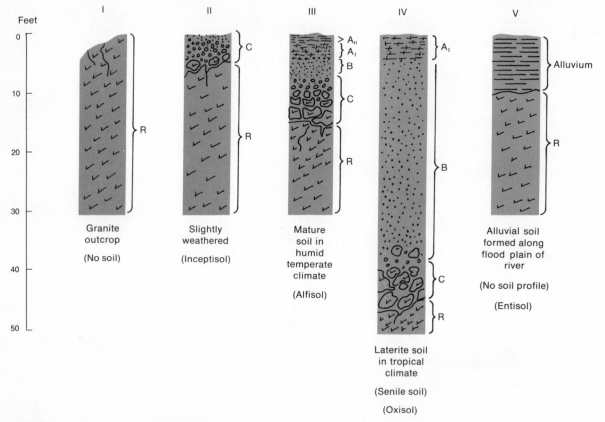

FIGURE 10.14   *Five soil situations developed in granite terrain. The corresponding soil descriptions in Table 10.4 are: II-(2), III-(6), IV-(10). and V-(1).*

As leaching continues, the upper layers become
nearly depleted of nutrients, and so plant growth
declines (column IV). The A_0 horizon fades because
decay exceeds accumulation. Given sufficient time
and a severe climate, the clay minerals decompose and
the silica in them dissolves. Aluminum oxide then accumulates in the B horizon with the iron oxides. This
condition is most common in the tropics. The resulting
infertile soils, rich in aluminum and iron, are called lat-
erites. Peak fertility has passed and the soil is old.

Column V of Fig. 10.14 portrays another situation in which a soil profile is absent. A valley is carved
in granite bedrock by a stream. Then stream-borne
sediment known as alluvium is deposited over the
bedrock. Floods every few years add a new alluvial
layer, and so weathering has little opportunity to develop a soil profile. Alluvial soils are very fertile even

if immature. Alluvium consists of a loose aggregate of
finely ground rock debris, clay, and organic matter that
readily absorbs moisture and provides plant nutrients.
Some of the world's best soils are alluvium, like those
along the Nile, Indus, Ganges, Mekong, and Mississippi Rivers.

Soils develop quite slowly. A rock may crumble in
a few decades, but a much longer time is required for a
fertile zoned soil to mature. Well-zoned soils have developed in about 1,000 years in Alaska, and over 30
centimeters of soil has formed on the ruins of a
Ukrainian fort that was abandoned in 1699. In the
American upper Midwest the present weathering
cycle began when the glaciers melted (10,000 to
15,000 years ago); yet the soils are nowhere near old
age. Soil development is faster in warm climates than
in cool ones and is encouraged by high precipitation.

Young, mature, and old soils do not necessarily reflect total years or even a sequence of events; rather, they reflect intensity and type of cumulative chemical alteration of the parent materials that developed into soil. An old-age soil can form only if chemical-environmental conditions are sufficient to cause leaching of silica from the clay minerals, regardless of the span of weathering time. In areas with less rigorous leaching, such as Canada or the northern United States, final attainment of equilibrium with environment probably will not produce an old soil. It is difficult to evaluate the ultimate effects of long periods of weathering in northern areas, since these areas had most of their original soil removed by glacial scraping during Ice Ages. The new soils are only post-glacial in age, a short span of time. The soil at any given locality is the result of all the past geologic and environmental history of the site.

Soil Classification

Over 7,000 varieties of soil have been named in the United States, and so this text will give only a broad outline of soil classification. Soils are named according to the type of profile developed. Table 10.4 summarizes a modification of the 10 orders recognized in the soil classification adopted by the U.S. Department of Agriculture in 1965. Figure 10.15 shows the geographic distribution of these soil orders in the United

TABLE 10.4   Names, Characteristics, and Environments of the Ten Soil Orders

Name	Soil characteristics and environments
1. Surface deposits (Entisol)	Soils lacking soil-layer horizons, as on stream alluvium, dune sand, or mountain slopes
2. Surface deposit with weakly developed profile (Inceptisol)	Soils that are usually moist, with soil-layer horizons of parent materials but not of accumulation, because of erosion or other preventive factors
3. Soils of deserts—semiarid regions (Aridisol)	Soils with soil layer, low in organic matter, and dry more than 6 months of the year in all horizons so that alkali and caliche, noxious to many plants, accumulate
4. Grassland soils (Mollisol)	Mature soils with nearly black, organic-rich horizons and high unleached calcium supply, under moderate to sparse rainfall on grasslands and forests
5. Podzol, brown podzol soils (Spodosol)	Soils with accumulation of amorphous materials consisting of organic matter and iron or aluminum oxides, with a bleached ashy-gray A horizon and conspicuous B horizon in forested areas under cool moist climates
6. Gray-brown podzols (Alfisol)	Soils rich in dark organic matter at top, underlain by bleached gray zone on a brown B horizon, with medium to high base supply, and subsurface horizons of clay and iron oxide accumulation; usually moist, but may be dry during warm season; characteristic of well-drained land in humid temperate forests
7. Bog and half-bog soils (Histosol)	Soils composed of plant debris in water-logged ground; peat, the first step toward coal; drained swamps well suited to truck farms
8. Soils with clays that crack and swell (Vertisol)	Dark soils with high content of swelling clays (montmorillonite) that can cause overturning or inversion in wet seasons and wide deep cracks in dry seasons; on volcanic tuff or clayey limestone or chalk
9. Red and yellow podzols (Ultisol)	Well-leached red-clay soils approaching old age; usually moist with horizon of clay accumulation and a low base supply, as in southeastern United States
10. Lateritic soil (Oxisol)	Old-age lateritic soil consisting of nearly pure aluminum or iron oxides, developed in tropical countries by extreme decomposition that removes silica and converts clay to bauxite, as in Caribbean countries, parts of Africa, India, and Southeast Asia

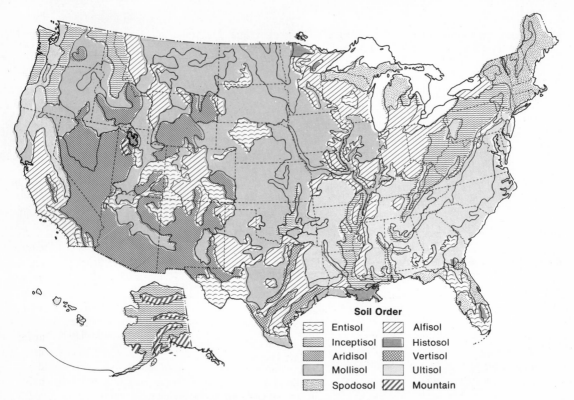

FIGURE 10.15 *Geographic distribution of soil orders in the United States. The soil orders in the legend (left column) correspond to descriptions 1 to 5 in Table 10.4. The orders in the right column, with the exception of the last order, correspond to descriptions 6 to 9.*

Soil Order

Entisol		Alfisol	
Inceptisol		Histosol	
Aridisol		Vertisol	
Mollisol		Ultisol	
Spodosol		Mountain	

States. The strong influence of precipitation on soil development is obvious when one compares annual precipitation with the geographic distribution of soil types. (Compare Fig. 9.5 and 10.15)

Soil Maintenance

Obviously, soil resources must be carefully preserved if the world is to cope successfully with its population explosion. Unfortunately, man has only recently developed an appreciation for conservation, and most agricultural areas have already suffered great damage, some of it almost permanent. Nature generally requires thousands of years to build a mature soil with peak fertility. Management practices can now control erosion (Fig. 10.16) and restore some chemicals extracted from the soil well enough to maintain a reasonable fertility or even improve it.

Erosion is controlled by terracing and contour plowing to decrease the slope and slow the runoff of precipitation. After the crops have been harvested and before the next growing season, the ground often lies bare for several months each year. Winter rainfall quickly erodes the unprotected ground, and so a cover crop should be planted, usually a legume or grass. Next spring the cover crop is plowed under before the new crop is planted, providing favorable physical conditions and some organic fertilizer for the new growth.

Soil aeration can be improved by plowing, which also breaks up the soil to aid root penetration. Exceptionally deep plowing every few years is beneficial for certain crops because the hardpan soil of the B horizon is broken.

Soil moisture can be controlled by terracing and contour plowing to increase water influx, by cultivating soon after a rain to aid evaporation from deep in the soil, and by irrigation that concentrates the water which fell over a much larger area. Irrigation waters can be used to dissolve harmful salts (especially excessive sodium) from certain arid soils creating productive soils out of desert and semidesert lands. Too much moisture can be reduced by draining swamplands.

The organic or humus content of the soil is easily eroded, since it is the topmost layer (Fig. 10.13). Hundreds of years were required to build up the A horizons, and they are not easily replaced. In many regions, including southeastern United States, farmers often burn the fields in the fall to get rid of weed seeds and pests and to produce fertilizer salts that can dissolve from the ashes. This practice increases the fertility of the top soil the next spring, but the long-range effect of field burning is to deplete soil nutrients. Winter rains wash away much of the ash, taking with it chemicals that growing plants extracted from deep in the soil and that would have been added to the topsoil by plant decay. Other nutrients are removed from the fields as plants are consumed by animals. Part of these can be reintroduced by adding organic fertilizer in the form of manure.

Chemical fertilizers are the newest way to improve soil fertility. Common fertilizer elements are calcium, potassium, phosphorus, and nitrogen. All are obtainable from localities where they are concentrated in the earth, usually as sedimentary rocks. Reserves of these essential elements are abundant, but their geographic distribution is very uneven—often scarce where they are needed most. Fortunately, this country is well supplied. Nitrogen can now be extracted from the air and formed into compounds that plants can utilize, at a cost cheaper than mining and transporting mineral nitrates. This has become the principal source of nitrogen fertilizer, replacing the limited supply of nitrate minerals formerly obtained from the deserts of Chile. Nitrogen is an ingredient of protein molecules. Developing adequate supplies of usable nitrogen compounds and improving the efficiency of plant utilization of soil nitrogen will be major steps in improving man's health.

Other elements in trace amounts that are essential to plant nutrition include iron, molybdenum, cobalt, copper, sulfur, zinc, boron, magnesium, and manganese. Crops must be watched for deficiencies, after several years of intensive farming, to see if a few pounds of trace-element fertilizer per acre are advisable.

Certain plants use small amounts of particular metallic ions, and such plants are naturally located on soils developed from rocks containing these metals. Identification of particular species by chemical analysis of the ash from selected plants may reflect the com-

FIGURE 10.16 *Sheet erosion by rainwash in a bean field near Torrance, New Mexico. Although erosion by rainwash is sporadic, its effects are appreciable. Topsoil, as in this photo, is washed away, and the fertility of the field is decreased. (Soil Conservation Service Photo.)*

position of rocks beneath the soil.

A cause of diminishing soil fertility is man's violation of the natural cycle that returns all plant material to the ground (or its natural return as manure). Now, man takes useful parts of plants to urban areas for garden fertilizers. These fertilizers ultimately enter the sewers and are washed to the sea.

Money spent for fertilizer is well invested. Tests of lime (Ca) fertilizer in northeastern Ohio resulted in an increased crop yield over several years worth 5 times the cost of the fertilizer. The C horizons here are rich in lime; yet the surface soil layers are lime-deficient. Such an example demonstrates the great promise of fertilizers to increase the world's crop productivity. Unfortunately, undernourished people are generally too poor to invest in fertilizers. Heavy government investment in fertilizer may be necessary in certain countries to overcome the threat of starvation.

REFERENCES

Blackwelder, E., 1925, *Exfoliation as a phase of rock weathering*, Journal of Geology, vol. 33, pp. 793-806.

Carroll, D., 1970, *Rock Weathering*, Plenum Press, New York.

Emery, K. O., 1960, *Weathering of the Great Pyramid,* Journal of Sedimentary Petrology, vol. 30, pp. 140-143.

Goldich, S. S., 1938, *A study of rock weathering*, Journal of Geology, vol. 46, pp. 17-58.

Hunt, C. B., 1972, *Geology of Soils*, W. H. Freeman and Company, San Francisco.

Keller, W. D., 1957, *The Principles of Chemical Weathering*, 2d ed., Lucas Bros. Publishers, Columbia, Mo.

———, 1966, *Chemistry in Introductory Geology*, Lucas Bros. Publishers, Columbia, Mo.

———, 1966, *Geochemical weathering of rocks: Source of raw materials for good living*, Journal of Geological Education, vol. 14, pp. 17-22.

McNeil, M., 1964, *Lateritic soils*, Scientific American, vol. 211, no. 5, pp. 96-102.

Millar, C. E., Turk, L. M., and Foth, H. D., 1965, *Fundamentals of Soil Science*, 4th ed., John Wiley & Sons, Inc., New York.

Ollier, C., 1969, *Weathering*, Oliver & Boyd Ltd., Edinburgh.

Reiche, P., 1950, *A Survey of Weathering Processes and Products*, The University of New Mexico Press, Albuquerque, New Mexico; University of New Mexico Publications in Geology no. 3.

Thompson, L. M., 1957, *Soils and Soil Fertility*, McGraw-Hill Book Company, New York.

United States Department of Agriculture, 1938, *Soil and Men*, Yearbook of Agriculture.

———, 1957, *Soils,* Yearbook of Agriculture.

———, Soil Survey Staff, 1960, *Soil Classification, A Comprehensive System*, 7th Approximation.

United States Geological Survey, 1969, *Distribution of Principal Kinds of Soils*, National Atlas Sheet 86, Scale 1: 7,500,000.

Winkler, E. M., 1965, *Weathering rates as exemplified by Cleopatra's needle in New York City,* Journal of Geological Education, vol. 13, pp. 50-52.

CHAPTER 11

SEDIMENTATION AND SEDIMENTARY ROCKS

Weathered rock transported by streams, glaciers, and the wind is eventually deposited as layers upon the underlying rocks of the earth's crust. Organic accumulations may also become buried with the rock-derived material. These layered unconsolidated materials are called sediments. With the passage of time, these layers of sediment may become firmly consolidated and lithified to form sedimentary rocks (Table 11.1). Three-fourths of the land area of our planet is covered by sediments and sedimentary rocks, and only one-fourth by igneous and metamorphic rocks. The bottoms of most lakes and streams are covered by a veneer of sediments, and the sea floor contains vast areas where sediments have accumulated for millions of years without appreciable interruption.

Sedimentary rocks contain mineral resources especially useful to man: oil, gas, and coal are examples. They also yield ceramic clay, building material, fertilizers, chemical raw materials, the ingredients of concrete, and ores of several metals, including iron, aluminum, manganese, and titanium. Sedimentary rocks are particularly valuable in studying the history of the earth. They contain the fossil record of life through the ages (Fig. 11.1). In contrast, igneous and metamorphic rocks contain very few fossils. The relative age of a sediment can be determined by its fossil content. Sedimentary rocks often contain structures that provide clues as to their environment of deposition. From such clues geologists can piece together a fairly complete geologic history of a region.

ORIGIN OF SEDIMENTS

Processes

Most sedimentary rocks are derived from the weathered and eroded products of older pre-existing rocks. Minor admixtures come from organic matter, volcanic ash, meteorites, and mineral-bearing waters. The decomposed and disintegrated rock, transported by gravity transfer, streams, wind, and glacial ice, is ultimately deposited as sediment. Most of it is in the form of solid particles, but part is in solution (Table 11.2). Every stream, whether a small brook or a large river, carries unconsolidated debris downstream. The coarse materials are rolled along the stream bed, finer materials are carried in suspension, and dissolved matter is transported in solution.

FIGURE 11.1   *A slab of fossiliferous limestone from western New York. (Hardin, U.S. Geological Survey.)*

TABLE 11.1   **Classification of Sedimentary Rocks**

Clastic varieties		Hybrid varieties	Chemical-organic varieties	
Texture (grain size)	**Varieties**		**Chemical or biochemical composition**	**Varieties**
Granule-, pebble-, cobble-, or boulder-size grains 2 mm or larger	Rounded fragments are called *conglomerates*: Quartz pebble conglomerate Angular fragments are called *breccias*: Quartz breccia	H y b r i d	Carbonate minerals e. g., calcite ($CaCO_3$)	Limestone
		v a r i e t i e s	Sulfate minerals	Gypsum
Sand-size grains 0.0625 to 2 mm	All varieties are called *sandstones*: Quartz sandstones		Iron-rich minerals	Hematite
Silt-size grains 0.0039 to 0.0625 mm	All varieties are called *siltstones*: Quartz siltstone		Siliceous minerals	Chert
Clay-size grains less than 0.0039 mm	All varieties are called *shales* or *claystones*: Red shale or claystone		Organic products	Coal

Millions of tons of sediment are carried to lakes and ocean basins every day. The Mississippi River alone transports and deposits in its delta nearly 3.2 cubic kilometers of sediment every year. Much of the sand and gravel transported by a stream is deposited periodically as sand bars, as islands, or as layers in the slack-water reservoirs of the stream channel. These temporary deposits may be picked up again during a flood stage, when both the volume and the velocity of the stream are greater.

Stratification is a feature of nearly all sedimentary rocks because conditions of sedimentation are not uniform, relatively coarse sediments may alternate with finer sized material. Likewise, fine sediment may be deposited between layers of chemical stratification. The walls of the Grand Canyon provide an excellent display of stratification (Fig. 1.1). An individual layer of sedimentary rock is called a stratum or, more commonly, a bed.

When sediments become deeply buried beneath younger layers of sediment deposited above them, they may become consolidated, or lithified, by pressure or compaction. Others are lithified by cementation due to chemical materials precipitated in the spaces between the grains.

In some areas, sedimentary rocks are thousands of meters thick, and most of the continental landscape is sculptured out of them. In the Grand Canyon and in the Niagara Gorge, sedimentary rocks occur in a nearly horizontal position. In many mountainous areas, however, the strata are commonly inclined at various angles (Fig. 1.4). Some of these regions, such as the Garden of the Gods or Red Rocks State Park, on the east slope of the Rocky Mountains in Colorado, have been carved into unbelievably beautiful places.

Sources

On the basis of their source, sediments may be classified as terrigenous, organic, volcanic, magmatic, and extraterrestrial. Each of these sources will be considered briefly.

Terrigenous   Included are all land-derived materials. Terrigenous sediments include solid particles, as well as chemical precipitates (calcium and magnesium carbonates, iron and manganese oxides, phosphates, sodium chloride, and nitrates). Some inorganic precipi-

tates are difficult to distinguish from those of biochemical origin. The terrigenous materials, both solids and precipitates, are the products of weathering and erosion upon rocks of all kinds.

Organic   Organic sediments are of two types. First are those made up of the protective and supporting structures of organisms, such as bones, teeth, shells, or tests (hard external covering of tiny invertebrates). These structures are composed of noncarbonaceous material and include phosphates, calcium and magnesium carbonates, iron oxides, and silica. Secondly,

TABLE 11.2   Constituents of Sedimentary Rocks

Primary constituents

Detrital	Chemical precipitates
Lithic fragments	Calcite, other carbonates
Quartzite	Opal, chalcedony (quartz)
Granite	Limonite
Gneiss	Hematite
Schist, phyllite, slate	Glauconite
Sandstone	Carbonaceous material
Coarse pyroclastics (volcanic bombs, blocks)	Halite
Glass shards, volcanic ash	Gypsum
Mineral grains	Anhydrite
Quartz	
Chalcedony, chert, jasper	
Feldspar	
Muscovite	
Magnetite, ilmenite	
Garnet	
Hornblende, pyroxene	
Clay minerals	

Secondary constituents

Introduced	Reorganization products
Opal, chalcedony	Quartz
Quartz	Hematite
Carbonates	Micaceous minerals
Gypsum	Chlorite
Hydrated iron oxides	Dolomite
	Anhydrite
	Pyrite
	Graphite
	Glauconite

plants manufacture various carbonaceous organic substances. Sediments of this origin are made up of partially decayed plant remains.

Volcanic   Sediments of volcanic origin include fragments ejected from volcanoes and deposited as sedimentary layers on land and in water. They consist of fine volcanic ash, dust, or coarser debris, and in some cases, stream-eroded fragments from volcanoes or lava flows. Volcanic-derived material often gets mixed with other terrigenous debris. Occasionally, nearly pure streaks of an altered volcanic ash, called bentonite, will be found interbedded with terrigenous sediments or in chemical sediments hundreds of kilometers from the volcanic source (Fig. 11.2).

FIGURE 11.2   *Clay bed at knee of geologist is Ordovician bentonite that can be traced for 645 kilometers in the southern Appalachians. Photo taken near Gadsden, Alabama.*

Magmatic   A sediment of magmatic origin represents a sediment formed at the surface of the earth from material in solution or suspension in hot magmatic waters. It may be possible for this material to reach the surface via hot springs, the waters of which may create deposits on the land surface as in Yellowstone National Park (Fig. 11.3).

Extraterrestrial   Extraterrestrial materials come from outer space and result largely from the burning up of meteoroids passing through the earth's atmosphere. These materials fall as very fine dust on land and sea. Larger fragments occasionally plunge to earth as meteorites, but they are extremely rare in sedimentary rocks. Extraterrestrial material has been detected in the snow accumulations of Antarctica and in sediments from deep oceans.

Natural objects called tektites are commonly found in deep-sea sediments. Tektites range in size from a sand grain to objects weighing over 1 kilogram. They are smooth, rounded, multicolored glass particles. The tektites are hypothesized as being extraterrestrial in origin, objects thrown from the lunar surface by meteorite impacts, or particles of terrestrial earth origin also a result of meteorite impact.

Figure 11.4 summarizes graphically the origin of sedimentary rocks.

CLASSIFICATION OF SEDIMENTARY MATERIALS

Sediment materials may be placed into three major categories: detrital sediments transported as solid particles, chemical sediments formed by precipitation from solution, and organic sediments derived from organisms buried with inorganic chemical and detrital sediments.

Detrital Sediments

The size of solid weathering products ranges from blocks of loosened rock down to minute clay-sized material. These solid fragments derived from pre-existing rocks that are transported by erosional agents are called detrital sediments (Table 11.1). Erosional agents move the largest sediments with difficulty, but the finest may be carried hundreds of kilometers beyond the land and deposited in the sea. The large fragments tend to lag behind where they are subjected

FIGURE 11.3 *The mound of siliceous material built up around Old Faithful Geyser in Yellowstone Park is a variety of opal called geyserite. It is a chemically precipitated sediment from the geyser waters. (J. Dennison photo.)*

FIGURE 11.4 *Derivation of sedimentary rocks.*

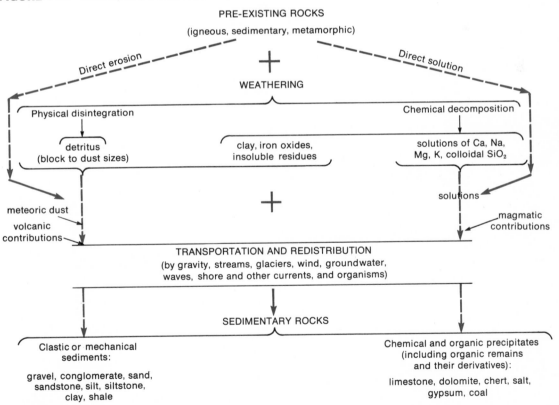

PRE-EXISTING ROCKS
(igneous, sedimentary, metamorphic)

Direct erosion

Direct solution

+

WEATHERING

Physical disintegration

Chemical decomposition

detritus
(block to dust sizes)

clay, iron oxides,
insoluble residues

solutions of Ca, Na,
Mg, K, colloidal SiO_2

+

meteoric dust

solutions

volcanic
contributions

magmatic
contributions

TRANSPORTATION AND REDISTRIBUTION
(by gravity, streams, glaciers, wind, groundwater,
waves, shore and other currents, and organisms)

SEDIMENTARY ROCKS

Clastic or mechanical
sediments:

gravel, conglomerate, sand,
sandstone, silt, siltstone,
clay, shale

Chemical and organic precipitates
(including organic remains
and their derivatives):

limestone, dolomite, chert, salt,
gypsum, coal

137

FIGURE 11.5 *Rounded sand-sized grains of quartz as a result of wear during transportation by an erosional agent. A sediment composed of fragmental grains of minerals consolidated together has a clastic texture.*

to renewed weathering. Consequently, detrital materials often become sorted into deposits with fairly uniform particle size.

Detrital particles are named according to their diameter, as shown in Table 11.3. Extensive transportation generally rounds the particles (Fig. 11.5), unless a new fracture produces sharp edges. Particle size and shape, plus degree of consolidation, are the bases of detrital-rock nomenclature, as shown in Table 11.1. Note that the names applied to the sedimentary rock reflect the names applied to the size of its component particles. Rock color is not important in the rock nomenclature, but mineral composition is often used to name special types of sandstones.

All detrital sedimentary rocks are characterized by a clastic, or broken, texture or by a grain to grain relationship. If mineral matter is precipitated around the grains or in the voids between the grains, the rock becomes lithified (Fig. 11.6). The fragmental origin of some seemingly chemically derived limestones is in-

dicated by the clastic texture of broken fossil material (Fig. 11.7).

During transportation, the fragments of gravel, sand, silt, and clay tend to be sorted on the basis of size. When a stream discharges into a body of quieter

TABLE 11.3   The Classification of Sedimentary Particles by Size

Name of fragment	Diameter (mm)	Sedimentary rock
Boulder	> 256	Conglomerate (if rounded)
Cobble	64–256	
Pebble	4–64	Breccia (if angular)
Granule	2–4	
Sand	$\frac{1}{16}$–2	Sandstone
Silt	$\frac{1}{256}$–$\frac{1}{16}$	Siltstone
Clay	< $\frac{1}{256}$	Shale or claystone (Mudstone)

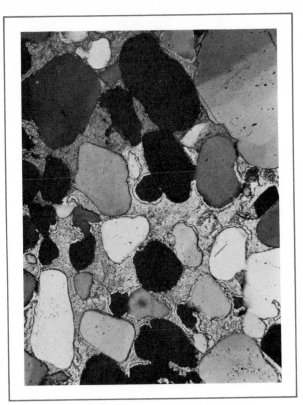

FIGURE 11.6 *In this calcareous sandstone seen in microscopic thin section, clastic grains of quartz became cemented together when calcite precipitated in the voids between the clastic grains. The clastic texture is still evident.*

FIGURE 11.7 *Clastic limestone from Tennessee made up chiefly of Cambrian trilobite, brachiopod, and echinoderm fragments. (William J. Byrd photo.)*

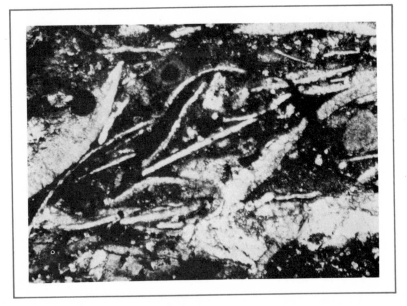

water, gravel may be deposited near shore, sand farther out, and mud (consisting of silt and clay) still farther out. Beyond the zone in which clastic material is deposited, deposits of $CaCO_3$ in the form of chemical precipitates or accumulations of shell material occur. Thus, successive belts of different sizes of sediments may occur roughly parallel to the shore.

However, the orderly processes of sedimentation may be, and is, modified by the variations in the amount of sediment deposited or the amount of wave action. Hence, the distribution of sediment is rarely uniform. Gravels generally contain sand, sands are mixed with silt or clay, and clay and silt often contain fine sand or some calcareous matter. Sorting by size is not perfect because of differences in particle shape and density, insufficient time for the natural sorting processes to finish, and variations in behavior of the erosional agent.

Types of Clastic Sedimentary Rocks

(*a*) *Breccia*   An aggregate of angular clastic fragments coarser than 2 millimeters in diameter (granules, pebbles, cobbles, and boulders) is a breccia

FIGURE 11.8   *A breccia consisting of angular fragments cemented together. The angularity of breccia fragments shows that they have undergone little, if any, abrasion during transportation.*

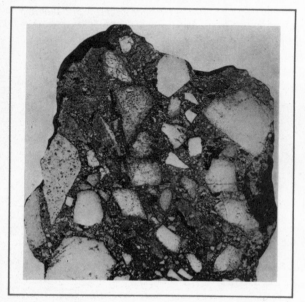

(Fig. 11.8). Breccia usually accumulates very near the original source of the particles. For example, the coarse particles that make up a breccia may be deposited at the base of a cliff, along a fault scarp, or along a small mountain stream. They may also originate by rock fracturing or collapse of a cave.

(*b*) *Conglomerate*   Unconsolidated rounded sedimentary fragments coarser than 2 millimeters in diameter are gravels. Lithified gravel, usually consolidated by cementation, is called conglomerate (Fig. 11.9). In many conglomerates the pebbles are commonly quartz, quartzite, or siliceous rocks since these are the materials that possess great resistance to disintegration and wear. Conglomerate pebbles may, however, consist of fragments of any kind of rock, such as a granite-pebble conglomerate or a limestone conglomerate.

Unusual gravel and sand deposits may contain fragments of valuable ore minerals, forming placer accumulations that can be mined profitably. Stream and beach gravels in South Africa contain commercial diamonds, and sapphires are produced from stream gravels in Ceylon.

(*c*) *Sandstone*   An aggregate of clastic grains $\frac{1}{16}$ to 2 millimeters in diameter is sand. Most sand grains are composed of the mineral quartz, but they may be composed of other minerals or of extremely fine-grained rock fragments. Such a consolidation or lithification of sand results in sandstone. Cementing material is commonly silica (quartz or chert), but it may be calcite (Fig. 11.6) or dolomite. Iron oxide cements are common and usually stain the rock a yellow, brown, or red color. Other sandstones may be coherent because the sand grains firmly adhere to clay compressed between the grains (lithified by compaction). Sand grains may be angular, subrounded, or rounded in proportion to increasing distances of transport (Fig. 11.5). The grain surfaces also may become pitted or frosted during transit.

Sandstones may grade in a rock exposure from a coarse-grained sandstone into a conglomerate or from a fine-grained sandstone into finer grained siltstones, shales, and mudstones. Sandstones that contain appreciable amounts of calcite are calcareous sandstones (Fig. 11.6); if they contain clay, they are argillaceous sandstones. Occasionally, a sandstone contains pebbles along a bedding plane or scattered randomly throughout the mass.

Several varieties of sand and sandstone are named

FIGURE 11.9 *Conglomerate, a sedimentary rock composed of gravel-sized sediment cemented together.*

according to the mineralogy of the grains. Sandstone with nearly all grains of quartz is an orthoquartzite: one composed of feldspar and quartz is an arkose. An arkose results from the partial chemical weathering of a granite or gneiss and a subsequent sorting during erosion and transportation. The red color of most arkoses is due to the pink color of potassium feldspar, or orthoclase. A graywacke is a sandstone that contains quartz, feldspar, and rock fragments set in a matrix of fine particles (more than 15 percent). Most graywackes are darker in color than other sandstones. They are associated with marine shales or mudstones, volcanic materials, chert, and submarine lavas. Their composition suggests a source area of moderately weathered igneous or metamorphic land masses with a high relief. They frequently exhibit graded bedding, indicating that they were deposited in their final environment by submarine flow mechanisms. Still another variety of sandstone is greensand, or glauconitic sandstone. Its green color results from the mineral glauconite (a complex iron silicate), which is deposited or formed along with grains of quartz and other minerals in a shallow marine environment.

(*d*) *Siltstone*   A consolidated sediment made up of silt-sized particles is called a siltstone. Grains of silt are too small ($\frac{1}{16}$ to $\frac{1}{256}$ millimeter) to see with the unaided eye, but they do exhibit a grittiness. The bedding of siltstone may be thin, and as a result, siltstones are often not rigorously distinguished from shale, claystone, or mudstones in a rock outcrop. Modern unconsolidated silt deposits along the flood plains of rivers form some of the most fertile soils in the world.

(*e*) *Shale, Claystone, and Mudstone*   The term claystone is used by many geologists for those rocks composed of clay-sized particles that do not exhibit a platy, or fissile, bedding. Other fine-grained rocks may contain unspecified amounts of silt and/or clay and are termed mudstones. As a rock term, shale refers to consolidated material dominated by clastic particles smaller than $\frac{1}{256}$ millimeter in diameter with a platy, or fissile, bedding. Mineralogically, most clastic particles smaller than $\frac{1}{256}$ millimeter consist of quartz with lesser amounts of feldspar, clay, and mica minerals. Common clay minerals are kaolinite, montmorillonite, and illite. The weathering of the minerals in a basalt, limestone, dolomite, or shale may produce clay minerals or clay-sized material without any significant sand-sized material. Clays accumulate as residual deposits on land or as deposits in quiet lakes or in the sea.

Thus, these rocks are composed of the finest products of rock decay. Fine particles are ordinarily transported farther out to sea than other clastic sediment. They exhibit a thin bedding, a fissile nature, a lack of bedding, or a graded bedding, all of which indicate frequent changes in the size-fineness of the mineral composing them. Such variations in grain size may result from seasonal changes, or they may represent differences in rainfall or some other factor that affected the amount and character of the sediment brought down by streams. Rocks that contain sand are arenaceous; those that contain calcium carbonate are calcareous; those containing iron are ferruginous; and those with large amounts of organic matter are carbonaceous. Carbonaceous rocks are usually black. Sometimes they grade into beds of coal. Some geologically ancient strata remain as beds of clay and differ from the original sediment only in that they have been compressed slightly (for example, the fire clays of Illinois, Ohio, and West Virginia).

Chemical Sediments

Whenever the concentration of certain ions in solution becomes sufficiently great, chemical precipitate may begin to form. Materials dissolved during weathering are transported in solution, usually reaching the sea before the concentration becomes great enough to cause precipitation. Consequently, the sea serves as a great storehouse of dissolved materials. Ultimately, some of these dissolved materials precipitate to form beds of chemical sediments.

As these sediments precipitate on the bottom of the sea or a lake, the grains usually grow in an interlocking manner, producing a crystalline texture (Fig. 11.10). The interlocking crystals are the primary cause of lithification for most chemical sedimentary rocks.

Many of the chemical compounds transported in solution are only slightly soluble in water; thus, changes in water chemistry (such as temperature changes, loss of carbon dioxide, or mixing of two water masses with different dissolved chemicals) can initiate precipitation of a chemical sediment. Many organisms extract calcium carbonate or silica from seawater to form shells and tests. Upon the death of the organism, these shells and tests are deposited on the sea floor. Some chemical compounds (for example, $NaCl$ and $MgCl_2$) are so soluble that they become concentrated to cause precipitation only when the water is evaporated; these sediments are termed evaporites.

Types of Chemical Sedimentary Accumulations   The chemical-sediment precipitates include $CaCO_3$ (limestone), dolomite, iron compounds, phosphate, manganese oxides, and silica in the form of chert and opal (Table 11.1). These chemical precipitates may form nearly pure beds, become cementing admixtures in other sediments, or occur as isolated lumps or nodules on the bottom of a lake or the sea.
(a) Limestone   This generally consists chiefly of calcium carbonate, mostly as calcite. Limestone may result from the inorganic chemical precipitation of calcite, from an accumulation of shells, or from a combination of these two processes. Limestone may contain abundant fossils (Figs. 11.1 and 11.7), or as in the case of lime muds that form by inorganic chemical precipitation, they may lack fossils. On the Grand Bahama Banks, shallow seawater becomes warmed, driving off the carbon dioxide, causing precipitation of tiny needle-shaped crystals of calcium carbonate:

$$\underset{\text{(dissolved)}}{Ca(HCO_3)_2} \overset{\text{precipitation}}{\underset{\text{weathering}}{\rightleftharpoons}} \underset{\text{(solid)}}{CaCO_3} + \underset{\text{(gas)}}{CO_2} + H_2O$$

Notice that in a weathering environment the reaction proceeds to the left.

Wave agitation in shallow water may keep the cal-

FIGURE 11.10   *A thin section of the Oneota Dolomite of Ordovician age. The rhombic-shaped crystals are dolomite*

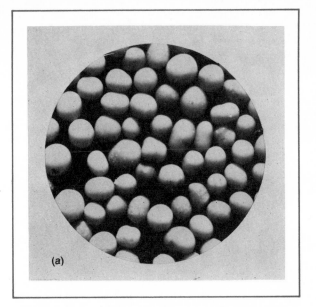

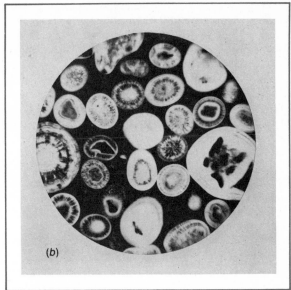

FIGURE 11.11 *Photomicrographs of oolites from Great Salt Lake, Utah. (a) Loose oolites from the northeast shore, enlarged 15 diameters; (b) thin section of cemented oolites enlarged about 35 diameters. Note the concentric and radial structures of successive layers in the oolite grains. (A.J. Eardley.)*

cite crystals in suspension, and small, concentric growths of calcite may form around the crystals. Tiny spheres formed this way are known as oolites (Fig. 11.11). Upon reaching a certain size, oolites settle to the bottom as clastic grains. They have an internal structure displaying their partially chemical origin.

Pure limestone accumulates in relatively clear, warm, shallow water adjacent to a low-lying land area. Life there is abundant, and neighboring land areas have too low an elevation or are too distant to contribute appreciable quantities of clastic sediment. Limestone may grade into a shale if clay is added to the calcareous sediments, or if sand is added, the limestone may grade into calcareous sandstone.

Other varieties of limestone include chalk, tufa, and marl. Chalk is a special type of limestone usually composed of tiny shells or shell fragments cemented together. Tests of foraminifera (small marine animals) and coccolithopores (marine planktonic algae) constitute a large part of the material, but small shells of other organisms are commonly present.

Tufas and travertines are crystalline calcite deposits often quarried as ornamental stones. Tufas form at the outlets of springs, and they are generally earthy, porous, or spongy. Many tufas are precipitated by calcareous algae. Travertine is more compact than tufas and will form around springs or within limestone caves.

Marls are porous accumulations of fine particles of calcite and clay on the bottoms of many fresh-water lakes. Marls most commonly contain abundant fine fragments of shells. The term marl is also used to designate certain marine sediments that contain a mixture of clay and finely divided shell fragments. Upon lithification, this material would be called a calcareous shale or an argillaceous limestone.

(b) Dolostone, or Dolomite   This is a deposit of the mineral dolomite, $CaMg(CO_3)_2$. The origin of dolostone has been an enigma for geologists for over a century. The puzzling feature about dolomite is that seawater is greatly supersaturated with dolomite, and it is the most stable Ca-Mg carbonate in seawater. For a long time, however, dolomite was not observed precipitating from seawater. In the early 1960s the actual formation of the mineral dolomite was observed for the first time. This mode of formation occurred when

spring tides and storms washed seawater onto flat-lying shores. Gradual drying concentrated the magnesium in solution so that previously deposited calcite reacted with the liquid to form dolomite. This phenomenon has been studied and observed in the Florida Keys, the Bahamas, the Leeward Islands of the Caribbean, and the Persian Gulf. Sediments of deep oceanic areas that were obtained by drilling often contain microscopic dolomite rhombs admixed with other sediments. Apparently, tropical and subtropical climates are necessary for dolomite deposition. This suggests that such areas as the Chicago, Illinois, area, where beds of dolomite exist, once had a much different climate. The area was the site of shallow tropical seas during the Silurian period and not the interior of a continental land mass as it is today. Eroded fragments of dolomite can be transported to the sea to form a detrital dolomite, with a clastic texture.

(c) *Iron Minerals*   Several iron minerals may accumulate as chemical sediments. Layers or nodules of limonite may form on lake bottoms or in swamps. Hematite forms beds several meters thick in some marine sedimentary rock deposits. Four formerly commercial rich beds of hematite occur at Birmingham, Alabama, where they accumulated some 420 million years ago. A portion of this hematite has an oolitic texture, while another part consists of calcareous shells that have been replaced by hematite. Iron carbonate (siderite) nodules deposited as lake sediments 250 million years ago were the original iron ores mined near Pittsburgh. These have long since been abandoned as commercial ores.

The iron sulfide mineral pyrite forms sedimentary nodules in sulfurous environments associated with organic decay. It occurs as nodules in black shales or as rare, thin beds, as in western New York. Pyrite lumps in coal are responsible for much air pollution in cities so that many urban areas now prohibit the burning of high-sulfur-bearing coal.

(d) *Phosphate Nodules*   These form at scattered sites on the sea floor, and phosphate minerals also occur as cement in other sediments. Limestones near Nashville, Tennessee, contain enough phosphate to be commercial fertilizer sources. The Phosphoria Formation of Wyoming, Utah, and Idaho contains vast quantities of phosphate minerals. Bone accumulations in Florida are also rich sources of phosphate.

Manganese oxide accumulates as nodules on the bottoms of lakes or in the sea. It has been estimated that on the Pacific Ocean floor the concentration of manganese nodules is 7,300 metric tons per square kilometer. Having a content of 24 percent manganese, 14 percent iron, and 1 percent nickel, they have a value upward of $2.35 million per square kilometer.*

(e) *Silica*   Silica occurs as nodules in limestones and dolomites, as distinct beds, and as a cementing material in sandstones. Silica nodules are very finely crystalline. If light in color they are called chert, while dark-gray to black varieties are known as flint. Chert colored red by small quantities of hematite is termed jasper. Chert strata over 30 meters thick have been found in Idaho, Texas, Arkansas, and West Virginia.

Chert had not been observed forming in nature until 1965, when it was found accumulating in desert lakes in Australia. Most sedimentary chert must have another origin, not yet understood, because chert nodules and beds are common in limestones that contain typical marine faunas. One of the more interesting results of the Deep Sea Drilling Project has been the discovery of chert beds buried deep within the oceanic sediments, or oozes.

(f) *Evaporites*   Seawater contains about 3.5 percent dissolved solids. If it is evaporated in an enclosed basin, a sequence of chemical precipitates forms until all the water is gone. The normal order of evaporite deposition from seawater is calcium carbonate, calcium sulfate, sodium chloride, and potash (K) salts. An arid climate is necessary for large-scale deposition of evaporite sediments.

Gypsum is commonly associated with dolomite and mud-cracked red shales. There are two calcium sulfate minerals—gypsum ($CaSO_4 \cdot 2H_2O$) and anhydrite ($CaSO_4$). Gypsum is the more common in rock exposures, but at temperatures above 42°C, anhydrite precipitates rather than gypsum. Gypsum and anhydrite are so soluble in a moist climate that bedrock outcrops of these materials are uncommon in those areas.

The mineral halite makes up extensive beds of rock salt. Evaporation of a shallow extension of the sea 400 million years ago produced a rock salt deposit tens to hundreds of meters thick under much of New York, Pennsylvania, West Virginia, Ohio, and Mich-

---

*A United States ship, the *Glomar Explorer,* built by Global Marine Co., is presently being outfitted for marine mining. It is designed to dredge manganese nodules from the sea floor in economic quantities beginning in 1974.

igan. Other salt deposits accumulate in saline lakes, as at Great Salt Lake and Bonneville Salt Flats in Utah and at Carson Sink, Nevada. The dome-shaped masses along the Texas and Louisiana coast are salt deposits that have resulted from upward flowage of salt layers from beds 6 kilometers below the surface. Other salt domes extend close enough to the ground surface to serve as commercial salt mines (Fig. 11.12).

Extreme evaporation of seawater yields potash salts (mostly the mineral sylvite, KCl) as a final precipitate. These sediments are very rare because the necessary extreme evaporative conditions are seldom found. Potash salts are used for fertilizer and chemicals. Major North American deposits of potash are found in western Texas and adjacent New Mexico, near Moab, Utah, and in Saskatchewan, Canada. Brines from Searles Lake, California, also yield potash when evaporated.

Certain desert lakes have dissolved salts in different proportions than those in the ocean. Brines from Searles Lake, California, are rich in borax, and older borax salts have been mined at several places in southern California, including Death Valley. The desert of Chile contains evaporite sediments rich in sodium nitrate.

Organic Sediments

These sedimentary materials develop from unusual concentrations of organic remains, such as coal, diatomaceous earth, and coral and algal reefs. Petroleum and natural gas, also of organic origin, occur in pore spaces in rocks. These fluid fuels will be treated in Chap. 12.

Coal Coal is derived from peat. Peat forms when organic materials accumulate faster in swamps and bogs than decay destroys them. Peat is porous and variable in texture, depending on its constituent plants and their degree of decomposition. In the tropics, plant decay is too rapid for major peat accumulations. Plant decay is slow in the tundra, but plant growth and resulting peat accumulation are also slow because of

FIGURE 11.12 *Salt mine 240 meters below ground at Winfield, Louisiana. This type of mined salt is called "rock salt," in contrast to salt obtained from the evaporation of naturally saline water. (Carey Salt Company.)*

the cold climate. Therefore, humid temperate climate is most favorable for peat accumulation. The Dismal Swamp in Virginia and the swamps of Louisiana, Texas, and the states around the Great Lakes are presently the major sites of peat deposition. Peat accumulations have also been found below the carbonate muds at Bimini Island in the Bahamas.

Burial of peat compacts the material and drives out moisture and volatile constituents, producing a lignite. Lignite still possesses remnants of the original woody structure and a brownish-black wood color. It is sometimes called brown coal. Vast lignite deposits occur in the northern Great Plains.

Deeper burial and time transform lignite into soft, black bituminous coal. Bituminous coal occurs interbedded with other consolidated sediments, as shown in Figs. 11.13 and 11.14. Most of the bituminous coal in North America and Europe formed during the Pennsylvanian period, about 280 million years ago. If bituminous coal is subjected to metamorphism, it is transformed into anthracite coal (hard coal), or even to graphite.

Diatomaceous Earth

Diatomaceous earth is an accumulation of siliceous tests formed by single-celled plants called diatoms. Diatoms grow abundantly in the sea and in fresh-water lakes. Upon death, their tests accumulate to form the diatomaceous earth occurring in gray to white layers associated with shale. Large diatomaceous earth deposits are found near Lompoc, California, where they are mined and used as insulating or filtration material.

Reef Deposits   These deposits represent wave-resistant mounds of calcareous organisms that built up from the sea floor toward the water surface. Calcareous algae are the most abundant organic constituent, with corals second. A host of other life forms live among the rigid parts of the reef, but these mobile organisms are less commonly preserved in the fossil record of a reef. Modern coral reefs form best where water temperatures are 21 to 26°C. Presumably, reefs were restricted to a similar warm climate in the geologic past. Reefs grow only where light penetrates to the sea floor, in water depths of less than 50 meters. In the United States, living coral reefs occur along the Florida Keys and Hawaii. The northernmost living reef in the world was discovered in 1969 off the North Carolina coast. Reefs are common on the fringes of many tropical Pacific islands. The largest modern reef is the Great Barrier Reef along the northeastern shore of Australia.

The rock of fossil reefs consists of limestone or dolomite. Ancient reefs are widespread in the Silurian dolomites of Indiana, Illinois, Wisconsin, and Iowa. A vast complex of reefs developed in the Permian seas of west Texas, as illustrated in Fig. 11.15. Cretaceous-age reefs have recently been found in the Kara Kum desert in the Turkmenia Province of Russia.

OTHER ASPECTS OF SEDIMENTARY ROCKS

Color of Sediments

Sedimentary rocks vary in color according to their composition. They may be white or light-colored from the presence of quartz, kaolinite, calcite, or other light-colored minerals; green from ferrous iron silicates; red from ferric oxides or pink due to orthoclase feldspar; yellow or brown from limonite; black from organic matter, black minerals, dark rock fragments, or finely divided iron sulfide; or gray from a mixture of light and dark ingredients. Color depends on such factors as composition, compositional purity, degree of oxidation of iron compounds, and amount of organic matter.

FIGURE 11.13 *Section showing typical relations of coal beds to associated strata in the Eastern Interior coal basin. (After Dunbar.)*

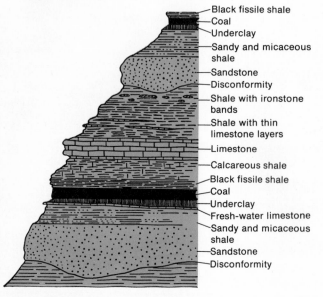

Black fissile shale
Coal
Underclay
Sandy and micaceous shale
Sandstone
Disconformity
Shale with ironstone bands
Shale with thin limestone layers
Limestone
Calcareous shale
Black fissile shale
Coal
Underclay
Fresh-water limestone
Sandy and micaceous shale
Sandstone
Disconformity

FIGURE 11.14   *A coal bed about 2 meters thick, with two clay partings, near Glendive, Montana. (U.S. Geological Survey.)*

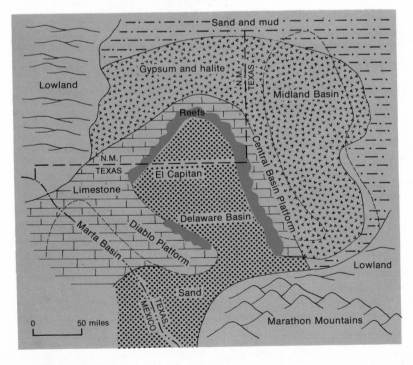

FIGURE 11.15   *Permian carbonate reef platforms and associated basins in west Texas and New Mexico.*

The color of a sedimentary bed is not commonly persistent over large areas. Also, color may differ with the degree of weathering a bed has undergone. For example, many dolomites that vary widely in original color weather to a tan color. Gray or black shales also may weather to a red clay, especially in the south. The red iron oxide (hematite) that colors the clay may be the oxidation product of pyrite or iron silicates, which was finely disseminated through shale or limestone.

Diagenesis

During and after the accumulation of a sediment, it may be affected by a number of low-temperature physical and chemical changes. The process including such changes is known as diagenesis. Mineral matter may precipitate between clastic grains, causing lithification. Some chert beds probably formed by diagenetic replacement of limestone. This is evidenced by the fact that within the silica, faint outlines of original shells and calcite grains are often preserved. The growth of new types of clay minerals in a shale may aid its lithification. Certain dolomites probably resulted from diagenetic modification of limestone.

With very deep burial, both temperature and pressure increase. New and sometimes larger mineral crystals form, such as the silicate minerals chlorite, muscovite, and biotite. These can develop from clays originally in the sediment. At the extreme of high temperatures and pressures, diagenesis grades into metamorphism (Fig. 8.12).

Relative Abundance of Sedimentary Rocks

Three types of rocks make up 99 percent of the volume of the sedimentary rocks present in continental areas. Shale and siltstone together form about 58 percent, sandstone 22 percent, and limestone 20 percent of all sedimentary rocks exposed on land (Pettijohn, 1957). The dominance of these three sedimentary types is one reason why coal, iron ore, gypsum, and fertilizer materials are rare and thus valuable. Deep-sea sediments are usually clays, lime, or siliceous oozes, and so the 22 percent for sandstone is excessive on a worldwide basis (land and sea together).

If the average igneous rock of the continental areas (the primary source of most sediments) theoretically were weathered completely and the sediments lithified, the resulting sedimentary rocks would be 70 percent shale, 16 percent sandstone, and 14 percent limestone. The discrepancy between these theoretical figures and the actual sediment proportions results from at least two factors: the incomplete weathering of clastic grains in sediments, and the uncertainties in the sampling program or method of sampling.

ENVIRONMENTS OF DEPOSITION

The products of weathering eventually find their way to the sea, although some temporarily become lodged on river flood plains or in lakes. The environments where sediments accumulate (Table 11.4) can be continental, marine, or transitional. Most environments of accumulation grade laterally into each other. This can be shown by the gradual transition between the waters of the intermediate and shallow depths in the ocean, between a flood plain and a delta, or between a delta and a shallow-water marine environment.

Continental Sediments

Continental sediments lack marine fossils but may contain plant and animal remains of fresh-water or of terrestrial origin. Sedimentary structures generally show evidence of transporting currents, and clastic sediments are dominant over chemical sediments.

Desert Deposits   Arid regions in the world presently cover approximately 28 million square kilometers. Sediments accumulate there by intermittent stream wash from upland slopes, by chemical deposition from the waters of playa and saline lakes, and by sedimentation of wind-blown materials.

TABLE 11.4   Environments of Sedimentary Deposits

Continental	Mixed continental and marine	Marine
Desert	Beach	Shallow depths
Glacial	Lagoon	Intermediate
Fluvial	Estuary	depths
Piedmont	Delta	Deep
Valley flat		
Lake		
Swamp		
Cave		

The valley and gully deposits are composed of coarse clastics that extend up valley into the highlands and down valley to alluvial fans. These fans coalesce laterally to form piedmont slopes.

Temporary playa lakes may precipitate evaporites. Stratification is conspicuous in the laminated clays of the lakes and playas. Most of the gravels on the piedmont slopes are poorly stratified. The eolian sediments are characterized by wedge-shaped, crossbedded units (see Figs. 11.19 and 24.18).

Glacial Deposits   Glacial sediments deposited directly by ice are unstratified, unsorted mixtures of both coarse and fine materials. Glacial meltwater deposits are better stratified and tend to be better sorted within a single stratum. The fine-grained sediments of glacial lakes may be varved.

Fluvial Sediments   Fluvial sediments of streams are deposited mainly on piedmonts and valley flats. Piedmont deposits at the base of mountains are the combined result of deposition by soil creep, rainwash, mudflows, and intermittent streams. The present extent of piedmont deposits in the Western states approximately equals the area of the mountains above them. In the Cucamonga district of California, piedmont deposits exceed 300 meters in thickness and consist of boulders, cobbles, pebbles, sand, and silt. The sizes are poorly sorted and indistinctly bedded.

Valley-flat sediments differ from those of the piedmonts by having better sorting and stratification, fewer large fragments, and more organic matter. Since most streams alternately erode and deposit, the sediments are in place only temporarily, and much of the bedding is lenticular. Many flood plains contain small lakes and swamps, both of which shift their positions as the streams move their channels over the valley flats. Fluvial sand and gravel thus are found interbedded with the clays and silt of lakes and with swamp mud or peat. Many of the sedimentary rocks of the Great Plains were originally deposited as fluvial sediments.

Lake Sediments   The origin and location of a lake basin greatly influences the type of sediment accumulating there.

Any geologic process that creates a closed surface depression or obstructs drainage may produce a lake basin. Most lake basins are the result of erosion, but

some are due to crustal movements or volcanic activity. Some lake basins have resulted from the combined effects of several processes.

Although the shape and capacity of a lake depend on the geologic events that formed the basin, the outline and depth do not remain the same. The existence of a lake is a dynamic process, with the lake-forming forces opposing the lake-destroying forces. A lake may be drained by downcutting of the outlet stream or dried up by climatic change or by a lowering of the water table. A lake basin may be filled by debris carried in by tributaries, debris derived from the shore by wave attack, or material derived by organisms (Fig. 11.16). Modification or destruction of a lake may involve a combination of all these factors.

Lakes in mountain valleys receive coarser material than those in the broad alluvial flats and delta regions. Lake sediments tend to have thinner and more continuous beds than stream deposits. In general, lake sediments consist of gravel, sand, silt, clay, marl, peat, iron and manganese oxides, or iron carbonate (siderite), and in saline lakes such evaporation products as salt and gypsum.

Swamp Deposits   Swamps and bogs are widespread and important features of the earth's surface. The terms marsh, swamp, and bog are often used interchangeably, as well as differently in different areas. For example, in the subarctic north, both swamps and bogs are called muskeg. A bog is one of several possible end stages in the life history of a lake. It usually represents a lake that has been completely covered and destroyed by a mat of plant life from shore to shore.

A swamp is vegetated land saturated with water and is characterized by a grassy or prairielike vegetation and by a few varieties of trees able to grow under wet conditions. The origins, characteristics, and physiographic environments of swamps vary greatly. They are most likely to occur wherever there is level, poorly drained land and an abundant supply of water. The upland swamps of northern Minnesota are examples. In the subarctic north, the impermeabilility is due to permanently frozen ground within a few inches to a few feet of the surface.

Coastal processes, with sea-level changes, may convert large areas along the seacoast into swamps or marshes. Irregularities of the sea floor near the shore

FIGURE 11.16 *Encroachment of vegetation on the margin of a lake. Bass Lake, Wolcottville, Indiana.*

may form poorly drained areas if the sea level falls. The swamps formed by coastal or shoreline processes vary from tidal flats with salty water to fresh-water swamps. Notable examples are the Dismal Swamp area in southern Virginia, the Atlantic coastal area of North Carolina, Okefinokee Swamp in southern Georgia, and essentially the whole coast of Louisiana.

The overflow from a river, carrying large quantities of fine sediment, may create a flood plain of low relief on an adjacent region. Often these flat areas are covered by sloughs, lakes, and secondary channels wandering through a maze of swampland.

Much of the extensive, poorly drained glacial deposits of the North Central states north to Hudson Bay are covered by a spongy, black soil made up of roots and their remains, partly decayed plant material, and fine rock waste saturated with water. This is called muskeg, which in some parts of Canada is developed directly over smooth, flat glaciated surfaces of relatively impermeable bedrock.

Just as ponds, lakes, and shallow embayments may be drained or gradually filled and converted to swamps or bogs, the further drainage or filling of swamps and bogs may form woodlands, meadows, prairies, or dry-land flats. If covered by later sedimentation, the sedimentary evidence of their former swamplike character will be preserved.

Swamp sediments are often dominated by peat or organic-rich muds. These may be interbedded with sand and silt deposits deposited by streams which pass through the swamp.

Cave Deposits   Caves are a sedimentation environment that is not on the earth's surface. Cave deposits are chemical sediments formed by precipitation from groundwater. These are mostly travertine deposits having a peculiar form. Some caves have small streams flowing through them, and these may deposit gravel, sand, and mud. Collapse of a cavern roof produces large breccia deposits. Organic sediments in caves include fossil bones and guano (bat, rat, and bird manure).

Coastal Sediments

The sediments that accumulate on or near the coastlines are a mixture of materials derived from both land and sea. They are deposited along the beach and in deltas, lagoons, and estuaries.

Beach Deposits   The beach environment (sometimes called the littoral zone) is that portion of the shore that is exposed between high and low tides. Wave ac-

FIGURE 11.17   *The calcareous shells of foraminifera (one-celled animals) in Cretaceous marl from New Jersey (greatly magnified).*

tion removes mud from the beach materials, and so most beach deposits are sand or gravel. Marine organisms are common fossil forms, but land plants may be washed in as driftwood and preserved as fossils.

Lagoon Deposits   Marginal lagoons are located between an offshore barrier island and the mainland. Lagoonal sediments are quite varied. The waters there are quieter than along the beach, and so lagoonal sediments tend to be finer grained than those on beaches. Land-derived clastic sediments may be brought to the lagoon by streams and wind; marine sediments are added by ocean currents; and in arid climates gypsum or salt may precipitate out of the water in the lagoon. Peat and organic-rich mud commonly form in shallow lagoons. Calcareous marls are formed by plants and invertebrate animals and to some extent by direct chemical action. In stagnant lagoons, bacterial activity forms hydrogen sulfide, which causes the precipitation of black sulfide in the sediments.

Estuary Deposits   An estuary is a broad mouth of a river where it enters the sea. Swift currents during the time of stream flood can deposit lenticular beds of sand and gravel. Incoming tidal currents rework the stream sediments. Marine fossils may occur in es-

tuarine deposits, although the water may have too low a salinity for most marine life forms.

Delta Deposits   Deltas form where sediment-laden streams supply clastics to the sea or a lake faster than the waves can distribute the sediment away from the river mouth. Consequently, the streams build extensions of land out into open water. Lenticular deposits of sand and silt are the rule; these may be interbedded with marine strata or with lake and swamp deposits.

Marine Sediments

The realms of present-day marine sedimentation include the shallow epicontinental seas, intermediate depths of the continental slopes, and the deep sea (Fig. 11.18, Table 11.5).

Shallow-Sea (Neritic) Deposits   The neritic zone extends from the low-tide line to depths of about 200 meters. This covers about 2 percent of the earth's surface, including the nearly flat continental shelves that border the continents in many areas. The neritic zone varies from less than a kilometer to several hundred kilometers wide, with an average width of about 60 kilometers.

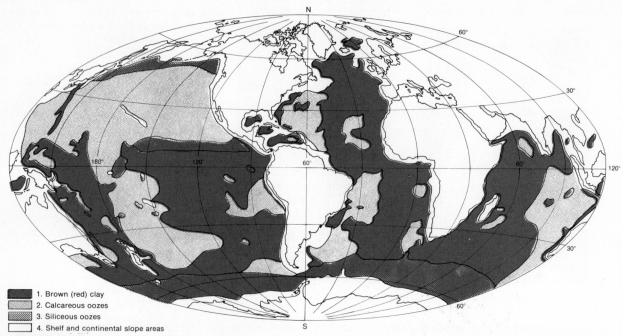

1. Brown (red) clay
2. Calcareous oozes
3. Siliceous oozes
4. Shelf and continental slope areas

FIGURE 11.18 *Distribution of marine sedimentary deposits throughout the world. (Modified from Dietrich, 1963.)*

Light generally penetrates about 50 meters into the upper neritic zone, and that is why marine organisms are more abundant in the neritic environment than in deeper water. Strata containing abundant and diverse shallow-water marine fossils are representative of deposition in a neritic environment. Perhaps half of the sedimentary rocks on the North American continent represent neritic accumulations formed when the continent was submerged beneath shallow epicontinental seas. Wave agitation commonly affects shallow neritic sediments. This usually occurs at depths of less than 25 meters but occasionally occurs to depths of 60 to 200 meters. Lenticular beds, ripple marks, cross bedding, and great variation in grain size from bed to bed are the result of wave action.

In general, coarse gravels occur near rocky points and sea cliffs or wherever strong currents affect the ocean floor. Sand prevails on the continental shelf off sandy shores or in areas of moderate currents. Fine silt and clay is deposited offshore of large rivers in sheltered areas near shore and in depressions on the continental shelf. Some chemical sediments show well-defined seasonal laminations.

The great bulk of shallow-sea sediments consists of land-derived detritus provided by streams, glaciers, shore erosion, dust and sand storms, and explosive volcanoes. Where the supply of detritus is small, sediment contributed by calcareous and siliceous diatoms and by chemical precipitation is the predominant sediment.

The rate of sedimentation in the neritic environment is difficult to determine, but it has been estimated that it would take approximately 1,470 years for a meter of sandstone to form, 2,950 years for the same amount of shale to form, and 7,380 years for a meter of limestone.

Intermediate-Depth Deposits   Beyond the continental shelf, the ocean floor descends rather abruptly to the deeper basin floor. This slope, called the continental slope, is the outer margin of the true continental mass. Sediments deposited in ocean depths of 200 to 2,000 meters are called bathyal sediments. The continental slope starts about 60 kilometers offshore and is generally covered by fine clastic sediment of both land and sea origin. Landward, the deposits grade into shallow-water deposits, and seaward, they pass into the oozes and clays of the deeper ocean depths. The latter have high calcium carbonate content as a result of accumulations of tiny calcareous organisms, especially foraminifera.

Land-derived vegetation has been found scattered on the sea floor several hundred kilometers from land. Off the coast of Central America, land-derived plant remains have been found in over 400 meters of water. The dredge hauls of the *Challenger* expedition (1872–1876) brought up fruits, seeds, leaves, twigs, branches, and parts of tree trunks.

Coarse sediments transported over the shelf edge may move downslope by turbidity currents. The layers of very fine sediment at the base of the continental slope may alternate with sandstone and siltstone beds and exhibit graded bedding.

Deep-Sea Deposits   With increasing distance from shore, the land-derived materials assume progressively less importance. In abyssal depths (greater than 2,000 meters), clastic sediments can be found that are of volcanic origin, are iceberg-rafted debris, or are turbidity-current sediments. Fine clay-sized material may extend outward some distance beyond the base of the continental slope.

Deep ocean-basin floors are not places where abundant life exists. Consequently, the principal organic sediments found there are bones, fish teeth, and especially oozes composed of the tests of tiny floating, planktonic organisms, such as foraminifera, diatoms, and radiolaria. Figure 11.18 shows the distribution of various kinds of oceanic sediments.

Calcareous *Globigerina* ooze is by far the most extensive organic sediment being deposited at present. An area of more than 128 million square kilometers of the deep-sea floor is composed of this type of ooze. Siliceous diatom ooze forms mainly in the cool waters of the North Pacific and in the waters of a belt surrounding Antarctica. Siliceous radiolaria are present in marine sediments deposited at all depths, but they become the dominant sediment only in areas with depths exceeding 4,000 meters. This is because the calcareous tests of planktonic organisms gradually dissolve as they sink through the deep water, leaving an insoluble organic residue of siliceous radiolaria.

TABLE 11.5   Classification of Oceanic Sediments

Pelagic
  Lithogenous brown (red) clay (having less than 30% biogenous material)
  Diagenetic deposits (consisting dominantly of minerals crystallized in seawater)
  Biogenous deposits (having more than 30% material derived from organisms)
    Foraminiferal ooze (having more than 30% calcareous biogenous, largely foraminifera)
    Diatom ooze (having more than 30% siliceous biogenous, largely diatoms)
    Nannofossil ooze (having more than 30% nannofossils—size less than 60μ)
    Radiolarian ooze (having more than 30% siliceous biogenous, largely radiolarians)
    Coral reef debris (derived from slumping around reefs)
      Coral sands
      Coral muds
Cosmogenous (extraterrestrial)
Terrigenous
  Terrigenous muds (having more than 30% of silt and sand of definite terrigenous origin)
  Turbidites (derived by turbidity currents from the lands or from submarine highs)
  Slide deposits (carried to deep water by slumping)
  Glacial marine (having a considerable percentage of particles derived from iceberg transportation)

SOURCE: Shepard, F. P., 1973, *Submarine Geology,* 3d ed., Harper & Row, Publishers, Incorporated, New York.

In other regions at great depths the ocean floor is covered with very-fine brown or red clay. The clay is composed of terrigenous clay minerals, insoluble portions of planktonic tests and other organic matter, volcanic ash, and meteoric dust. This abyssal red clay covers about 100 million square kilometers of the sea floor, mostly in the Pacific Ocean, and has a very slow sedimentation rate.

Because of slow sedimentation in the deep ocean, it is possible to sample the continous accumulations of the past 160 million years. Deep-sea drilling begun in 1968 by the research vessel *Glomar Challenger* has in many places penetrated the sedimentary sequence overlying the basaltic layer of the abyssal ocean floor. The total sediment thickness in abyssal depths is generally less than 600 meters thick and locally less than 100 meters.

Ericson, Ewing, and Wollin (1964) estimate the *Globigerina* oozes in the Atlantic Ocean have accumulated at an average rate of 2.5 centimeters per thousand years over the past 175,000 years. The average thickness of Pleistocene sediments in cores studied was 38 meters. This allowed an extrapolated estimate of 1.5 million years for the beginning of the Pleistocene epoch, which marked the onset of the ice ages. Within the cores, four intervals are recorded by cold-water fauna, corresponding to the four ice ages that affected much of the earth's land area.

Sediments apparently of abyssal origin are now exposed on the Mediterranean island of Malta, on Barbados in the Caribbean, and on Christmas Island off Java in the Indian Ocean. Radiolarian cherts of Mesozoic age that occur on Borneo and Timor are probably of abyssal origin. Geologists think these are lithified equivalents of radiolarian oozes and abyssal clay.

SEDIMENTARY STRUCTURES

Sedimentary structures are geometric patterns in the arrangement of the grains making up the sediment. These may be patterns resulting from size differences or from differences in mineralogy. Sedimentary structures are either primary structures, formed at the time of sediment deposition, or secondary structures, developed after deposition and burial by overlying rocks.

Primary Structures

Stratification This is the arrangement of sediments in layers, or strata (Fig. 11.13). Sediments can be called stratified when they readily separate into layers along bedding planes. Stratification results from different-sized material, the kinds of material, or an interruption in deposition. The difference in character of sediments may result from variations in currents, seasonal changes, climatic changes over a period of years, fluctuations of sea level, or marked changes in the types or numbers of organisms. The individual layers range in thickness from a small fraction of a centimeter in some muds to many meters in coarser sediments. Thus sedimentary rocks can be thin bedded or thick bedded.

When the beds are deposited, they are generally nearly parallel to the surface over which they are deposited. As a rule, they are approximately horizontal. In many places, however, the surfaces of deposition are uneven, and inclined stratification results. Sediments may accumulate in an orderly sequence upon surfaces inclined as much as 30°. The steepest depositional slopes are found in small bodies of water and protected bays, where there is slight agitation and very limited spreading of sediments. In most sedimentary environments, however, currents spread the sediments over a broad area so that strata accumulate as nearly flat layers. This rule of action is known as the principle of original horizontality. If strata are observed with the bedding steeply inclined over a large area, one can infer that the original horizontality of the sediments has been disturbed by deforming forces within the earth.

Stratification is the most distinctive feature of sedimentary rocks.

Cross Bedding Cross-bedded sediments are those that show layering inclined obliquely to the main bedding planes of general stratification (Fig. 11.19). Coarse sediments, such as gravel and sand, are likely to show cross bedding, although some limestones and siltstones are also cross bedded.

Whenever steep slopes are produced by the rapid deposition of sediments, inclined sedimentation occurs. This may take place at the front of a delta, on an offshore bar, where sand waves move along the bottom of a river, in an alluvial fan, or in a sand dune. Succeeding beds again may assume a horizontal position, and so the cross-bedded layers may be interstratified with horizontal layers of master bedding. These features are shown in Fig. 11.19.

Graded Bedding   When a turbulent mixture of particles of different sizes, shapes, and densities is brought to the site of sedimentation, the coarser, heavier, and more nearly spherical grains tend to settle more rapid-

ly than the others. The lighter, smaller, and more angular ones follow in a more or less progressive series. The bed of sediment finally accumulated tends to show a segregation of the particles determined by their relative rates of settling. Such an arrangement is called graded bedding (Fig. 11.20). Repetition of the process may develop graded bedding within each of the succeeding layers. Graded bedding is typical of marine deposits found at the base of the continental slopes or

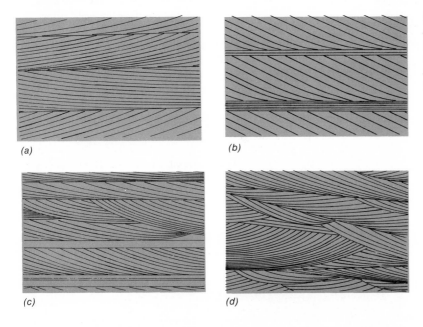

(a)

(b)

(c)

(d)

(e)

FIGURE 11.19   *Varieties of cross bedding. (a) Ordinary marine near-shore type; (b) common stream-current type; (c) torrential-stream type; (d) eolian type, or dune structure; (e) eolian cross bedding. (Gregory, U.S. Geological Survey.)*

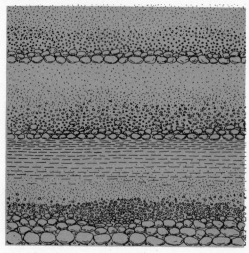

Fine silt
and clay

Fine sand

Coarse
sand

Gravel

FIGURE 11.20   *Graded bedding.*

at other sites where currents move mixed sediments down subaqueous slopes. The deposits left by a receding flood on land may also exhibit graded bedding. Where sedimentary rocks have been tilted steeply by earth movements, or even overturned, graded bedding can be used by geologists to distinguish the original tops and bottoms of the beds.

Lenticular Beds   Some individual beds of sediment can be traced for many kilometers, but others change thickness or disappear in much shorter distances. Massive beds, such as sandstone or certain limestones, at many places gradually decrease in thickness and may be seen to "pinch out" when traced along the face of a cliff. This type of bedding is called lenticular bedding. It is especially common in deposits made at the outlets of rivers or in stream deposits.

FIGURE 11.21   *Mud cracks in clayey sediment on the floor of a dry stream bed, caused by shrinkage in drying. (U.S. Geological Survey.)*

Mud Cracks   These structures are caused by shrinkage of mud as it dries. River flood plains (Fig. 11.21) and the floors of playa lakes (temporary desert lakes) are favorable sites. Some mud cracks are several centimeters wide and 10 times as deep. When the mud is thoroughly dried, such cracks may remain open for months and even years. When water again covers them and sedimentation is renewed, they may be filled by younger sediment. In that event, the sediment, when thoroughly indurated, preserves the mud cracks indefinitely in solid rock.

Ripple Marks   Ripples are formed by the drag of waves and currents of water or wind over sediment. They consist of a series of small, almost equally spaced ridges and troughs of sand or other fine sediment (Fig. 11.22). They may be symmetrical or asymmetrical, depending on the fluid motion that formed them (Fig. 11.23). Oscillatory or undulatory waves give rise to symmetrical ripples (equal slope to trough on both sides of ripple crest). These generally form in shallow water less than 15 meters deep. Symmetrical ripples have sharp crests and rounded troughs. They can be used to distinguish the tops of beds in deformed strata. Currents of wind or water produce asymmetrical ripples with the steep slope on the side toward

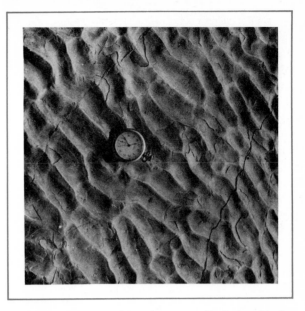

FIGURE 11.22   *Ripple marks preserved in the basal beds of the Beekmantown Dolomite, deposited nearly 500 million years ago at Perth, Ontario. (Geological Survey of Canada, Ottawa, No. 73593).*

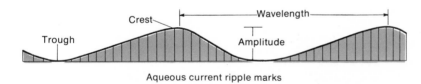

Aqueous current ripple marks

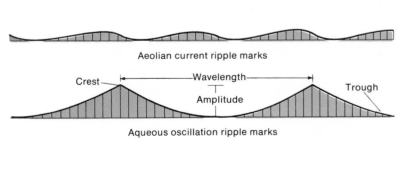

Aeolian current ripple marks

Aqueous oscillation ripple marks

Aqueous oscillation ripple marks

FIGURE 11.23   *Profiles of current and oscillation ripple marks. (After Shrock, 1948.)*

which the current is flowing. Current ripples can form at shallow water depths in streams or at depths of thousands of meters where density currents move across the ocean floor.

Sand waves, or giant ripple marks, may be formed by flooding streams or tidal currents in narrow straits. Their crests may be 5 to 10 meters apart and have a height of 1 meter.

The wind also forms ripple marks on sand dunes, and these resemble some of those formed in water. They are destroyed before burial beneath younger sand layers and are not preserved in rocks.

Fossils   Many sedimentary rocks contain remains or traces of ancient life known as fossils. The term is restricted to organic remains or direct evidence of organisms preserved in the rocks of the earth's crust.

Fossils are of the following types:

1. Actual remains of organisms (soft or hard parts)

2. Impressions

3. Replacements of original substances

4. Tracks, trails, and burrows

Plants have a softer organic structure and decay more quickly; consequently, plants are not so well represented in the fossil record. However, some excellent plant fossils occur as carbonized remains or as impressions of leaves or stems in shale or sandstone. Others have their woody fibers gradually filled in by silica, producing specimens of silicified wood, such as in the Petrified Forest National Monument. Still other plants are preserved in coal.

The bones, teeth, shells, and general skeletal matter of animals are more likely to be preserved as molds or casts (Figs. 11.1, 11.7, and 11.17); but the tracks, trails, burrows, or other impressions of animals may also form fossils. In some cases, the forms of entire animals have been preserved. The best known of these are the fossil insects in amber from the Baltic region and the wooly mammoths frozen in gravels in Siberia.

Fossils record the development of living things over the last 3 billion years of the earth's history. Not

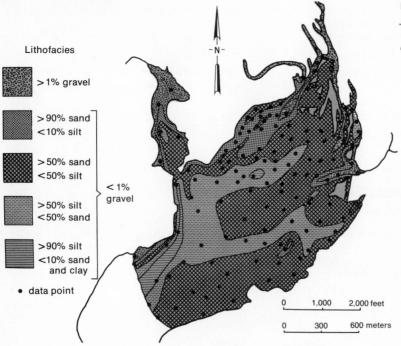

FIGURE 11.24   *Facies patterns (litho-facies-grain size) in a delta built into Watts Bar Lake, Tennessee. (T. Worsley and J. Dennison.)*

Lithofacies

> 1% gravel

> 90% sand
< 10% silt

> 50% sand
< 50% silt

> 50% silt
< 50% sand

> 90% silt
< 10% sand
and clay

< 1% gravel

• data point

| 0 | 1,000 | 2,000 feet |
| 0 | 300 | 600 meters |

only do they reveal the nature of primitive life, but they also show the stages leading to the development of modern life forms and indicate trends of migration from place to place throughout the world.

Since fossils show the progressive development of life, a fossiliferous layer of rock may be dated by the character of the life existing when that rock was deposited.

STRATIGRAPHIC RELATIONS OF SEDIMENTS

The study of the geometric and age relations of strata over large areas is the subscience of stratigraphy. Several stratigraphic patterns will be briefly considered.

Facies

Facies are different sediment types that were deposited simultaneously in adjacent areas. Figure 11.24 illustrates facies in modern sediments. The different sedimentary rock types grade into each other by what is known as a facies change. Frequently, the facies change is a series of intertonguing strata, as illustrated in the stratigraphic cross section of Fig. 4.2. In the Book Cliffs of Utah, facies changes may be observed directly by tracing exposures for several kilometers. Intertonguing of strata demonstrates equivalent ages of two rock bodies.

Conformity and Unconformity
If deposition of strata has been more or less continuous through time, one bed is said to lie on another with conformity. If, however, there is an erosion surface between any two beds, there has been a major interruption in deposition. A region of deposition may be uplifted, and thus be converted into a site of nondeposition or erosion. After a time lapse, the same area may be depressed so that sediments are again deposited on the old surface. Such a buried erosion surface or surface of nondeposition is called an unconformity. If the strata above and below the erosion surface are parallel, the contact is called a disconformity (Fig. 11.25b and c). If the beds below the erosion surface are tilted so that they make an angle with the beds lying on top of it, the contact is called an angular unconformity (Figs. 11.25a and

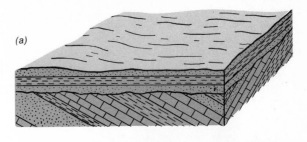

(a)

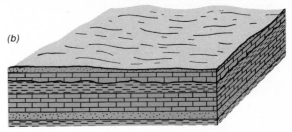

(b)

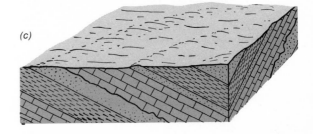

(c)

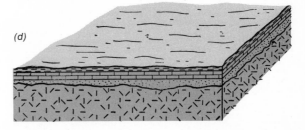

(d)

FIGURE 11.25 *Diagrams illustrating unconformable relations of sedimentary rocks. (a) An angular unconformity with a discordance in bedding; (b) an erosional unconformity, or a disconformity—neither series of strata is tilted or folded; (c) a disconformity with both series tilted; (d) a nonconformity, where sedimentary beds rest on the eroded surface of plutonic igneous rocks.*

FIGURE 11.26   *Angular unconformity between the Laramie Sandstone and the overlying Wasatch Conglomerate. (Fisher, U.S. Geological Survey.)*

FIGURE 11.27   *Diagram showing an overlap resulting from progressive submergence of a land area or a progressive transgression of the sea. (a) Beach sands are deposited on an old land surface; (b) muds that later form shale cover much of the sand as the shoreline moves inland; (c) calcareous oozes that form limestone are deposited over the previous muds as the sites of deposition of sands and muds shift landward.*

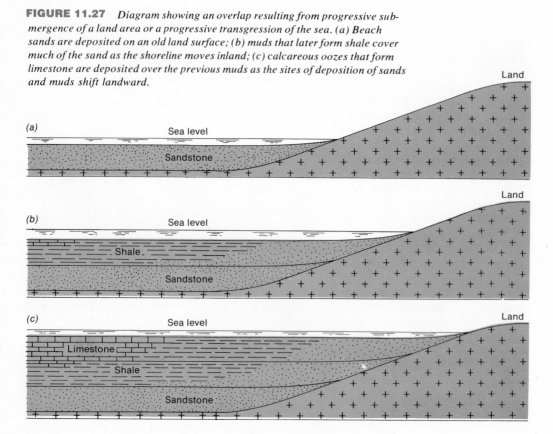

11.26). A nonconformity occurs where strata rest un-conformably on igneous or metamorphic rock (Fig. 11.25*d*).

Overlap and Offlap   In the normal course of events, such as rise in sea level or the sinking of a continent, the sea may gradually encroach upon the land. The deposition of the coarsest sediments follows the shifting coastline, as the sea inundates the land surface, to form an unconformity. The resulting unconformable contact is known as an overlap, or onlap (Fig. 11.27).

The reverse takes place, however, when the land is rising: coarse sediments are deposited farther and farther seaward on top of finer sediments previously deposited. This type of contact between sediments is called an offlap (Fig. 11.28). Alternate onlap and offlap by migrations of the shoreline is one mechanism for producing the intertonguing of facies shown in Fig. 4.2.

INTERPRETING THE HISTORY OF SEDIMENTARY ROCKS

Sedimentary rocks have many characteristics that reflect their environment of origin. Sandstones show ripple marks, cross bedding, and tracks and burrows, just as do sands of the beaches today. Other kinds of sedimentary rock, by their composition, structure, texture, fossils, and other features, also carry telltale signs of the past. Ancient sediments are interpreted in the light of our knowledge of what is taking place today. The present is the key to the past.

Ancient epicontinental shallow seas, former ice sheets, extinct lakes, and other features of the past are

FIGURE 11.28   *Diagram showing an offlap resulting from progressive emergence of a land area as the sea recedes. This stratigraphic cross section shows Fig. 11.27 modified by an offlap occurrence.*

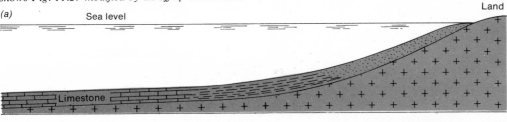

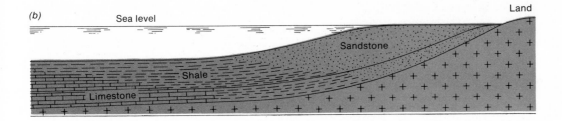

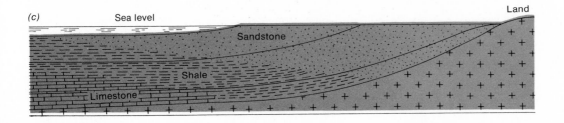

recorded by their sediments. Such sediments reveal not only the environment of their deposition but also something about their derivation, the relief of the land from which they came, the climate under which they were weathered, and other geologic conditions of the past (Table 11.6).

Since sediments have been accumulating in an orderly succession throughout geologic time, they constitute the main documents of geologic history. They also contain the direct evidence of evolutionary history of life on the earth.

TABLE 11.6   Genetic Factors in Sedimentary Rocks

Factors	Properties
Source area	
Relief	These factors control
Stability	weathering, erosion, and
Climate	transport
Source material	
Composition	Composition, grain size, color
Texture	
Structure	
Coherence	
Weathering, fragmentation	Grain size, composition, color
Type	
Degree	
Erosion and transport	
Agency (wind, ice, water)	Grain size
Time and distance	Rounding
Velocity	Sorting
Depositional environment (subaerial deposits)	
Topography	
Drainage	Composition
Climate	Grain size
Position of water table	
Stability (whether subsiding)	
Depositional environment (water-laid deposits)	Sorting
Depth	Color
Stability	Stratification
Temperature	Organic content
Salinity	Porosity
Circulation (waves, currents)	Crystallinity
Aeration	
Subsequent history	
Burial, pressure	Porosity
Exposure (to air or fresh or salt water)	Induration (compaction)
Movement	Cementing
Fracturing	Crystallinity
Compaction	Concretions
	Color

REFERENCES

Blatt, H., Middleton, G., and Murray, R., 1972, *Origin of Sedimentary Rocks*, Prentice-Hall, Inc., Englewood, Cliffs, N. J.

Degens, E. T., 1965, *Geochemistry of Sediments*, Prentice-Hall, Inc., Englewood Cliffs, N. J.

Dietrich, G., 1963, *General Oceanography*, John Wiley & Sons, Inc., New York.

Dunbar, C. O., and Rodgers, J., 1957, *Principles of Stratigraphy*, John Wiley & Sons, Inc., New York.

Garrels, R. M., and Mackenzie, F. T., 1971, *Evolution of Sedimentary Rocks*, W. W. Norton & Company, Inc., New York.

Hutchinson, G. E., 1957, *A Treatise of Limnology*, John Wiley & Sons, Inc., New York.

Kuenen, Ph. H., 1950, *Marine Geology*, John Wiley & Sons, Inc., New York.

Langbein, W. B., 1961, *Salinity and Hydrology of Closed Lakes,* United States Geological Survey Professional Paper 412.

Pettijohn, F. J., 1957, *Sedimentary Rocks,* 2d ed., Harper & Brothers, New York.

——, Potter, P. E., and Siever, R., 1972, *Sand and Sandstone*, Springer-Verlag, New York.

Shepard, F. P., 1973, *Submarine Geology*, 3d ed., Harper & Row, Publishers, Incorporated, New York.

Shrock, R. R., 1948, *Sequence in Layered Rocks*, McGraw-Hill Book Company, New York.

Zumberge, J. H., and Ayers, J. C., 1964, *Hydrology of lakes and swamps,* in Chow, Ven Te (ed.), *Handbook of Applied Hydrology,* pp. 1-33, McGraw-Hill Book Company, New York.

CHAPTER 12
MINERAL RESOURCES

Mineral resources play an important part in providing man with a myriad of necessities and conveniences. Mineral resources include mineral fuels, mainly petroleum, coal, and natural gas; metallic ores; and various nonmetallic minerals and rocks, other than fuels. The mineral fuels account for about two-thirds of the dollar value of the total mineral production of the United States (Fig. 12. 1). Copper and iron are the leading metals. Building stone, cement, and sand and gravel are the principal nonmetals. The relative values of the mineral production of the United States for 1972 are listed in Table 12.1

MINERAL FUELS

Petroleum

Composition   Petroleum is a naturally occurring, complex mixture of liquid hydrocarbons. Hydrocarbon compounds consist chiefly of carbon and hydrogen, with minor amounts of nitrogen, oxygen, and sulfur. Most petroleum contains 82 to 87 percent carbon by weight and 11 to 15 percent hydrogen. Distillation of the light fractions of crude petroleum leaves a residue of light-colored paraffin wax, a dark tarry asphalt, or a mixture of the two. For this reason crude oils sometimes are classified as paraffin-base, asphalt-base, or mixed-base oils. Asphalt-base and mixed-base oils are most abundant.

Origin   Geologists and chemists agree that petroleum originates from plant and animal remains that accumulate on the sea floor along with the sediment that forms sedimentary rocks. The evidence for organic origin seems conclusive: Petroleum compounds have certain optical properties that are peculiar to organic compounds; they contain pigments called porphyrins (compounds like the chlorophyll of plants and the hemoglobin of animal blood); more than 99 percent of all occurrences of crude petroleum are associated with marine sedimentary beds. Oil is rarely found in igneous or metamorphic rocks or in fresh-water sediments, and then only where these sources are associated with marine sedimentary beds. Oil found in such rocks presumably has migrated from marine rocks nearby; hydrocarbons resembling petroleum are found at shallow depths in modern marine sediments.

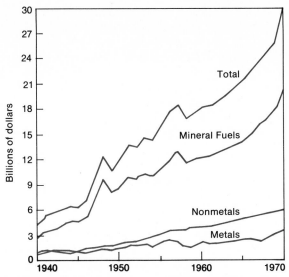

FIGURE 12.1 *Value of the mineral production in the United States to 1970. Mineral fuels account for over two-thirds of the $30-billion total. Stone, portland cement, and sand and gravel are the principal nonmetals other than fuels. Copper and iron are the leading metals. (U.S. Bureau of Mines.)*

The processes by which the parent organic matter is converted into petroleum are not understood. The contributing factors are thought to be bacterial action, shearing pressure during compaction, heat and natural distillation at depth, possible addition of hydrogen from deep-seated sources, presence of catalysts, and time.

Anaerobic bacteria can liberate oxygen, nitrogen, phosphorus, and sulfur from organic matter, and in this manner they increase the relative percentages of carbon and hydrogen remaining. Some bacteria produce the natural gas methane, CH_4, but methane's role in the origin of petroleum has not been clearly defined. The time factor also is not well established. It is not known whether the transformation of organic matter to petroleum takes place at the time of deposition of the associated sediments, at some time after burial but before or during consolidation, or as a later post-depositional process. Many geologists favor an early origin.

Source Beds Source beds are the rocks in which the petroleum (or natural gas) originated. In many oil-bearing districts, thick layers of dark-gray, chocolate-

TABLE 12.1 **Mineral Production in the United States in 1972**
Values in Millions of Dollars

Mineral fuels		Nonmetals (other than fuels)	
Petroleum	11,700	Stone	1,475
Coal	4,489	Portland cement	1,269
Natural gas	4,500	Sand and gravel	1,200
Others	1,355	Salt	304
Total mineral fuels	22,044	Lime	286
Metals		Clays	268
Copper	1,700	Phosphate rock	200
Iron ore	888	Sulfur	193
Molybdenum	190	Potash salts	98
Lead	188	Boron minerals	87
Zinc	174.8	Magnesium salts	62
Uranium ore	158.5	Bromine	61
Silver	63	Sodium carbonate	56
Gold	84	Gypsum	35
Vanadium	35	Barite	13
Bauxite (Al ore)	27	Asbestos	11
Tungsten	22	Others*	215
Titanium	19	Total nonmetals	5,833
Mercury	1.4		
Other metals	58		
Total metals	3,608.7	Grand total	31,485.7

* Abrasives, diatomite, feldspar, fluorspar, perlite, slate, talc, and vermiculite.

brown, bluish-gray, or black shale, all of which have been darkened by organic matter, are thought to have served as the source beds. In other areas calcareous shales, siltstones, fine sandstones, or limestones apparently were sources. Recent marine sediments contain an average of 2.5 percent by weight of organic matter. More organic matter is preserved in fine-grained sediments than in coarse-grained ones. Can you think of a reason why this is so? It is for this reason that fine-textured dark shales and limestones seem to be especially favorable source beds for petroleum.

Reservoir Rocks   Reservoir rocks are the permeable rocks in which oil accumulates underground. Even though petroleum may originate in shale, it cannot ordinarily be recovered in commercial quantities from shales because they are not sufficiently permeable. Very recently, because of the need for petroleum resources, oil shales are being considered as a potential reserve. The cost and the techniques of recovery

are slowing development. Good reservoir rocks must have high porosity and high permeability to provide both storage capacity and freedom of discharge.

Oil generally accumulates in open-textured sandstones or in porous limestone or dolomite. In producing areas, these rocks will have porosities of more than 10 percent, and they are tens of meters thick. Some sandstone bodies that serve as reservoirs are extensive sheets, while others are small lenses representing former beaches, or they are narrow stringers ("shoe-string sands") that were once offshore bars. Limestone reefs and the solution cavities within the reefs also make excellent reservoirs. In a few areas, highly fractured rocks of almost any type, including shale, basalt, chert, and schist, may yield petroleum.

Apparently, when a column of sedimentary rocks is compacted by the weight of beds deposited at a later time, petroleum, water, and natural gas are squeezed out of the soft source muds into the more permeable reservoir rocks, where the fluids then can migrate freely. During compaction, the source muds will have

FIGURE 12.2   *Diagram showing the types of traps that can allow the accumulation of oil and gas. (After Heiland, 1946.)*

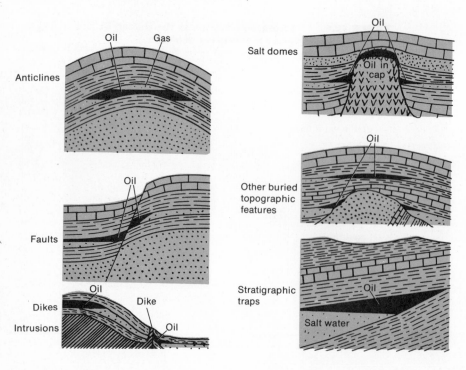

their porosity reduced from as much as 90 percent to 35 percent or less, and they become almost impermeable because of the fineness of the openings between the particles.

Migration of Petroleum When petroleum is expelled from the source rock into a reservoir rock, it may migrate laterally or vertically, depending on the structure of the reservoir rock. Its movement is facilitated by the water on which it floats. If the reservoir rock is tilted, the oil held under the reservoir roof floats up the tilted surface until it is trapped and held in a structural trap. Oil accumulations found near the edges of certain structural basins may have moved laterally several tens of kilometers or more before being trapped.

Oil Traps An oil trap is a body of reservoir rock surrounded by impervious rock in such a way that the two form a naturally closed container in which the oil collects (Fig. 12.2). The oil is held beneath an impermeable cap rock. Some oil traps are formed by structural features, such as upfolded or domed strata (Fig. 12.3). Some are stratigraphic traps resulting from differences in permeability within sandstone lenses and stringers, variations in cementation, facies changes, and limestone reefs. Other traps may represent combinations of structural and stratigraphic conditions.

Natural gas and salt water are nearly always associated with petroleum. In the oil traps, the gas, oil, and water are separated into layers, since oil is insoluble in water and gas is only slightly soluble. The separation occurs in order of increasing density, with gas (the lightest) on top, oil in the middle, and water (the heaviest) on the bottom. The gas may be present in small or large quantities, and free gas appears above the oil only when the oil is saturated with gas. The salt water is believed to be seawater that filled the openings in the sediments when they were deposited in a marine environment.

Oil Pool and Oil Fields An oil pool is any accumulation of oil in reservoir rock that yields petroleum on drilling. Since the petroleum is distributed throughout the rock between the mineral grains, it is not a pool in the ordinary sense. An oil field is a geographic district containing underground accumulations of economic value. A single oil field may draw upon

FIGURE 12.3 *Aerial photograph of an oil-producing anticline in Wyoming. This arched type of structure favors entrapment of oil. (U.S. Geological Survey.)*

several different oil pools located in different stratigraphic and structural settings. The major oil fields of the world occur in the United States, Canada, Mexico, South America, the Caribbean, the Middle East, Russia, Eastern Europe, and Indonesia. In the United States, Texas, Louisiana, California, Oklahoma, Kansas, New Mexico, Pennsylvania, Wyoming, Michigan, and Arkansas are the leading producing areas.

Uses, Production, and Reserves Petroleum has three main uses—fuel, lubrication, and petrochemicals. It furnishes about one-half of the fuel used to produce energy. The principal refinery products, in order of bulk, are gasoline, 50 percent, distillate fuel oil, 25 percent, and decreasing percentages of residual fuel oil, kerosene, asphalt, jet fuel, and petroleum coke. Minor products (by volume) are lubricating oil, liquefied refinery gases, grease, road oil, and wax. In addition, petroleum yields thousands of chemical compounds, called petrochemicals. These compounds are used in making such products as synthetic rubber, synthetic fiber, plastics, paints, solvents, dyes, detergents, resins, fertilizers, pesticides, and various drugs. The total production of crude petroleum in the United

States up to 1972 is estimated to have been about 4 billion barrels, an increase of 1 billion barrels since 1960. The world's supply of proved oil reserves is judged to be about 634 billion barrels, with five-sixths of this in land reserves and one-sixth in offshore reserves on the continental shelf. The estimated total potential oil resources is between 100 billion and 1,000 billion barrels.

The principal proved reserves by percentage are 62 percent in the Middle East, 11 percent in the U.S.S.R. and other Communist countries, 7 percent in the United States, 6 percent in Libya, 3 percent in Venezuela, 2 percent in Indonesia, 2 percent in Canada, 1 percent in Mexico and in Nigeria, and 5 percent in all other areas. The relative ranks of these countries may be greatly changed by additional discoveries, such as on the North Slope of Alaska or in the North Sea.

Natural Gases

Natural gas occurs either with oil, occupying the same reservoirs and structural traps, or separately (Fig. 12.2). Where oil is absent, the gas lies directly above water in the reservoir rocks. Most natural gas consists of colorless, odorless, and highly flammable methane (CH_4). In oil fields, ethane, propane, butane, pentane, and vapors of gasoline may also be present. Butane and related gases, which can be liquefied under pressure and marketed in tanks, are widely used as bottled gas. Natural gas is a highly valued fuel, which now is piped to nearly all parts of the country. Some of it is reduced to carbon black for use in printer's ink and rubber tires. Texas, Louisiana, Oklahoma, New Mexico, California, and Kansas lead in natural-gas production in the United States.

Other Natural Gases

Among the important but noncombustible gases encountered in the earth are nitrogen and carbon dioxide. In some mixtures of natural gases, nitrogen and carbon dioxide exceed the quantity of flammable gases. Certain oil wells in Colorado yield such large quantities of admixed carbon dioxide that rapid release of pressure freezes the gas to dry ice and converts the accompanying oil to a stiff slush. New Mexico, Colorado, Utah, and California lead in the production of carbon dioxide.

Helium is a light gas, and because it is not flammable, it is very desirable for spacecraft fuel and for many other technical uses. Nearly all the world's supply of helium comes from a few wells in Kansas, Texas, Oklahoma, New Mexico, and Arizona. Some noncommercial gas comes from swamps (marsh gas) and coal beds.

Coal

Distribution   Coal beds are widely distributed in the United States, Canada, Europe, Russia, China, and Australia. The principal coal fields in the United States are the anthracite field near Scranton, Pennsylvania; the Allegheny-Appalachian field, extending from Pennsylvania to Alabama; the Illinois field, reaching into western Indiana and western Kentucky; the Mid-Continent field in parts of the states from Iowa to Texas; and the Rocky Mountain–Great Plains field, extending southward from Montana and North Dakota to New Mexico (Fig. 12.4).

The Pennsylvania anthracite field is about the only American source of anthracite. It has an area of less than 1,200 square kilometers, and much of the field has already been worked out. The coal in the Allegheny-Appalachian, Illinois, and Mid-Continent fields is mostly bituminous. The coal in the Rocky Mountain–Great Plains field ranges from lignite in the Dakotas to generally subbituminous and bituminous nearer the Rockies. The type of coal depends on the former depth of burial and the amount of structural deformation the coal and associated rocks have experienced. Some coal in Colorado and Arkansas approaches anthracite in grade.

The fuel value of lignite is only 7000 to 8000 British thermal units (Btu) per pound. (One Btu will raise the temperature of 1 pound of water 1 degree Fahrenheit.) High-grade bituminous and anthracite coals yield 13,000 to 15,000 Btu per pound. Impurities in the coal lower the fuel value and add to air pollution.

Methods Used in Mining Coal   The methods of mining have to be adapted to the structure of the coal beds and their topographic setting. For example, in the Pennsylvania anthracite field, the sedimentary rocks in which the coal occurs are steeply dipping. For this reason, inclined and vertical shafts or deep open pits must be used to reach the coal.

In the stream-dissected Allegheny Plateau, where

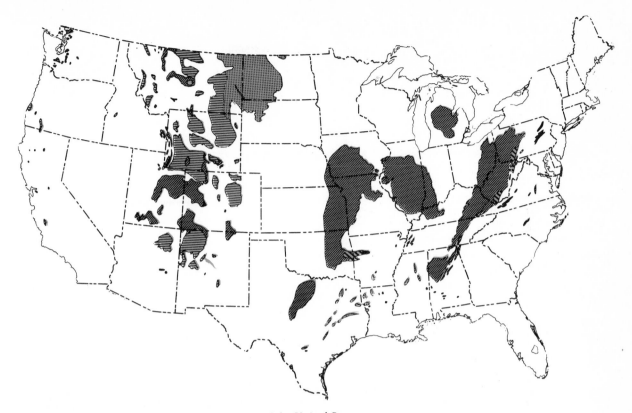

FIGURE 12.4 *The major coal-producing areas of the United States.*

many flat-lying to gently tilted bituminous coal beds are exposed along the sides of valleys, the coal is mined by digging adits (single-opening entrances) into the hillsides along the bedding plane. In the nearly flat-lying coal beds of central United States, both shafts and open-pit mines are used extensively. Where the overburden is thin, open-pit mining is utilized.

Uses, Production, and Reserves Coal is used primarily as fuel for domestic and industrial purposes. Most coal is used directly, but some is converted to coke before it is used as fuel. Coke is made by heating coal in special ovens, where in the absence of air the volatile constituents are distilled out. The residue is carbon-rich coke. Coke is valuable as a fuel and reducing agent for smelting iron ore in blast furnaces. The by-products driven off by the coking process are combustible gases and tars, from which a host of chemical products are obtained.

Large coal-burning electric-generating plants are now being constructed in or near the coal fields. The electricity is then transmitted to the areas of principal consumption. As a result of this trend, large-scale use of nuclear energy for generating electric power could be delayed. If the air pollution problems from coal begin to outweigh the popular fear of atomic installations and their thermal pollution problem, the trend could be reversed.

Reserves of coal are difficult to estimate because of changing technology and economic conditions, but by conservative standards the present annual production of coal in the United States is at least 453.5 million metric tons. This quantity is ample to last about 2,000 years at present rates of consumption. The recoverable reserves of the world, 7,619 billion metric tons, are located mainly in North America (about one-third), the U. S. S. R., China, Germany, and the United Kingdom.

METALLIC ORES

Definitions

An ore deposit is a concentration of one or more metallic minerals sufficiently rich in some metal to make
its mining profitable, while an ore mineral is one
that contains a valuable metal. Some metals, such as
gold and platinum, are found in an elemental state, but
more commonly the metallic elements occur in mineral
compounds with other elements in chemical compounds as sulfides, oxides, and carbonates.

In most ore deposits the ore minerals are associated with large quantities of gangue and country
rock. Gangue is the valueless and usually nonmetallic
material deposited along with the ore. Country rock is
the rock that encloses the ore. In many deposits the
ore does not have sharp boundaries. Instead, the ore
grades into the country rock, much of which is removed in mining the ore.

Many ores contain two or more metals. Thus the
copper ores of Butte, Montana, carry important
amounts of silver and gold, as do practically all the copper deposits of the West. Lead and zinc, or lead and
silver, are also common associates in the same deposit.
These extra metals found in an ore, especially gold and
silver, make profitable the extraction of otherwise sub-
standard grades of ore.

Genetic Classes

According to their mode of origin, valuable mineral deposits may be classified into one of eight categories
(Table 12.2).

Magmatic segregations are formed by the accumulation of differentiates separated from a magma,
by the settling of early-crystallized and heavy ore minerals. Most ores of this type occur at the bottoms of
sills or in dikes. Pegmatites are also magmatic-dif-
ferentiation products, but they represent the light
rather than the heavy components of a magma. Generally, they are composed of large crystals and are sometimes called "giant granites." Some have crystals of
tourmaline (a boron mineral) and apatite, which contains fluorine and more rarely chlorine. Boron,
fluorine, and chlorine are thought to facilitate the
growth of the large crystals. Gases and steam tend to
keep a magma liquid, thus promoting the development
of unusually large crystals. Pegmatites generally fill
dikes in or near the roofs of batholiths.

Ore deposits in contact-metamorphic zones are
replacements of invaded rocks by solutions that are
extruded from an intrusive magma. Although they are

TABLE 12.2 **Genetic Classes of Mineral Deposits**

Genetic type	Representative examples of products
Magmatic segregations	Magnetite (Kiruna area, north-ern Sweden)
	Chromite (eastern Cuba)
	Ilmenite (Ti ore) (Allard Lake, Quebec)
	Platinum (Ural Mountains, U.S.S.R.)
Pegmatites	Nonmetals: mica, lithium min-erals, feldspar
	Metals: beryllium
Contact-metamorphic deposits	Tungsten, tin, gold, copper, iron, zinc, silver, lead
Deep, high-temperature deposits	Gold (Homestake mine, Lead, South Dakota; Kirkland Lake and Hollinger areas, Ontario)
	Tin (Cornwall, England)
Intermediate-depth and -temperature deposits	Copper (Bingham Canyon, Utah; Butte, Montana)
	Lead-zinc-silver (Coeur d'Alene, Idaho)
Shallow, low-temper-ature deposits	Gold (Cripple Creek, Colorado)
	Silver-gold (Comstock Lode, Nevada)
	Silver (Pachuca, Mexico, fabu-lously rich)
	Antimony (Hunan, China—world's largest supply)
	Mercury (Almaden, Spain)
Near-surface, low-temperature deposits	Lead-zinc (Tri-State area, Okla-homa-Missouri-Kansas)
	Zinc (Tennessee)
Sedimentary deposits	Uranium-vanadium (Utah, Co-lorado, New Mexico, Ari-zona)
	Residuals: bauxite, iron, manga-nese, gold, platinum, native copper, nickel (Cuba, New Caledonia), barite, mercury (cinnabar), clays
	Placers: gold (Sierra Nevada foothills, California; Nome and Fairbanks, Alaska; Wit-watersrand, South Africa)
	Chemical precipitates: metals (ores of iron and manganese); nonmetals (phosphate, salt, gypsum, and potash salts)

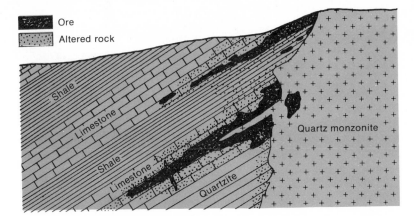

Ore

Altered rock

FIGURE 12.5 *Section of a contact-metamorphic ore deposit. The alteration (mostly in limestone) was caused and the ore introduced by the intrusion of the magma that formed the quartz monzonite.*

found in the contact zones of both sedimentary and igneous rocks, they occur mainly in sedimentary rocks, particularly in limestones and calcareous shales (Fig. 12.5). The deposits are usually adjacent to the intruding igneous rock, rarely more than 0.5 kilometer away. Sulfide minerals, such as pyrite and chalcopyrite; oxides, such as hematite and magnetite; and gangue minerals are mutually intergrown, having formed at about the same time.

Some contact-metamorphic ore deposits are parallel to the sedimentary layering, while others cut across the beds. Because ore-bearing solutions come from intruding magma masses, they have high temper-

atures and are under strong pressures. As a result of these physical characteristics, they are able to penetrate minute joints and the cleavage planes of minerals.

Metalliferous deposits formed at high temperatures (300 to 500°C) and great depths are found in and around granitic stocks and batholiths. Some deposits are tabular fissure fillings, while others are selective replacements of favorable host rocks (Fig. 12.5). Other deposits (Fig. 12.6) will form at intermediate temperatures (200 to 300°C) and moderate depths (1 kilometer). The country rock is usually thoroughly altered hydrothermally.

Deposits formed by magmatic solutions at low

FIGURE 12.6 *The Bingham Canyon, Utah, copper deposit. (Spence Air Photos.)*

temperatures (100 to 200°C) and at depths of perhaps 0.8 kilometer or less, are usually found in geologically young volcanic areas. The wall rocks of the deposits are extensively altered. The ores usually occur as fissure fillings in breccias, as complex vein networks, and as replacement masses.

Some metalliferous deposits are formed by hydrothermal solutions at shallow depths and low temperatures and at a long distance from an inferred but unproved magmatic source. They may form in part from cold meteoric waters. The generally flat-lying host rocks are altered very little, but the association of the minerals fluorite and barite, with sphalerite and galena in some places, suggests a magmatic contribution to the ore-forming fluids. The lead-zinc deposits of the Tri-State area (Oklahoma-Missouri-Kansas) are fracture fillings and replacements of adjacent limestone. The sedimentary uranium and vanadium minerals of the Colorado Plateau occur in the following forms: disseminations or small lenses in sandstones and conglomerates, thin incrustations along fractures and bedding planes, thin sheets a few hundred meters wide, and elongate stringers in buried stream channels. In the stream-channel deposits, uranium and vanadium minerals have replaced wood and fossil bones.

Supergene Enrichment of Sulfide Ores

Supergene enrichment is the natural upgrading of sulfide ore deposits. It is formed by a solution process in the acidic oxidizing zone near the surface and a reprecipitation in greater concentration in the alkaline zone just below the water table.

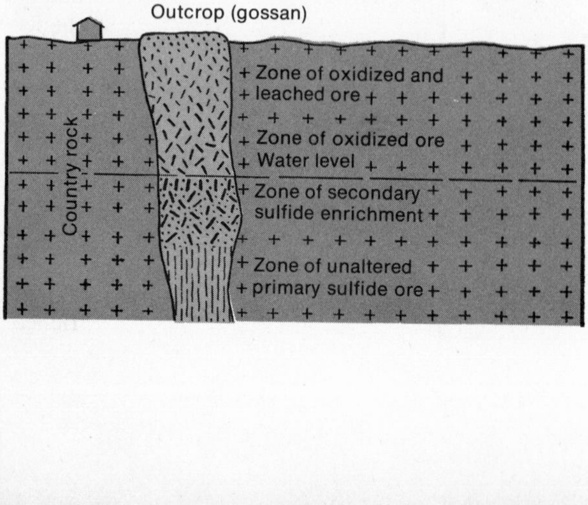

Nearly all veins bearing ores of copper, silver, lead, zinc, and other metals contain sulfides. Where the veins are exposed at the surface of the earth, they are attacked by air and water and undergo a series of weathering changes. For example, pyrite is oxidized, forming sulfuric acid and iron sulfate:

$$2FeS_2 + 16H_2O + 7O_2 \rightarrow 2H_2SO_4 +$$

pyrite   water   oxygen   sulfuric acid

$$2FeSO_4 \cdot 7H_2O$$

iron sulfate

The sulfuric acid that is formed dissolves copper, zinc, and other metals, which are carried downward. The iron sulfate is further oxidized and breaks down to form limonite. The limonite is insoluble and remains at the surface, forming a rusty-stained outcrop called gossan.

The copper in sulfide ore deposits is soluble in the presence of air and sulfuric acid. The copper is carried downward as copper and sulfate ions, and so the outcrops of copper sulfide veins carry very little copper. When the water containing sulfuric acid and dissolved copper sulfate reaches the water table, it enters an alkaline, low-oxygen environment. The solution reacts with pyrite or other primary sulfides and precipitates copper sulfide as a replacement. The secondary copper sulfide just below the water table makes a rich ore. The vertical sequence is illustrated in Fig. 12.7.

Deposits of ores of other metals also show changes that are brought about by surface alteration, but each metal behaves in its own way, depending upon its chemical properties. Silver ores frequently are enriched, lead ores form insoluble lead sulfate, and zinc may be removed in solution.

Sedimentary Metallic Ore Deposits

Sedimentary ore deposits include products concentrated as residuals from weathering, those concentrated as placers by sediment transport processes, and chemical precipitates.

Residuals One of the principal residual deposits is bauxite, the main ore of aluminum. Some bauxite deposits are decomposition products of low-silica syenite, but others are thought to have been derived by the

FIGURE 12.7 *Diagram showing secondary sulfide enrichment of a low-grade, sulfide-bearing vein. Because of weathering, valuable metals such as copper are dissolved in the acid zone near the surface and reprecipitated as an enriched high-grade zone of secondary sulfides in the alkaline environment just below the groundwater level.*

Outcrop (gossan)

Country rock

Zone of oxidized and leached ore
Zone of oxidized ore
Water level
Zone of secondary sulfide enrichment
Zone of unaltered primary sulfide ore

slow leaching and alteration of clay minerals occurring in impure limestones. The waters leaching the limestones must have just the right acidity to convert the clay minerals into bauxite.

More than 90 percent of the production of bauxite in the United States comes from central Arkansas, where the deposits were produced by the alteration of nepheline syenite, a coarse-grained igneous rock composed largely of feldspars and no quartz. Extensive leaching under tropical conditions changed the nepheline ($NaAlSiO_4$) and feldspars into clay minerals. With further leaching and removal of the silica, the clays were converted to bauxite. In tropical countries, bauxitic laterite soils are widespread, but large high-grade bauxite deposits are uncommon. Jamaica, in the Caribbean, is the leading producer. Large deposits also occur in Guiana and Surinam in South America, in the U.S.S.R., in Europe, and in the Near East. Within the past few years large new deposits have been discovered in Australia.

Placers   All ore minerals that have a high specific gravity and are not easily dissolved are likely to accumulate in placer deposits. Figure 12.8 illustrates the formation of a diamond placer deposit by the erosion of diamond-bearing material from a peridotite intrusive. Where gold-bearing veins are exposed at the surface, the mineral may be washed away by running water. Particles of gold are strung out along the ground below the vein cropping and are washed into the beds of streams. The method of mining gold occurring in this setting is called placer mining, and the gold-bearing gravels are placer deposits.

Chemical Precipitates   The main metallic chemical precipitates are ores of iron and manganese, but other nonmetallic substances of economic importance, such as limestone, phosphate, salt, gypsum, and potash salts, are also sedimentary in origin.

Most of the United States supply of iron ore comes from sedimentary beds of hematite in the Great Lakes states (Fig. 12.9). Some is derived from magnetite of contact-metamorphic origin, as in New York, Pennsylvania, Utah, and California. Minnesota and Michigan supply more than three-fourths of the United States production.

The great iron-bearing deposits of the Lake Superior districts originally were chemical precipitates. The initial sediment formed an iron formation that con-

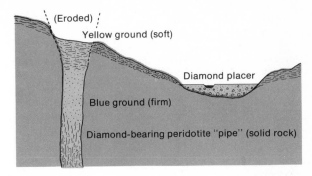

FIGURE 12.8   *A diamond-bearing pipe of peridotite and its associated placer deposits. Diamonds removed from the weathered surface are washed downhill and deposited in gravels and sands along a stream near the pipe.*

tained chert, hematite, magnetite, siderite, and a hydrous iron silicate. The unaltered rock in the iron formation (taconite) contains 33 percent magnetite and 25 to 35 percent iron. Where taconite was leached by weathering and groundwater, much of the original silica was removed and iron was left behind. Thus the iron content was increased to 50 percent or more, the usual shipping grade of untreated commercial ore. In recent years, methods have been devised to concentrate and pelletize the iron minerals from taconite. Pellets containing more than 60 percent iron make excellent blast-furnace feed and, due to their small size, increase the capacity of existing furnaces reducing the cost of pig iron.

In the Mesabi district of Minnesota, the iron ore is mined in enormous open pits along an outcrop of gently tilted beds (Fig. 12.10). In other iron districts of Wisconsin and Michigan, where the beds are tilted more steeply, underground mines are used. A newly developed iron-mining district lies in eastern Canada near the Quebec-Labrador boundary. The ore is hematite, enriched by natural leaching.

Beds of oolitic hematite of marine origin are mined or were formerly mined for iron in Newfoundland, Alabama, France, and England. Near Birmingham, Alabama, one bed, the "Big Seam," has 5 to 7 meters of ore. The primary rock contains about 30 to 35 percent iron and a large quantity of calcium carbonate. The hematite is partly a replacement of oolitic limestone and calcareous shell fragments and partly a colloidal precipitate. The outcrops have been enriched

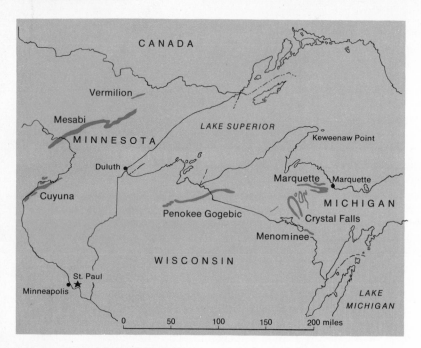

FIGURE 12.9 *The distribution of iron ranges of the Lake Superior region. (After Leith, Lund, and Leith, 1935.)*

FIGURE 12.10 *Monroe open-pit iron mine, Mesabi Range, Minnesota. Weathering removed silica from ferruginous chert (taconite) along the outcrop of the iron-bearing formation. The iron was concentrated residually to form commercially valuable iron ore. (Oliver Iron Mining Company.)*

by leaching. The steel mills in Birmingham now operate on ore shipped in from other areas, since the oolitic iron ores can no longer compete with processed ore.

NONMETALLIC MINERAL RESOURCES (OTHER THAN FUELS)

Nonmetallic mineral resources (other than fuels) include a wide variety of rocks and minerals. Data on some of the main resources are presented in Table 12.3. In addition to the major mineral resources, there are many other commercially valuable nonmetallic substances. These are summarized in Table 12.4.

POLLUTION CONSIDERATION

A student sensitive to environmental issues realizes that mineral fuels create problems. Pollution can arise from the time of drilling for oil or the mining of coal to the time of their actual use. A list of a few problem issues will show the scope and magnitude.

Strip Mining of Coal

In the Appalachian Region of the United States, this has converted a beautiful natural landscape into a wasteland of open pits, piles of debris, and barren land surfaces. Other waste products are discarded during the mining, quarrying, and beneficiation processes. Rainwash often carries part of these wastes into streams, with serious effects on aquatic life and the stream regimen. The safe disposal of polluted water from mines and oil wells without polluting surface water or groundwater is especially difficult.

Drilling for Offshore Oil

This has led to major oil-pollution problems off the coast of California and the Gulf of Mexico.

Transportation of Fuel

The discovery of petroleum reserves on the North Slope of Alaska was a boon to those who predicted a serious fuel shortage. But the transportation of this fuel to markets either by supertankers, such as *The Manhattan,* or by a cross-tundra pipeline, has alarmed environmentalists.

TABLE 12.3   Major Nonmetallic Mineral Resources (Other than Fuels)

Material	Origin	Use
Limestone	Sedimentary	Building stone, cement, crushed-stone aggregate, blast-furnace flux
Crushed stone	Various	Concrete aggregate, macadam roads, general fill
Dimension stone	Various	Building stone
Sand and gravel	Sedimentary	Concrete aggregate (general construction, pavement, building blocks), glass sand, fiberglass, foundry sands
Phosphate rock	Sedimentary	Fertilizers, chemicals, detergents
Salt	Evaporite	Source of chlorine and soda ash
Clays	Residual and transported	Brick, tile, pottery, firebrick, fillers, decolorizers
Sulfur	Volcanic, oxidation of H_2S, and reduction of sulfates	Sulfite papermaking, chemicals, fertilizer, bleaches
Potassium salts	Evaporite	Fertilizers, chemicals, munitions
Gypsum	Evaporite	Wallboard, lath, plaster, retarder in portland cement

TABLE 12.4   Minor Nonmetallic Mineral Resources (Other than Fuels)

Material	Use	Geologic Source
Abrasives	Grinding, scouring, polishing	Sandstone, quartzite, flint, volcanic ash, hard minerals (garnet)
Asbestos	Fireproof fabrics, paper, brake linings, shingles	Veinlets in serpentinite and serpentinized dolomite
Asphalt	Paving, roofing, waterproofing	Petroleum bitumen, natural asphaltite
Barite	Oil-well drilling muds, glass, paint	Veins, beds, replacements, concretions, residuals
Bentonite	Oil-well muds, foundry molds, oil bleaching	Altered volcanic ash
Borates	Flux, glass, detergents, chemicals	Salt lakes, bedded evaporites
Bromine	Gasoline additives	Brines, seawater
Diamond	Drills, abrasives, gems	Placers, crystals in peridotite pipes
Diatomite	Filters, insulation, absorbent, abrasive	Marine and lake beds
Feldspar	Glass, porcelain, enamel, tile, glazes, abrasives	Pegmatites
Fluorite	Acid, flux, glass, enamel, chemicals	Veins and replacements
Garnet	Abrasives, gems, watch jewels	Metamorphic rocks, pegmatites, placers
Graphite	Foundry facings, crucibles, lubricants	Altered organic sediments, veins, lenses in metamorphic rocks
Magnesite	Refractory brick	Hydrothermal alteration of serpentinite
Mica	Radio tubes, electrical insulation	Pegmatites, washed clays
Olivine	Refractories, gems	Peridotites
Perlite	Insulating plaster, light-weight aggregate	Obsidian flows
Pumice	Light-weight concrete, insulation, abrasive	Volcanic ejecta
Quartz	Broadcast frequency control, silica glass	Vugs in pegmatites and veins, placers
Sodium	Kraft paper, glass, detergents, chemicals	Evaporites, brines
Talc	Filler in paint, rubber, ceramics, roofing	Altered soapstone or dolomite
Vermiculite	Sound insulation in plaster and loose fill, plastics	Hydrothermal alteration of pyroxenite

Atmospheric Pollution

The pollution of our atmosphere by the burning of fossil fuels, either from power plants or from automobiles, has reached serious proportions. The discharge of sulfur compounds from the smokestacks of sulfide-ore smelters and from coal- and oil-burning power plants pollutes the air, water, and soil. Partial abatement of this problem can be accomplished by the use of low-sulfur fuels, the extraction of sulfur before combustion, or the recovery of sulfur from the stack gases.

Fluorides emitted from aluminum plants are also harmful to native plants, crops, and livestock. Improved technology despite increased costs is necessary if man is to prevent irreparable damage to the earth.

What are the solutions? Some answers might be: stop the strip mining of coal, outlaw the offshore production of petroleum, find safer methods to transport the petroleum from known oil fields, and stop burning fossil fuels and use an alternate power source.

FUTURE OUTLOOK FOR MINERAL RESOURCES

The supplies of mineral resources in the accessible parts of the earth's crust are limited and (except for water) nonrenewable. Some of these materials can be recycled, for example, scrap iron or lead batteries, but many valuable metals are scattered, lost, or wasted. For instance, much of the iron eventually goes into land-fill, and the lead used in ethyl gasoline is irretrievably dispersed as a pollutant in the earth's air and soil. Lead, zinc, and titanium used in paint are also dissipated.

Mineable concentrations of metallic ores are distributed very unevenly over the globe (Table 12.5). No

TABLE 12.5   Continental Distribution of Mineral Resources

Continent	Mineral resource
Africa	Gold, platinum, cobalt, diamonds, phosphate, petroleum
Asia	Tin, tungsten, antimony, cobalt, chromium, manganese, coal, petroleum, sulfur
Australia	Gold, lead, zinc, iron, manganese, coal
Europe	Lead, zinc, mercury, manganese, bauxite, potash, sulfur, coal, natural gas, petroleum
North America	Iron, nickel, molybdenum, gold, silver, lead, zinc, tungsten, phosphate, sulfur, potash, coal, natural gas, petroleum
South America	Tin, copper, iron, manganese, bauxite, petroleum

nation, not even the United States, is self-sufficient in its mineral resources, and every industrial nation is increasingly dependent upon foreign sources of supply. The United States is fairly well supplied with part of its requirements (molybdenum, magnesium, sulfur, coal, potash, phosphate) but lacks others, such as tin, almost completely. The United States is increasing its imports of many resources, such as petroleum, natural gas, iron ore, copper, manganese ore, bauxite, nickel, chromite, tungsten, mercury, lead, zinc, antimony, platinum, graphite, asbestos, mica, fluorspar, and industrial diamonds.

Known reserves of mineral resources are being rapidly depleted, as the rates of production accelerate a few percent every year. The population and industrial expansion create exponential demands on the world's mineral supply far in excess of any addition to the reserves by new discoveries. Except for iron, aluminum, and magnesium (obtained mainly from seawater), the reserves of commercial-grade ores are approaching exhaustion within a few decades. Meeting the crisis will require new discoveries, utilization of lower grade ores, increased secondary recovery and recycling, improved technology of mining and metallurgy, increased freedom of international trade, increased use of substitutes, and reduction of waste.

REFERENCES

Bates, R. L., 1960, *Geology of the Industrial Rocks and Minerals*, Harper & Brothers, New York.

Flawn, P. T., 1966, *Mineral Resources: Geology, Engineering, Economics, Politics, Law*, Rand McNally & Company, Chicago.

Frashche, D. F., 1962, *Mineral Resources*, National Research Council Publication 1000-C, Washington, D.C.

Gillson, J. L. (ed.), 1960, *Industrial Minerals and Rocks*, 3d ed., American Institute of Mining Metallurgical, and Petroleum Engineers, New York.

Heiland, C. A., 1946, *Geophysical Exploration*, Prentice-Hall, Inc., Englewood Cliffs, N. J.

Hubbert, M. K., 1962, *Energy Resources*, National Research Council Publication 1000-D, Washington, D. C.

Leith, C. K., Lund, R. J., and Leith, A., 1935, *Precambrian Rocks of the Lake Superior Region*, United States Geological Survey Professional Paper 184.

Levorsen, A. I., 1967, *Geology of Petroleum*, 2d ed., W. H. Freeman and Company, San Francisco.

Mouzon, O. T., 1966, *Resources and Industries of the United States*, Appleton Century Crofts, New York.

Park, C. F., Jr., and MacDiarmid, R. A., 1970, *Ore Deposits*, 2d ed., W. H. Freeman and Company, San Francisco.

Riley, C. M., 1959, *Our Mineral Resources*, John Wiley & Sons, Inc., New York.

Skinner, B. J., 1969, *Earth Resources*, Prentice-Hall Inc., Englewood Cliffs, N. J.

Smith, H. (ed.), 1971, *Conservation of Natural Resources*, 4th ed., John Wiley & Sons, Inc., New York.

United States Bureau of Mines, 1970, *Mineral Facts and Problems*, Bulletin 650.

———, *Minerals Yearbook*, published annually.

The Deep Sea Drilling Project is a part of the National Science Foundation's national Ocean Sediment Coring Program, a program to explore the crust of our own planet. Its objective is to learn about the origin and history of the earth through the study of samples obtained from previously inaccessible sites.

The immediate aim of the Deep Sea Drilling Project is to increase our understanding of the suboceanic crust of the earth by studying long cores of sediments taken from the ocean bottom. The ridges, rifts, and great abyssal plains that comprise the ocean basins are being systematically examined, and samples from deep under the sea floor are being obtained for careful study.

Ocean-bottom sediments come from two sources. One, the land, contributes such products of weathering and erosion as sand, silt, and volcanic ash—carried to the sea by streams and rivers, by the winds, or by glacial ice. The other is the steady "rain" of shells from microscopic animals and plants that live near the surface of the ocean. Through the ages, as they have died, their debris has dropped to the ocean floor, there to become part of the record of the earth's history.

From a study of these sediments, scientists are obtaining information about the behavior of oceanic crust. When this is combined with knowledge about the ancient history of the continents and their migration and development during the last 250 million years, investigators can better reconstruct the history of the origin, growth, and movement of the earth's crust. Besides telling us how our planet evolved and what we should expect in the future, such knowledge should help make possible better use of oceanic resources for the good of all mankind.

EARTHQUAKES

Earthquakes represent the passage of a series of waves through the rocks of the earth. They have been known and feared by man since ancient times because they are among the most destructive of natural phenomena. They may be more awe-inspiring than volcanic eruptions; perhaps, because earthquake shocks move through the ground surface that we have been conditioned to regard as terra firma.

The seismic waves originating within the earth radiate outward in all directions from the source (Fig. 13.1), just as the sound of an explosion spreads through the air. Their paths are modified by differences in the materials traversed; their persistence is proportional to their initial energy. Some go completely around or through the earth, whereas other, weaker, ones die out in short distances. Even when strong seismic waves become weakened by traveling great distances, they can still be detected and recorded by suitable instruments.

CAUSES OF EARTHQUAKES

The earth is an elastic medium capable of transmitting energy vibrations through and around its body. Such vibrations are characteristic of any disturbed elastic substance. An elastic substance is one that can be deformed and yet return to its original shape after the deforming forces have been removed.

The basic causes of earthquakes are probably disturbances that have been or are originating deep in the earth's interior; however, the more immediate cause of the earthquakes is the sudden breaking of rocks that have been exposed to forces beyond the limit of their strength. The process is called faulting.

Faulting

Faulting is a type of earth deformation. If a fracture or break develops, there will be subsequent movement of the rocks on either side of the fracture. The San Andreas fault of California (Fig. 1.5) is an example. Many Californians have seen evidences of movement along this fault.

The direct association of some earthquakes with known faults is well established. Several large shocks have been accompanied by vertical or horizontal displacements of the earth's surface. These displacements have ranged from 1 to 2 meters to several tens of meters, extending for distances of many kilometers

181

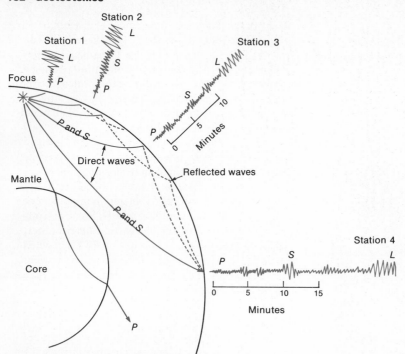

FIGURE 13.1 *A cross section through a portion of the earth, showing the paths of a few of the many earthquake waves and the records they leave on the seismogram at four stations. Note the reflected waves. (After Gilluly, 1963.)*

along the trend of the faults. In many other instances, no faulting has been apparent, but it may be reasonably suspected. Although thousands of faults are known, only a small number now seem to be active sources of earthquakes; the others are at least temporarily inactive. The San Fernando Valley, California, earthquake of February 1971 resulted from movement along a small branch fault of a larger fault zone, the San Gabriel fault (Fig. 13.2). However, the area where the fault occurred had been considered by many to be a nonactive area (Fig. 13.3).

The faults that are active sources of earthquakes are carefully watched by seismologists, who will often give a hint of possible earthquakes in these areas by simply saying, "it is not a case of if the quake will occur but simply a case of when will it occur." Quite simply, this indicates geologists' awareness of the tremendous strains developing along many fault zones.

The association between earthquakes and faulting is explained by the elastic-rebound theory, developed by H. F. Reid (1910) and others after studying the California earthquake of 1906. According to this theory, the rocks beside a fault, in close contact with

one another, are able to undergo gradually accumulating strain by changing shape, until their elastic limit is finally reached. They suddenly break, and much of their stored-up elastic energy is released in earthquake waves. The rocks resume their original shapes, but they have been faulted and have different relative positions on either side of the break.

The process of elastic rebound is illustrated diagrammatically in Fig. 13.4. The point O lies on a fault, and A and B are points on opposite sides. In stage 1, before any strain has accumulated, the line AOB is straight. In stage 2, after strain has accumulated but before an earthquake, A and B have been displaced in opposite directions, and the line AOB is bent. The situation then is somewhat analogous to that of a bending steel spring. In stage 3, after an earthquake, the points formerly adjacent to O have been shifted to O' and O", faulting has occurred, the segments AO' and O"B are straight, and the strain is relieved.

Careful surveying many years prior to, and just after, the California earthquake of 1906 showed that points on the northeast side of the San Andreas fault were displaced 1 to 2 meters to the southeast (the distance displaced was dependent upon the nearness to the fault) and that points on the southwest side had shifted 2 to 3 meters to the northwest. Subsequent surveys have shown that a differential displacement of the same nature is taking place presently at a rate up to 4 centimeters per year.

The movements are now under careful study. Scientists are using a number of different instruments, such as strain gauges, tilt meters, gravimeters, laser beams, and magnetometers, to find clues that might be useful in forecasting shocks. For example, local changes in the earth's geomagnetic field have been recorded by magnetometers up to 10 hours preceding a small shock on the San Andreas fault.

Vulcanism

Nontectonic causes are responsible for only local and mostly minor tremors. Most volcanic eruptions, such as Krakatoa in 1883, the relatively quiet Hawaiian eruptions, or the eruption of Helgafjell near Iceland in 1973, have generated many small earthquakes (Fig.

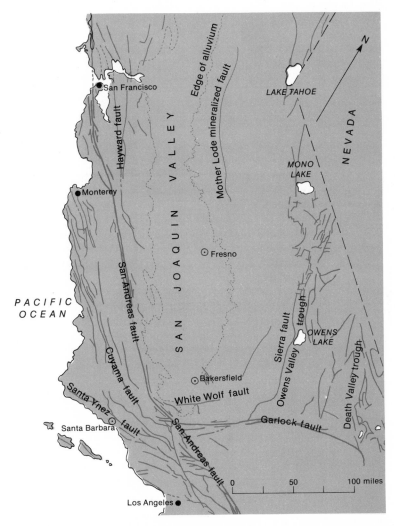

FIGURE 13.2 *Map of the San Andreas and certain other faults in California. (After Crowell, 1962.)*

FIGURE 13.4 *Elastic rebound. (1) Before strain has accumulated; (2) after strain has accumulated but before an earthquake; (3) after an earthquake, with strain relieved. (After Reid, 1910.)*

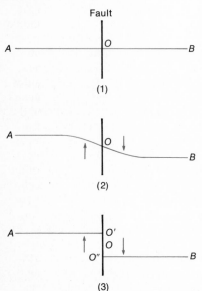

13.5). These earthquakes generally have only small, local, and shallow effects. The following description of the Kilauea, Hawaii, volcanic events in 1959 and 1960 will show the relationship between vulcanism and earthquakes.

In 1959, a domal rising occurred around the Kilauea caldera. This was caused by the rise of magma from a depth of about 48 to 64 kilometers. The swelling stopped on February 19, 1959, following several earthquakes. Subsurface fractures apparently had diverted the upward flow. After a few months of partial subsidence of the Kilauea summit, another series of earthquakes from August 14 to 19, originating at the 56-kilometer depth, signaled the release of additional magma. The rise of magma from August to October could be followed (1) by means of mild earthquakes like those known to be associated with lava extrusions at the surface, and (2) by the resumption of swelling of the subsurface reservoir.

Additional, but small, earthquakes originating at a shallow depth (less than 60 km) began in September 1959 and increased to more than 1,000 per day in early November. After a 5-hour crescendo of earthquakes on November 14, presumably caused by fracturing of near-surface rocks, lava broke out from a 0.8-kilometer-long fissure on the wall of Kilauea Iki, a sub-

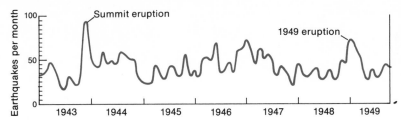

FIGURE 13.5 *Earthquake frequency on the volcanic island of Hawaii, from January 1943 to September 1949. During the period shown, earthquakes averaged several tens per month and had a maximum of about 90. Unrest continues between eruptions. (After MacDonald and Orr, 1950.)*

sidiary crater just east of Kilauea. The eruption, mostly from one major lava fountain, lasted until November 21.

During late December 1959 and early January 1960, swarms of earthquakes of shallow depth of origin occurred in the area 41 kilometers to the east of Kilauea and later near Kapoho, 50 kilometers east of Kilauea. On January 13, 1960, a trough 3 kilometers long and 0.8 kilometer wide at Kapoho subsided several meters, and lava broke through a fissure 1.2 kilometers long in the middle of the trough. The sharp earthquakes abruptly changed to minor tremors of the type that is characteristic of times of lava expulsion.

Four days after this lateral outbreak, the top of Kilauea began to subside rapidly, apparently because of the transfer of magma from the reservoir under the mountain to the site of the flank eruption at Kapoho. The settling of the brittle rocks of the Kilauea dome generated small earthquakes. The total collapse was about 2 meters vertically and equal in volume to the lava concurrently erupted elsewhere. The flank eruption ceased in February 1960, and the settling stopped soon afterward.

Studies of a similar nature are presently going on near other active volcanos, especially those in the Caribbean. The studies are directed toward predicting volcanic events by observation of seismic activity that may precede such an event.

Ultimate Cause

While faulting and vulcanism may be the important immediate causes of earthquakes, the ultimate cause lies deeper in the earth's interior structure. For example, local differences in heat accumulation below the crust of the earth may be a cause. Conceivably, these heat differences may necessitate adjustments within the earth because heat, as well as mineral phase changes controlled by temperature, cause direct volume

changes. Such changes in volume, whether due to swelling or shrinking, lead to movements that may include faulting. Deep-seated earthquakes (deeper than 300 kilometers) are attributed by some seismologists to a volume collapse with or without associated faulting. The study of the deep-seated origins of earthquakes has increased in recent years. Some geologists believe a pattern does exist and believe they understand the cause of the deep earthquakes originating within the earth. This work has resulted in definite clues to patterns of movement and formation of continents and ocean basins.

CHARACTERISTICS OF EARTHQUAKES

Geographic Distribution

Although earthquake waves are transmitted over the entire earth, the actual earthquakes originate in limited belts. The largest number and the more violent ones take place in two main belts (Fig. 13.6): the circum-Pacific belt, extending from Chile, Central America, through a loop in the Caribbean-Antilles area, California, the Queen Charlotte and Aleutian Islands, Japan, the Philippines, Indonesia, and New Zealand; and the Alpine-Mediterranean-trans-Asiatic belt, including northern Africa, Spain, Italy, Yugoslavia, Greece, Turkey, Iran, northern India, and Burma. The fact that these two belts coincide with either the volcanic or the mountainous belts of the earth suggests that both phenomena may be related to a common cause. As you look at the distribution of all earthquakes (Fig. 13.6), other coincidences or correlations should come to mind.

The circum-Pacific belt is the source of 80 to 90 percent of the earthquake shocks. Many of the Pacific border earthquakes are generated along a deep shear

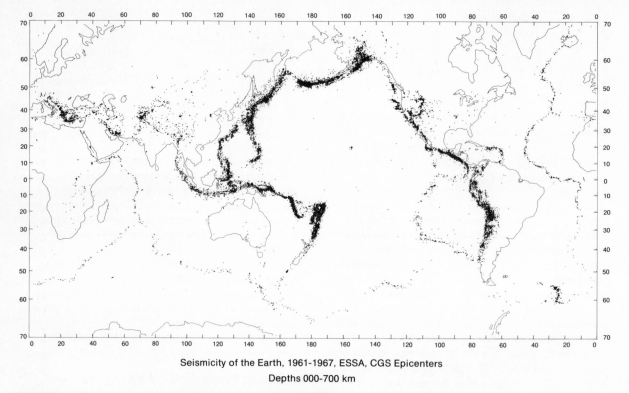

Seismicity of the Earth, 1961-1967, ESSA, CGS Epicenters

Depths 000-700 km

FIGURE 13.6 *Distribution of earthquakes, 1961 to 1967. Note their concentration around the Pacific Ocean and along mid-oceanic ridges. (From M. Barazangi and J. Dorman, 1969.)*

or fault zone that slants down under the continent-ocean border. About 5 percent of earthquakes, all of shallow origin, occur along the submarine mountain rise or ridge systems and their offsetting fracture zones in the Arctic, Atlantic, eastern Pacific, and Indian Oceans. A few occur in the fault zones of eastern Africa and east-central Siberia.

The ocean-continent boundaries and the volcanic island arc systems are zones of special weakness, having the greatest and most deep-seated movement. In contrast, the interiors of the geologically old continental cores or shields of east-central Canada, Brazil, Scandinavia, part of Siberia, much of Africa, southern India, and western Australia, except in fault zones, are comparatively free of earthquakes. The center of the Pacific Ocean Basin, aside from the Hawaiian Islands, the fracture zones, and the East Pacific mountainous rise, is nearly inactive.

Not all earthquakes, however, are in the great deformation belts of the earth (Fig. 13.6). Some originate in areas remote from recent mountain building and present-day volcanoes, such as the 3.7-magnitude earthquake that occurred in Lake, Lee, and Ogle Counties in Illinois on September 15, 1972. The earthquake, of intensity IV, is the first instrument-recorded shock since 1912. To the south, the New Madrid, Missouri, area had a rather severe earthquake in 1811. The mid-continent region or the Mississippi Valley seem to be unlikely places for earthquakes, but the area is classified as a major seismic-risk area. The United States Geological Survey reports that more than 1,000 earthquakes have been reported east of the Mississippi River since 1700.

Some minor earthquakes occur in areas of delta building because of concentration loading by the great amount of sediments. Others occur in areas of recent continental glaciation, presumably because removal of the ice sheets allows the continental area to rise, or uplift, vertically due to decreased weight.

Frequency

On the average, about 15 to 20 large and destructive earthquakes occur in a year, but comparatively harmless small earthquakes number up to 1 million per year. This figure includes large numbers of minor shocks associated with volcanic activity, such as on the island of Hawaii (Fig. 13.5). Minor shocks are so numerous that a tremor is taking place somewhere on the earth every few minutes. In 1970, there were 37 earthquakes with a magnitude of 6.75 or greater. The earth may be said to be in a state of perpetual tremor. Adjustments in the earth's crust or its deeper structure are in progress almost continuously, and so earthquakes are a common and frequent occurrence. Many of them are so slight as to be negligible.

Recurrence

Earthquakes repeat in the same place but apparently not in any regular cycle. Elastic strain has been accumulating along the San Andreas fault in California since 1906. The results have been only minor episodes of local relief and slow creep along the fault. Geologists cannot predict the year, much less the day or hour, when another major shock may be expected, but they can predict that one will occur. The fanatical prediction of the plunging of western California beneath the Pacific Ocean in 1969 or a disastrous earthquake for San Francisco on Janurary 4, 1973, did not materialize, but given another 50 million years, who knows!

There have been attempts to attribute earthquake trigger action to moon phases, high tides, heavy rainfall, sharp changes in barometric pressures, and especially, other earthquakes elsewhere. Some even blame our Apollo space program, perhaps assuming that the weight of men and spacecraft leaving earth or landing on the moon is the cause. Earlier, in Chap. 1, we noted that Ruben Greenspan, a predictor of earthquakes for 38 years, was retiring. Mr. Greenspan's record is worth enlarging upon here. In 1935, he accurately predicted the relative times for future earthquakes in Chile (1939), Peru (1970), and India (1935), as well as foretelling a volcanic eruption for Krakatoa. In the late 1960s, Greenspan correctly for-

saw the date (February 9) and time (6:00 A.M.) of the San Fernando, California, earthquake of 1971. His predictions are based on relative positions of the sun and moon, which apparently cause pressures on the earth's surface. However, his latest prediction, in 1971, told of a disastrous earthquake that would strike San Francisco at 9:20 A.M. on January 4, 1973. In December 1972, he withdrew the prediction, saying he needed to verify his data. No earthquake occurred, but in prior years he has been remarkably accurate. Despite Greenspan's success, detailed studies by competent seismologists have failed to make useful predictions possible. Similar studies on earthquake prediction are now being pursued vigorously in California and Japan. Some new and different approaches to the problem are being reached. Early in 1973, Russian scientists reported that a "glow in the sky," occurring hours before the actual earthquake, could be used to predict an earthquake for an area. The glow is caused by electrical charges in the atmosphere-earth system.

Also in 1973, the United States Geological Survey set its first earthquake prediction. Hollister, California, less than 160 kilometers southeast of San Francisco, was predicted to experience a 4.5-magnitude earthquake. The prediction was based on observations of a strain build-up on the San Andreas fault, which occurs in the Hollister area. As of October 1973, an earthquake whose epicenter could be considered to be in the "Hollister area" did not occur. However, Hollister has experienced tremors from quakes along the San Andreas fault, including a 3.1-magnitude quake on August 2, at 36°N, 121°W.

Man's activities also generate earthquakes, ranging from dynamite explosions to underground nuclear tests. The weight of water during the filling of reservoirs, such as Lake Mead behind Hoover Dam, has caused a series of minor tremors in many instances. Injection of radioactive contaminated water during the years 1962 to 1970 into deep wells drilled at the Rocky Mountain Arsenal near Denver, Colorado, set off more than 700 small earthquakes centered around the wells. The increased fluid pressure in the highly fractured rock at the bottom apparently lubricated the faults, allowing greater ease of movement. The earthquake frequency corresponded closely to the volume and pressure of water injected. The hypothesis is strengthened by the fact that the tremors ceased when the fluid injection ceased. Some of these man-generated earthquakes are being restudied. The studies are attempting to discover whether man-induced earthquakes might release stresses along active fault zones, decreasing the potential of a severe shock. However, there is opposition to this approach, as witnessed by the widespread concern over the nuclear blast at Amchitka Island in the Aleutians.

Intensity

The intensity of an earthquake is measured by its effects on man, its damage to buildings and other structures, and the changes it produces in rock and soil at the earth's surface. The intensity differs not only from place to place but also at the same place because of differences in the stability of foundations, the manner of construction, and other variables. The destruction wrought by an earthquake depends on the geological characteristics of the ground surface and on the size and shape of the man-made structures, including their design, workmanship, and construction materials. Also important are the characteristics of the seismic waves, such as the acceleration, period, velocity, and duration.

In a severe earthquake, buildings are leveled (Fig. 13.7), cornices fall off, chimneys topple, bridges and piers collapse (Fig. 13.3), pavement is broken, roads and railroads are buckled and twisted, telephone and telegraph wires and cables are broken, power lines are downed, fires are started, water towers give way, oil tanks burst, and utility lines are severed.

To measure the severity of an earthquake the Mercalli intensity scale was developed and modified by Wood and Neumann in 1931. This intensity scale is graded in steps from I to XII in order of increasing severity (Table 13.1).

The intensities at different places from a single earthquake are rated by trained seismologists on the basis of in-the-field surveys and postcard questionnaires distributed by the U.S. Coast and Geodetic Survey. The intensity at each locality is plotted on a map, and isoseismal lines are then drawn at the boundaries (Fig. 13.8). Such an isoseismal map shows the area of maximum damage, surrounded by more or less concentric belts of progressively less damage. Its principal value is to call attention to areas with poor geologic foundations, unstable methods of construction, or other hazards. Better planning can then reduce destruction in future earthquakes.

The geologic effects of an earthquake include ground fissuring (Fig. 13.9), sunken ground, sag ponds, raised hummocks, slumps, earthslides (Fig. 13.10), mudflows, eruptions of water and sand, seismic sea waves (Fig. 13.11), and seiches (oscillation waves) on lakes. In addition, there may be disturbances in the circulation of groundwater, which will cause the water

FIGURE 13.7 *Damage to stores (ground floor) and collapse of public auditorium (above), resulting from an earthquake near Bakersfield, California, July 21, 1952. (National Ocean Survey, NOAA.)*

TABLE 13.1   Abridged Wood-Neumann Scale of Earthquake Intensities

I.   Not felt, except by a very few who are favorably situated

II.   Felt only on upper floors by a few people at rest. Swinging of some suspended objects

III.   Quite noticeable indoors, especially on upper floors, but many people fail to recognize it as an earthquake; standing automobiles may sway; vibrations feel like those of a passing truck

IV.   Felt indoors by many during day, outdoors by few; if at night, awakens some; dishes, windows, and doors rattle, walls creak; standing cars may rock noticeably. Sensation like heavy truck striking a building

V.   Felt by nearly all; many wakened; some fragile objects broken and unstable objects overturned; a little cracked plaster; trees and poles notably disturbed; pendulum clocks may stop

VI.   Felt by all; many run outdoors; slight damage; heavy furniture moved; some fallen plaster

VII.   Nearly everyone runs outdoors; slight damage to moderately well-built structures, negligible to those substantially built, but considerable to those poorly built; some chimneys broken; noticed by automobile drivers

VIII.   Damage slight in well-built structures; considerable in ordinary substantial buildings, with some collapse; great in poor structures. Panels thrown out of line in frame structures; chimneys, monuments, factory stacks thrown down; heavy furniture overturned; some sand and mud ejected, wells disturbed; automobile drivers disturbed

IX.   Damage considerable even in well-designed buildings; frame structures thrown out of plumb; substantial buildings greatly damaged, shifted off foundations; partial collapse; conspicuous ground cracks; buried pipes broken

X.   Some well-built wooden structures destroyed; most masonry and frame structures destroyed or knocked off their foundations; rails bent, ground cracked; landslides on steep slopes and river banks; water slopped over from tanks and rivers

XI.   Few if any masonry structures left standing; bridges destroyed; underground pipes completely out of service, rails bent greatly; broad cracks in ground and earth slumps and landslides in soft ground

XII.   Damage total; waves left in ground surface, and lines of sight disturbed; objects thrown upward into the air

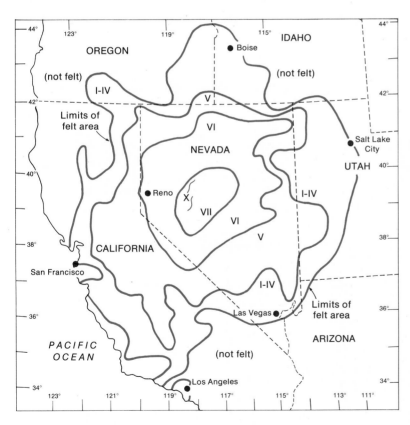

FIGURE 13.8 *Map showing isoseismal lines of an earthquake centered in west-central Nevada, December 16, 1954. This earthquake was felt over an area of about 518,000 square kilometers in five states. Its maximum intensity, X, was reached near the source, where fault displacements reached nearly 6 meters vertically and 4 meters horizontally. (After National Ocean Survey, NOAA.)*

FIGURE 13.9   *Alluvium cracked by an earthquake near Arvin, California, July 21, 1952. The house was shifted off its foundation and severely damaged structurally. (California Division of Mines and Geology.)*

FIGURE 13.10   *Landslide scars in Sycamore Canyon, Kern County, California, after the earthquake of July 21, 1952. The White Wolf fault zone, along which movement took place, causing the earthquake lies a fraction of a kilometer to the left (northwest). (Robert C. Frampton.)*

in wells and springs to become muddy, stop some springs, and initiate others. Yellowstone National Park, after the 1959 Montana earthquake, illustrated many changes in the hot spring and geyser system. Similar changes to hot springs and geysers have occurred in Iceland due to earthquakes. Earthquakes may also start avalanches of snow or release icebergs from tidewater glaciers.

These diverse effects are relatively superficial, largely local, and generally of minor geological consequence. Despite their destructiveness, earthquakes themselves are comparatively unimportant as geologic agents, though earth movements over long periods of time in seismic areas may make major changes in the bedrock or geologic features.

SCIENTIFIC STUDIES OF EARTHQUAKES

The instruments used to determine and record earthquakes are called seismographs. The seismograph is simply a vibrating system designed to measure and record movement of parts of the earth. The vibrating element must be enclosed and mounted on solid rock so that it moves with the ground. It is usually damped (i. e., diminished in amplitude). The design of a seismograph varies considerably. One variety uses a pendulum mass that is mounted horizontally, similar to a sagging gate. The pendulum mass is free to move horizontally in a fixed direction, north-south or east-west (Fig. 13.12), but gravity tends to restore it to its original position. An inverted pendulum mounted vertically on springs may be used to measure the vertical motion. Since the mass of a pendulum may be either very heavy or relatively light, and either damped or undamped, the natural periods, sensitivities, and magnifications of these instruments differ substantially.

The recorder has mechanical, optical, or electromagnetic devices, or combinations of them, to transfer the vibrations to heat- or light-sensitive paper on a revolving drum driven by a synchronous motor. The earliest recorders used a pen on a clock-driven drum. The movement of the drum in all cases is such that, in the absence of an earthquake, the light or heat source describes an unvarying fine line. As soon as the earth vibrates beneath the steady mass, the previously smooth line changes to a series of zigzag or crosswise loops. The result is a seismogram (Fig. 13.13).

Instruments vary in the degree of magnification they provide, but those commonly used give mag-

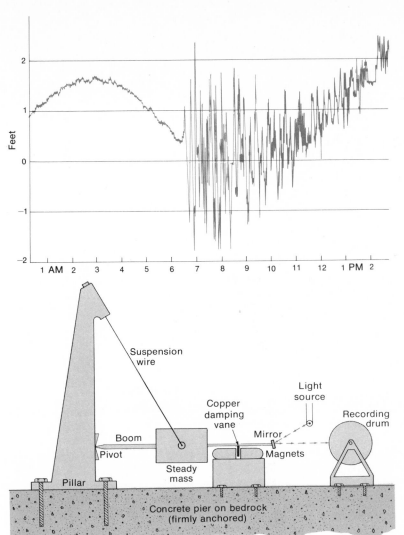

FIGURE 13.12 *Sketch of the main elements of a horizontal-pendulum seismograph. The suspension of the steady mass resembles a sagging gate, free to swing horizontally as the earth moves beneath it. It is damped magnetically and restored by gravity. The swing of the mirror attached to the end of the boom is recorded photographically on a revolving drum. Many variations and adjustments of this scheme are possible.*

FIGURE 13.13 *Seismogram (middle line) of an earthquake nearly 5,000 kilometers from the seismograph that recorded it. A (above) shows the undisturbed line 1 hour before the earthquake; B (below) is a record of minor waves still arriving 1 hour after the earthquake. (Harvard University Press.)*

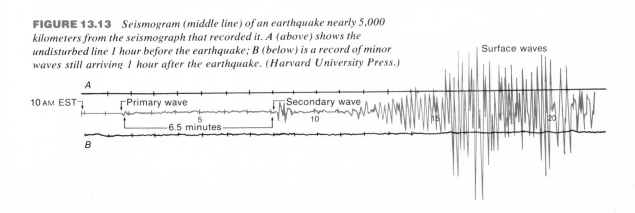

nifications in the range of a few hundred to tens of thousands times the actual movement of the ground. The maximum feasible is about 4 million. Background of cultural "noise" usually sets a practical limit of about 100,000.

Exact timing of the record to small fractions of seconds is very important; thus, radio time signals, such as those from the U.S. Naval Observatories, indicating the time appear on the recording.

There is no universal seismograph. Instruments differ with need and location. A strong-motion accelerometer, for instance, is triggered only by a movement that has a large, predetermined amplitude. It continues to operate for only a fixed period of, perhaps, 60 to 70 seconds, unless the motion lasts longer. It will supply a record in a strong-earthquake area when more delicate instruments may have been thrown out of commission by a heavy shock.

Instrumentation has been undergoing rapid changes in recent years. Sensitive laser-beam units mounted in abandoned tunnels represent the extreme in sophisticated instruments. Laser units are also mounted on opposite sides of major surface-fault zones in order to detect the slightest creep or movement. Other developments include the establishment of seismograph networks. A World Wide Standard Seismograph Network, composed of standardized instruments, sponsored by the United States and Canada, has recently been established. Earthquake-prone California has its own seismograph network. Other improvements include the use of magnetic tape recorders and direct digital recording on tape so that electronic computers can be used for rapid analysis of the data.

Earthquake Waves

Seismograms show that earthquakes consist of the passage of a series of waves. Near the source of an earthquake, the amplitudes of these waves may be large, but at a distance they become so small as to require magnification. In the Imperial Valley earthquake of May 18, 1940, at El Centro, California, the maximum horizontal vibration was about 37 centimeters in somewhat less than 3.5 seconds, and the maximum vertical motion was about 10 centimeters in 2.5 seconds—exceptionally large ranges of movement.

The patterns recorded on seismograms show that earthquake waves are of three main types: (1) P, or

primary, waves, which are transmitted by alternate push-pull changes of volume or by alternate compression and rarefaction in the direction of propagation. An example of this motion can be seen in the spring "Slinky," a popular toy. (2) S, or secondary, waves are vertical waves, or waves in which the particle motion is transverse to the direction of movement. (3) L, or long, waves, which are complex sinuous or undulatory gravity waves that travel mainly along or near the surface of the earth (Figs. 13.13 and 13.1). The depth of penetrations of these L waves depends upon their frequency.

The P waves resemble sound waves, but they are far below audible frequencies. Most P waves vibrate only once in 2 or 3 seconds, and the longest ones only once in 25 seconds or more. P waves pass through gases, liquids, and solids.

The S waves, or transverse waves, involve a shearing of the material of the earth. Volume change in the material is not associated with S waves, but shape changes are.

The L waves include both the Rayleigh waves, waves that vibrate vertically and those that vibrate horizontally. Their periods range from several seconds to several hundred seconds.

The P waves travel at velocities of about 5.0 to 13.8 kilometers per second, increasing with depth; the S waves, about 3.2 to 7.3 kilometers per second; and the L waves, about 4.0 to 4.5 kilometers per second. Because of their different velocities and different routes of travel, the three sets of waves arrive at a seismographic station at different times. A simple record of an earthquake in a 1,600- to 11,000-kilometer-distance range shows a threefold set of signals well separated in time (Fig. 13.13). If there is an earthquake, a seismograph at a station will first record the time of arrival of the faster P wave, then the time of arrival of the slower S wave, and finally, the time of arrival of the L wave that has passed around the earth. Since the interval between the arrivals of the P and S waves increases with distance traveled, the time lag (S minus P) may be used to compute the distance to the source (Fig. 13.13).

When the travel times of the different waves are plotted against the surface distances from the source, we find that the travel times of the L waves are almost directly proportional to the surface distances—hence, the inference that they tend to follow the surface. They travel at slightly different rates in areas of different kinds of rock and somewhat faster in rocks beneath the oceans than they do under the continents. On the other hand, the velocities of P and S waves increase as the surface distances from the source become greater. Therefore, it is inferred that these waves do not travel along surface arcs or along straight subsurface chords, but along curved paths concave upward within the earth, and that their velocities increase (within limits) as their paths reach greater depths (Fig. 13.1). Seismograms indicate that the travel of waves within the earth is greatly complicated by refraction, reflection, diffraction, and dispersion, often causing very complex wave patterns.

Focus and Epicenter

The point of origin of an earthquake is its focus, and the point on the surface directly above the focus is its epicenter. The depth of focus is classified in three categories: (1) shallow, up to 60 kilometers beneath the surface, (2) intermediate, 60 to 300 kilometers, and (3) deep, 300 to 720 kilometers, the maximum known depth. The deepest focus recorded for an earthquake was 720 kilometers and was associated with the island arc zones of the Pacific. Interestingly, however, a deep-focus earthquake of 600 kilometers occurred beneath southeastern Spain. Since the intensity of an earthquake dies out inversely as the square of the distance from the source, the depth of focus can be computed from the records of intensity at different places. At least 85 percent of all earthquakes, some seismologists say more than 90 percent, are of the shallow type. The depth of focus of most of these is less than 8 kilometers.

The existence of deep-focus earthquakes was discovered from the early arrival of P waves on the opposite side of the earth and from discrepancies in the arrival times of P and S waves, which made epicentral distances disagree, often by hundreds of kilometers. The weakness or lack of surface L waves is also an indication of a deep-source earthquake. The waves from a deep-focus tremor not only follow the usual direct courses through the earth but also travel up to the surface, from which they are reflected back to the receiving station. The extra time, ranging from a few seconds to about a minute, required by such a roundabout route is a measure of the depth of the focus. Earth-

quakes originating at great depth show that the rocks affected are highly rigid, undergo elastic strain, and ultimately change, by fracture or recrystallization, into minerals of greater density.

The geographic location of distant epicenters may be determined by the three-point-positioning method (Fig. 13.14). First, the epicentral distances from three stations are calculated from the intervals between the arrival times of P and S waves at each station; then a circle is drawn about each station, with radius appropriate to each. The point at which the three circles intersect indicates the location of the epicenter.

The direction to the source also may be indicated at a single station if the record is clear enough to show the first P wave. This direction and the epicentral dis-

FIGURE 13.14   *Map showing the three-circle method of locating the epicenter of an earthquake from seismograph records at three stations. The distances from San Francisco, St. Louis, and Washington are determined from the time elapsed between the arrivals of the first (P) and the second (S) seismic waves. Using these calculated distances as radii, circles are drawn about the stations (preferably on a globe) to find their intersection—in this instance, off the coast of Lower California.*

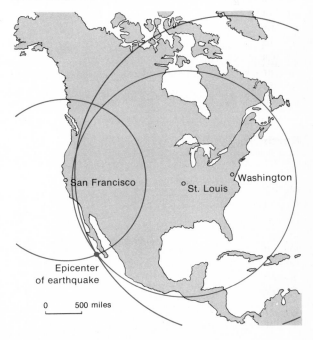

tance obtained from the *S*-minus-*P* arrival times give at least an approximate location of the epicenter.

Scale of Magnitudes

The actual energy of earthquakes at the source is rated in terms of a standard magnitude scale originated by Richter in 1935. This scale is based on the measurement of the maximum amplitudes of the swings traced on a seismogram by a standard seismograph of specified free period, magnification, and damping, at a distance of 100 kilometers from the epicenter. Only rarely is this type of seismograph exactly 100 kilometers from the epicenter, but calculations can be applied to a particular seismograph to find the desired equivalent for other distances and other seismographs. Similar calculations for several other stations generally agree and hence confirm the calculated results.

This quantitative rating system in terms of magnitude is obviously more scientific than the subjective judgments required to apply an intensity scale to superficial earthquake effects. The intensity scale involves complex variables of personal reactions, construction standards, and stability of foundations.

The Richter scale is logarithmic, and so it encompasses an extremely wide range of magnitudes. An earthquake of magnitude 3.0 has ten times the energy of one with a magnitude of 2.0. The very faint earthquakes are rated little more than 0, whereas the most severe shocks recorded run to 8.6 (Colombia, South America, in 1906 and Assam, India, in 1950). An earthquake of magnitude 2.5 can be felt near the source; one of 4.5 may cause minor damage; and one of 6.0, considerable damage.

The magnitudes of many artificial earthquakes caused by explosion of bombs underground in Nevada and elsewhere have been determined. An explosion equivalent to 1 million tons of TNT generates an earthquake of about 6.0 magnitude. The 8.5 magnitude of the 1964 Alaskan earthquake was equal to the energy of 320 million tons of TNT.

From 1904 to 1956, 26 earthquakes had magnitudes of 7.0 or more. The first large shock of the May 21, 1960, earthquake in Chile had a magnitude of 7.5, and the principal earthquake of the five strong aftershocks on the following day, 8.5. Hundreds of aftershocks were characteristically low in magnitude, but others ranged as high as 7.5 to 8.0. Ten of the af-

tershocks of the 8.5 Alaskan earthquake of 1964 exceeded a 6.0 magnitude.

Microseisms

The records of many sensitive seismographs of high magnification commonly show tiny quivers from earth waves of small and irregular motion. These waves are called microseisms. They come and go, but they usually continue for hours or even days at a time.

They have been ascribed to the pounding of surf on seacoasts; to waterfalls; to hurricanes, typhoons, and other storms; to strong winds, monsoons, and trade winds; and to surges of lava beneath volcanoes during an eruption (volcanic tremors). Microseisms recorded at specially equipped stations enable us to locate a hurricane at sea.

NOTEWORTHY EARTHQUAKE CASE HISTORIES

Managua, Nicaragua, 1972-1973

An earthquake of magnitude 6.25 occurred on December 23, 1972, in the Nicaraguan capital of Managua. The capital was at or near the earthquake's epicenter, with the focus located 14.5 kilometers below the surface. Over 80 percent of the city was destroyed, with an estimated 3,000 to 6,000 deaths. The damage was described as "complete devastation."

Most of the damage was caused by the strongest of six tremors over a 12-hour period on December 22-23. This was the fourth earthquake to strike Managua: the first in 1885; the second, of magnitude 6.0, in 1931 caused $70 million damage and 1,400 deaths; and the third, in 1968, had a magnitude of 4.5.

The 1972 earthquake can best be described by outlining the following points. The country is near a major seismic zone extending from Tierra del Fuego (Cape Horn) to the Aleutians; it is located along an active volcanic belt extending into Mexico. Perhaps unrelated to the Nicaraguan earthquake, but notable for the timing, was the eruption of the volcano Acatenango in Guatemala in November 1972, the first eruption since 1927; minor tremors had been felt by many a week prior to the December 23 earthquake; and aftershocks up to a magnitude of 5.0 were felt a week after the earthquake. Three strong aftershocks were felt in March 1973. One was strong enough to cause a previously damaged church to cave in completely. In addition to those killed, 50,000 were injured and over 200,000 left homeless. Most homes of adobe construction were reduced to rubble. Over three-fourths of the buildings in the city were destroyed by the earthquake. The energy was equivalent to 1 million tons of TNT.

Water, electricity, and other services were completely disrupted; thousands were seeking shelter, medical help, and food; looting became prevalent; and uncontrollable fires swept through the city.

Over 20 countries air-lifted food, personnel, medical supplies, and temporary housing. Although there was speculation as to whether the city would be rebuilt, it seems as though it will be. It will be rebuilt on the same volcanic tuff foundation, using earthquake-hazard construction guidelines.

Prince William Sound, Alaska, 1964

At twilight on Good Friday, March 27, 1964, in Anchorage, a violent earthquake struck (Fig. 13.15). The initial tremor lasted 3 to 5 minutes. The ground pitched like waves at sea, paved streets buckled in waves, and gaping cracks opened up, causing destruction of man-made structures in the city. In the Turnagain Heights residential section of Anchorage, huge landslides and subsidence devastated the area, as slice after slice of earth slid toward Cook Inlet. What was left was a jumble of open chasms, tilted street segments, and ruined houses. The destruction of water, gas, electric, telephone, and sewer lines was virtually complete.

The damage was not confined to Anchorage. The waterfronts and docks at Valdez and Seward slumped into the sea. This slumping caused huge waves that headed back to the shore to heights of 10 to 15 meters. Seismic sea waves, generated beneath the Gulf of Alaska, destroyed waterfront property everywhere along the coast. These waves spread across the Pacific to Hawaii, Japan, and Antarctica, inflicting the worst damage on 27 business blocks at Crescent City, California.

The total property damage from this earthquake

FIGURE 13.15 *Map of the Alaska earthquake area of 1964.*

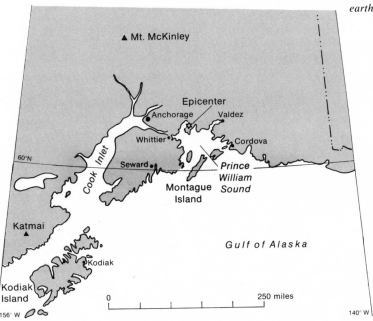

exceeded $300 million. Deaths, mainly by drowning, numbered 114 in Alaska, 12 in California, and 4 in Oregon.

The earthquake set up oscillations of water levels in wells throughout much of Canada and the United States, sloshed water in swimming pools in Texas, caused surges to be recorded on tide records on the Gulf of Mexico. These incidents indicated that a large segment of the earth was vibrating like a tuning fork.

The source of the initial earthquake was at a depth of 20 to 50 kilometers near the head of Prince William Sound (Fig. 13.15). The first tremor was followed by 28 aftershocks during the first 24 hours, 10 of them of damaging size, and by tens of thousands of other aftershocks over a 2-month period. Minor tremors continued for a year and a half. These aftershocks came from a belt extending 730 kilometers northeast-southwest from Prince William Sound to an area off Kodiak Island. The primary fault responsible for the earthquake can be traced into the crust to depths of 100 to 200 kilometers. The elevated side of the fault moved about 8 meters toward the continent.

Large changes in land level accompanied the earthquake. An area of about 67,000 to 90,000 square kilometers along the earthquake belt was uplifted, and an area of about 100,000 square kilometers northwest of the uplifted area subsided an average of 1 meter. The maximum rise of the sea floor was estimated to be more than 16 meters, the largest displacement ever observed in historical time. This submarine uplift was responsible for the seismic sea waves.

Chile, 1960

A succession of several hundred earthquakes began in Chile on May 21, 1960, and continued for many months afterward. These shocks caused 2,000 deaths, made a half million homeless, and damaged property estimated at $500 million. An early shock of magnitude 7.5 to 8.0 was followed the next day by five severe shocks of comparable magnitude. These represented a relatively sudden release of enormous energy.

Ordinarily, the earth experiences only a few exceptionally strong earthquakes over a period of years and few very severe shocks in an entire year. For example, from 1904 to 1946, only 14 were considered great (with magnitudes of 7.0 to 8.6). Several tens of the Chilean earthquakes during the first 10 days were classified as potentially destructive (having magni-

tudes of 6.0 to 7.0) or major (having magnitudes of 7.0 to 7.7).

The principal shocks of the Chilean earthquake began near the coast or just offshore at depths between 41 and 66 kilometers beneath the surface. Their centers of origin clustered near an offshore fault line between 37 and 41° south latitude and spread as far south as 51°, a distance of about 1,600 kilometers.

The effects of these shocks included the shattering of brick and adobe buildings, collapse of their roofs, cracking and spalling of concrete structures, bending of steel beams, collapse of bridges (mainly from failure of their abutments). Geologic effects included slumping of river banks, landslides, mudflows, eruption of the volcano Puyehue, local uplift, and general subsidence and submergence of an agricultural lowland between Valdivia and Puerto Montt. Lake Rinihue, dammed by landslides, rose an extra 33 meters and for a time posed a flood threat to the city of Valdivia downstream.

The strong Chilean earthquakes of 1960 set up body oscillations of the entire earth, like the ringing of a bell, that have been of special interest to geophysicists studying the interior of the earth.

California, 1906

A major earthquake struck California on April 18, 1906. Buildings disintegrated, chimneys crumbled, pavements buckled, piers collapsed, water mains broke, and landslides plunged down the hillsides. Structures on filled land along San Francisco Bay were damaged most. Fires sweeping through the wreckage subsequently burned over about 11.6 square kilometers of San Francisco. The loss was about $400 million, about four-fifths by fire. About 700 persons were killed and 250,000 left homeless. This damage was due partially to the earthquake and partially to groups in the city who were attempting to block fire routes.

This earthquake originated along the San Andreas fault, a nearly vertical rift zone that cuts diagonally across western California (Fig. 1.5). The movement along the fault in 1906 was largely horizontal and ranged up to 7 meters near Tomales Bay. The southwest side moved northwesterly with reference to the northeast side (Fig. 13.2). The vertical displacement was generally less than 1 meter.

TABLE 13.2A   Earthquake Activity during January 1972 to November 1972*

Date	Location	Remarks
January 1972	Osima Island, Japan	Swarm; included 27 shocks felt
	Torres Island, New Hebrides	Magnitude 7.4; shallow (50–60 kilometers)
	Taiwan	Magnitude 7.3; shallow
February 1972	Hachijo Jima, Japan	Magnitude 7.3; shallow
April 1972	Jahrom, Iran	Magnitude 7.0; shallow; intensity VIII; 10 aftershocks; focus 33 km
	Luzon, Philippines	Magnitude 7.3; shallow
	Taiwan	Magnitude 7.2; shallow
June 1972	Sangihe, Indonesia	Magnitude 7.4; focus 300 kilometers
July 1972	Sitka, Baronof Islands, Alaska	Magnitude 7.3–7.8; shallow
August 1972	New Ireland, South Pacific	Magnitude 7.4; shallow
September 1972	Santa Cruz Island, South Pacific	Magnitude 7.2; shallow
	Banda Sea, Indonesia	Magnitude 6.9; shallow
November 1972	Tanna Island, New Hebrides	Magnitude 7.5; shallow

*As reported on Smithsonian Institution Earth Science Event Cards.

Other Earthquakes

Earthquakes of great intensity have occurred periodically for eons. By one estimate they have killed at least 13 million persons during the past 4,000 years. Some noteworthy earthquakes of the past and those that occurred during the period January 1972 to November 1972 are listed in Tables 13.2A and 13.2B.

Seismic Sea Waves

Seismic sea waves, or tsunami, often accompany large coastal earthquakes. They originate as the energy from an earthquake is transferred to the water as well as the sea floor. They are characterized by high velocity and large wavelength; yet they attain a height of only a few feet on the open sea. A ship at sea would scarcely detect their passage. Reaching shallow water, they can be very destructive. Each wave may rise to heights of several meters.

Seismic sea waves occurred after the Lisbon earthquake of 1755 and the Chilean earthquake in 1868. On April 1, 1946, a tsunami originating in the Aleutian area destroyed a lighthouse at Dutch Cap, Alaska, situated 15 meters above sea level, and it crossed 3,800 kilometers of ocean to the Hawaiian Islands at an average speed of 780 kilometers per hour (Fig. 13.11). At sea, the waves were only a few meters high and 150 kilometers in wavelength, but near shore

TABLE 13.2B   Noteworthy Earthquakes of the Past

Year	Place	Magnitude	Remarks
1755	Portugal	8.7	Tremors felt in eastern United States; 60,000 deaths
1811	New Madrid, Missouri		Affected the Mississippi River's course and flood plain; felt over a 5-million-square-kilometer area
1897	Assam, India	8.7	Destruction over a 390,000-square-kilometer area; felt over a 2-million-square-kilometer area
1923	Tokyo, Japan	8.3	250,000 dead; 1,100 aftershocks
1959	Hebgen, Montana	7.0	Affected geyser activity in Yellowstone Park; prominent landscape changes
1970	Chimbote, Peru	7.8	Debris avalanche on Nevados Huascaran; buried two towns and killed 17,000 people
1971	San Fernando, California	6.6	Intensity of VIII to XI; over $1 billion damage

they rose 3 to 6 meters in height. In constricted places they were funneled to elevations of 10 to 15 meters above sea level. These waves, acting like walls of water, did great damage to buildings, roads, railroads, bridges, piers, breakwaters, and ships and caused 159 deaths. The total property damage in Hawaii was estimated at $25 million. Along the California coast the water rose as much as 4 meters. An international warning system has since been set up to alert communities around the Pacific whenever tsunami danger threatens.

In 1960, giant sea waves originating on the coast of Chile reached the Hawaiian Islands, 11,000 kilometers away, in about 15 hours. The tide gauge at Hilo alternately rose and fell at about 30-minute intervals. Despite warnings, these waves killed about 60 persons and caused $75 million property damage at Hilo and elsewhere in the Hawaiian Islands. Reaching Japan 8 hours later, they again smashed waterfront structures and killed 180 persons. Additional deaths and property damage occurred in the Philippine Islands, New Zealand, and other parts of the Pacific rim.

PROTECTION FROM EARTHQUAKE HAZARDS

When a large shock occurs in a heavily populated area, homes, stores, and a variety of other buildings will suffer damage or be destroyed. Poor construction is the main reason. The destructive effects result because of unstable ground, the use of adobe and loose stonework, the falling of roofs and chimneys, and the breaking of foundations and walls.

Potentially dangerous are heavy cornices, parapet walls, and unnecessary ornaments. Old mortar, unanchored roofs and roof trusses, unstiffened elevator shafts and frames, unreinforced stairwells, and common walls in adjoining buildings of different size are also hazardous. In-ground construction, such as water, gas, steam, and sewer pipes and similar piping, are subject to rupture by differential movements.

To minimize damage, builders should consider pertinent geological factors. Solid bedrock is the ideal foundation for large structures. Construction on soft ground, steep slopes, and filled land should be avoided. Sea cliffs, river bluffs, sites near deep excavations, and areas of high water level in loose sediments are also undesirable.

Bridges and tall buildings require special consid-

eration because of their weight, resistance to horizontal forces, and internal balance. Reinforced-concrete buildings have proved to be relatively stable, but wood, steel, and reinforced-masonry structures can be made earthquake-resistant if they are well designed and well built. They need adequate reinforcements and appropriate ties, bracing, struts, and anchor bolts. The most secure structure is one that moves as a unit and not as separate components which may batter each other.

Earthquake-resistant construction is imperative in earthquake-prone areas. The additional cost is estimated by engineers to be less than 10 percent if met at the design stage. Architects along the Pacific Coast now routinely design structures to reduce earthquake hazards. Voters in California are repeatedly asked to support bond issues to rebuild schools that do not meet earthquake standards.

In the context of this discussion, it will be advantageous to summarize some results of the engineering studies following the 1971 San Fernando, California, earthquake.

1. The most practical approach in building design is for earthquake resistancy, and the protection level should be in excess of building code provisions. With this approach as a basis, more than 20,000 buildings, as far as 50 kilometers from the epicenter, should be reinforced or razed. Interestingly, the typical one-story wood-frame house withstood the strongest shaking.

2. Architectural damage without structural damage was so common as to warrant new codes for architectural elements such as cornices, pillars, and overhangs.

3. Some of the newest and tallest buildings in the Los Angeles area never were subjected to a shaking that would test their ultimate strength.

4. Earthquake standards for the freeway bridges and overpasses are considered inadequate. Reservoir dams in the area showed substantial damage.

An enlightened government can supply additional protection for the public, whereas a laissez-faire policy allows possible disaster (Fig. 13.16). Zoning laws may be used to control land use and types of structures in high-risk zones, such as unstable land fill areas. Building codes can dictate standards of construction. Classification of various areas for their different risks according to their geologic setting already is common practice by some insurance companies, following a 1969 map of seismic-risk areas prepared by the National Ocean Survey. Earthquake insurance in some areas is expensive. Control by zoning, building codes, and hazard classification is especially appropriate in Pacific-border states in order to prevent death or catastrophe in the future.

PRACTICAL USES OF SEISMOLOGY

Seismic Prospecting

Small artificial earthquakes are used to probe the structure of the outer part of the earth's crust, especially in the search for petroleum. Small charges of explosives are detonated in shallow holes. The resulting waves are picked up and recorded by a series of geophones and recorders carefully spaced along a line of traverse (Fig. 13.17).

The rates of travel of underground waves differ in different types of rock, and so the travel times suggest the general character of the underlying rocks. Emphasis is placed on the reflections and refractions that the waves undergo. These changes in the directions of wave travel serve to indicate the position of underground contacts between different bodies of rock, especially layered ones. From a series of explosions it is possible to compute the depth to a reflecting or refracting layer at different places, thus working out the structure of the subsurface rocks.

Sonic Sounding of Sediments

An adaptation of sonic waves is used to penetrate sediments on the ocean floor, enabling geologists and oceanographers to determine their thickness and other characteristics. A continuous profile showing multiple reflections may be obtained as a ship proceeds on its course (Fig. 23.2).

FIGURE 13.16 *Housing development along the San Andreas fault, South San Francisco, California. Many homes are directly over the fault rift, which runs from top to bottom through the middle of the view. (U.S. Geological Survey.)*

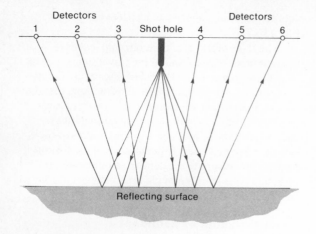

Detectors

1 2 3 Shot hole 4 5 6

Detectors

Reflecting surface

FIGURE 13.17 *Diagrammatic sketch of the reflection method of seismic prospecting. Waves spreading out from a shot point are reflected from an underground surface (where velocities change abruptly) back up to a series of carefully spaced detectors at the surface. Comparisons of the records of the first arrivals by direct horizontal routes and of the reflections received later, plotted as time-distance curves, indicate the depth and slope (if any) of the reflecting surface.*

REFERENCES

Barazangi, M., and Dorman, J., 1969, *World seismicity maps compiled from ESSA, Coast and Geodetic Survey, epicenter data 1961-1967*, Seismological Society of American Bulletin, vol. 59, no. 1, pp. 369-380.

Crowell, J. C., 1962, *Displacement Along the San Andreas Fault, California*, Geological Society of America Special Paper 71.

Duke, C. M., 1960, *The Chilean earthquake of May 1960*, Science, vol. 132, pp. 1797-1802.

Eckel, E. B., et al., 1972, *The Great Alaskan Earthquake of 1964: Geology,* National Academy of Sciences, Washington, D.C.

Gilluly, J., 1963, *The tectonic evolution of the western United States,* Quarterly Journal, Geological Society of London, vol. 119, pp. 133-174.

Hansen, W. R., et al., 1966, *The Alaska Earthquake, March 27, 1964: Field Investigations and Reconstructive Effort,* United States Geological Survey Professional Paper 541.

Hodgson, J. H., 1964, *Earthquakes and Earth Structure,* Prentice-Hall, Inc., Englewood Cliffs, N. J.

MacDonald, G.A., and Orr, J.B., 1950, *The 1949 Summit Eruption of Mauna Loa, Hawaii*, United States Geological Survey Bulletin 974-A.

Malloy, R. J., 1964, *Crustal uplift southwest of Montague Island, Alaska*, Science, vol. 146, pp. 1048-1049.

Plafker, G., 1965, *Tectonic deformation associated with the 1964 Alaska earthquake*, Science, vol. 148, pp. 1675-1687.

Reid, H. F., 1910, *The Mechanics of the Earthquake, the California Earthquake of April 18, 1906, Report of the State Investigation Commission,* Carnegie Institution of Washington, Volume 2.

Richter, C. F., 1958, *Elementary Seismology,* W. H. Freeman and Company, San Francisco.

———, 1969, *Earthquakes*, Natural History, vol. 78, no. 10, pp. 36-45.

Robertson, E. C., 1966, *The Interior of the Earth—An Elementary Description*, United States Geological Survey Circular 532.

Stacey, F. D., 1969, *Physics of the Earth*, John Wiley & Sons, Inc., New York.

Takeuchi, H., Uyeda, S. and Kanamori, H., 1970, *Debate about the Earth,* 2d ed., Freeman, Cooper and Co., San Francisco.

Thomas, G., and Witts, M. M., 1971, *The San Francisco Earthquake*, Dell Publishing Co., Inc., New York.

United States Geological Survey, 1971, *The San Fernando, California, Earthquake of February 9, 1971*, Professional Paper 733.

Weigel, R. L. (ed.), 1970, *Earthquake Engineering,* Prentice-Hall, Inc., Englewood Cliffs, N. J.

CHAPTER 14
THE INTERIOR OF THE EARTH

For decades, seismic waves have supplied information concerning the interior of the earth. In recent years, improved instrumentation, a worldwide network of standard seismic stations, information from underground explosions, free oscillations from large earthquakes, and studies of long-period surface waves have greatly increased that knowledge. These seismic studies have been supplemented by ultrasonic, shock-wave, and high-compression tests on minerals, rocks, and metals. Additional experimental data have been obtained in laboratories about the stability fields of various minerals and rocks at pressure-temperature conditions equivalent to a depth of 700 kilometers. As seismology, solid-state physics, and experimental petrology join forces, new information emerges. This chapter will summarize the present knowledge of the earth's interior based on these findings.

CHARACTERISTICS OF EARTH-WAVE TRAVEL

Meeting different zones within the earth, earthquake waves undergo refraction and partial or complete reflection (Fig. 14.1). The energy in the P waves just grazing the earth's core is also spread by diffraction as they pass its edge. Diffraction is a spread of energy parallel to the wave crest. Refraction abruptly changes the direction of propagation, just as light rays are refracted at the boundary between air and water. Reflection sends part of the wave energy back to the surface, where its multiple signals add complications to the seismographic records. Both P and S waves are reflected from the earth's surface and from the outer boundary of the core, and P waves are also reflected off the inside of the core boundary.

Still another distinct feature of seismic-wave travel within the earth is that, at epicentral distances of 103 to 143° of arc (11,000 to 16,000 kilometers), both P and S waves are blacked out. The S waves cannot penetrate the core, and the P waves are doubly refracted inward at the core boundary. The net result is a "blind," or "shadow," zone about 4,300 kilometers wide (Fig. 14.1). This shadow zone for any given earthquake has the form of a doughnut-shaped ring. Within the "hole-in-the-doughnut" the P waves are received again, but even these have been modified and delayed in transit.

Refraction, reflection, and sharp increases in the

201

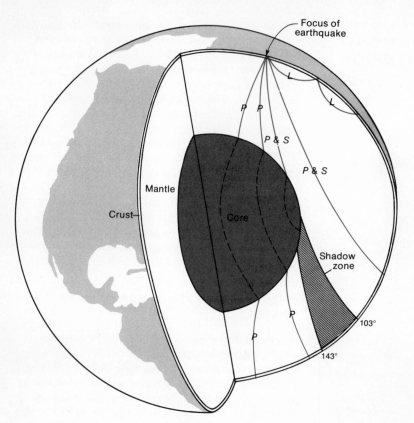

FIGURE 14.1 *Diagram showing refraction and reflection of earthquake waves within the earth. A ray just grazing the core is diffracted; another ray is strongly refracted twice at the mantle-core boundary; and a third ray is reflected repeatedly at the surface.*

TABLE 14.1   Characteristics of the Interior of the Earth

Layer	Thickness (km)	Depth (km)	Volume (%)	Seismic velocities (km/sec)		Estimated density (g/cm³)
				P waves	S waves	
Crust			1.5	5.8–7.6	3.2–4.3	2.8 (avg.)
—————————32*————— 33*————————. Mohorovičić discontinuity ———————————————						
Mantle	2,848		82.3	7.9–8.2	4.3–4.6	3.3–3.4
				13.6	7.3	5.5–5.8
——————————————— 2,898† ————— Core boundary (Gutenberg-Wiechert discontinuity) —————————						
Outer core	2,208		15.4	8.1	None	9.4–10.0
						10–11 (avg.)
				10.4	—	12(?)
——————————————— 5,120 ————————— Outer core–inner core boundary ‡ ————————————————						
Inner core	1,243		0.8	11.1	—	
						13.3–13.7 (avg.)
——————————————— 6,371 ——————————————— 11.3 ——————————————————						

*Usually considered to be 5 kilometers (3 miles) beneath ocean basins and 40 kilometers (25 miles) beneath continents.
†Diffraction studies suggest that the boundary may be 64 kilometers higher.
‡ Depths between 4,730 and 5,150 kilometers may be a transition zone from the liquid outer to the solid inner core.

velocities of *P* and *S* waves occur at certain distinct depths within the earth. The boundaries of these different depth zones, based primarily on seismic-velocity changes, are called discontinuities. Two major discontinuities separate the interior of the earth into three zones: (1) crust, (2) mantle, and (3) core. Their relative volumes are approximately 1.5, 82.3, and 16.2 percent. Other data are shown in Table 14.1.

INTERNAL EARTH STRUCTURE AND COMPOSITION

The Earth's Core

Within the doughnut-shaped shadow ring, the *P* waves arrive up to a few minutes late because their velocity slows to about 8 kilometers per second (Fig. 14.2). This delay, coupled with the fact that *S* waves which travel only through solids and are completely lacking within the shadow ring, may indicate that the outer part of the core is liquid. Another possible discontinuity occurs within the core itself at a depth of about 5,120 kilometers. The *P* waves again increase their velocity, an indication that the inner core is solid.

The molten outer core is generally considered to be a nickel-iron melt, with possibility of some associated lighter elements such as silicon and sulfur.

FIGURE 14.2   *Velocities of P and S waves within the earth. The S waves do not enter the core. The velocity of P waves drops back from 13.6 to 8 kilometers per second at the mantle-core boundary, but it jumps from about 10 to more than 11 kilometers per second at the inner core boundary. The hooks on the curves at the left mark the low-velocity zone in the upper mantle. (After Anderson, Sammis, and Jordan, 1971.)*

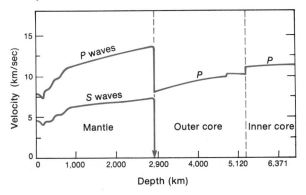

These lighter elements are theoretically included to improve experimental density and seismic-velocity calculations.

The composition and structure of the inner core is still a subject for continued discussion and investigation. Densities reach 13 grams per cubic centimeter (Fig. 14.3), and pressures exceed 218,000 kilograms per square centimeter.

The Mantle

Earthquake-wave velocities generally increase sharply at the crust-mantle boundary. This discontinuity is the Mohorovičić discontinuity, and is called the M, or "Moho," for short. A second important discontinuity in the mantle is the Gutenberg-Wiechert discontinuity that occurs at the mantle-core boundary.

The wave velocities in the mantle immediately below the Moho under the Colorado Plateau, the Basin and Range province, and the mid-oceanic ridges are a little lower than normal, probably because of higher than average subsurface temperatures there. This may also be due to an altered mantle material. Elsewhere, the *P*-wave velocities are 8.0 kilometers per second or more.

Seismic velocities suggest at least a threefold division of the mantle into the upper mantle, a transition zone, and the lower mantle. Velocity differences in the mantle between different depths or between different zones may be due to differences in temperature and pressure, local and partial melting, changes in mineral composition, density differences resulting from phase changes between minerals, transitions from minerals to noncrystalline glass, or some combination of these causes.

The upper mantle consists of (1) a high-velocity zone from 0 to 50 kilometers thick, (2) a middle zone of low velocity about 100 kilometers thick, and (3) a progressively homogeneous but denser zone about 250 kilometers thick, to a depth of 400 kilometers. The solid upper zone and the overlying earth's crust are the main sources of earthquakes.

The characteristics of the upper mantle have a direct bearing on the geotectonic processes of continent and ocean-basin motions. Two zones of major interest are the 0-to-50-kilometer-thick, high-velocity zone directly below the earth's crust and the underlying 100-kilometer-thick, low-velocity zone.

The upper zone is a brittle solid and, together

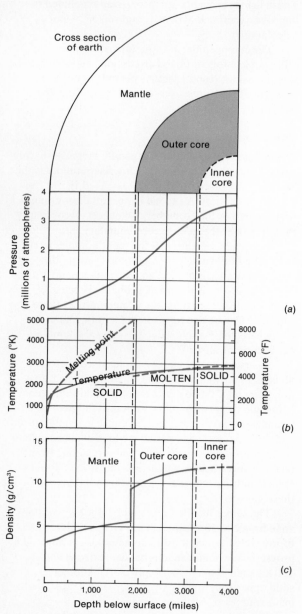

FIGURE 14.3 *Distribution of pressure, temperature, and density within the earth. (Data from J. Verhoogen, 1960,* American Scientist, *vol. 48; and K. E. Bullen, 1963,* An Introduction to the Theory of Seismology, *3d ed., Cambridge University Press. From Strahler, 1971,* The Earth Sciences, *2d ed., copyright 1963, 1971 by Arthur N. Strahler. By permission of Harper and Row Publishers, Inc.)*

with the earth's crust, constitutes a strong outer layer of the earth termed the lithosphere. The important minerals making up this upper zone are the low-pressure phases of olivine, pyroxene, and garnet, with accessory spinel and amphibole.

The low-velocity zone has also been termed the asthenosphere, or plastic layer (Fig. 14.4). The low seismic velocities cannot be explained adequately by mineral or temperature gradients. This layer, which has very sharp boundaries, can best be explained by partial melting, dehydration, or a small amount of water in a mineral assemblage. Investigations have found that partial melting of olivine, pyroxene, plagioclase, garnet, and minor minerals can begin at suitably low temperatures and pressures to explain this low velocity zone.

The transition zone, at a depth of 400 to 800 kilometers, shows several abrupt increases in seismic velocities. These are attributed to recrystallization of the component crystals to more compact minerals with higher specific gravity under the influence of pressure.

The Earth's Crust

Continental Crust The continental crust consists of two layers separated by an ill-defined discontinuity. The layers have been defined on the basis of seismic velocities and densities. In the upper layer the velocity of seismic waves corresponds to the velocity found by experiment to be characteristic of granite. This fact combined with the abundance of granite, schists, and gneisses at the earth's surface is the basis for calling an upper layer the granitic, or sialic, layer. The lower layer corresponds to basalt by its density and seismic velocities and is called the basaltic, or simatic, substratum. The discontinuity, or boundary, is not sharp between the granitic and basaltic parts of the continental crust, and so this two-layer concept may be oversimplified.

The *P* waves traveling in the basaltic substratum from an epicenter in western California are delayed 2 or 3 seconds on seismographs in the Owens Valley, immediately east of the Sierra Nevada. The Sierra Nevada must have a deep-seated "root," or downward extension of lower velocity granitoid rocks, that intercepts these waves (Fig. 14.5). The existence of the

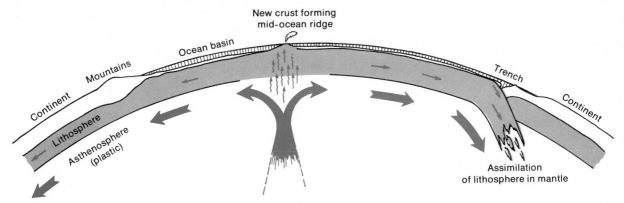

FIGURE 14.4 *Schematic view of the asthenosphere, its motions and relationship to the earth's crust. (From Menard, 1964.)*

mountain root has been confirmed by gravity studies and by the dispersion of surface seismic waves.

The granitic layer has been found to be thicker beneath mountains and plateaus than it is under low-lying plains, but not without exceptions. Variations in crustal thicknesses are well established across the United States (Table 14.2). The maximum known crustal thickness of 70 kilometers underlies the Himalayas and the Tibet Plateau.

Oceanic Crust   The earth's crust beneath the oceans consists of (1) a low-velocity layer of deep-sea sediments about 300 to 400 meters thick in the Pacific Ocean and 600 to 700 meters thick in the Atlantic, (2) a layer of intermediate velocity called the basement, about 0.8 kilometer thick, probably composed of compacted and indurated sediments and lava flows, and (3)

a third layer called the oceanic layer, about 4.1 to 5.8 kilometers thick and of uncertain composition but with the density and velocity of basalt. Sialic rocks like those of the continents are lacking. This three-layered oceanic crust is generally 5 to 8 kilometers thick.

Physical Properties of the Earth

Density   The specific gravity, or density, of the earth is 5.52 (Fig. 14.3). Most crustal rocks have a density of 2.6 to 3.0. These surface rocks are not sufficiently compressible, even under tremendous pressures in the earth's interior, to yield an average specific gravity of 5.52 over all depths. So the material at depth within the earth must be different in composition.

FIGURE 14.5 *Seismic refraction section across the California Coast Range, Great Valley, Sierra Nevada, and part of the Great Basin province. The numbers are seismic velocities in kilometers per second. The vertical exaggeration is about 2 times. A thickened crust under the Sierra Nevada is evident. (After Bateman and Eaton, 1967.)*

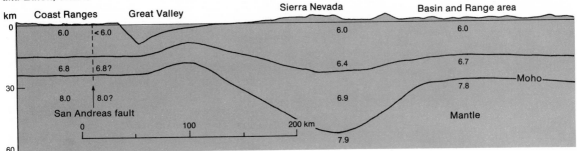

TABLE 14.2   Crustal Thicknesses under Areas of the United States

Area	Thickness (km)
Atlantic Coastal Plain	32–39
Central Lowland	37–43
Upper Great Plains	50
Rocky Mountains	35–60
Colorado Plateau	27–42
Basin and Range Province	22–34
Sierra Nevada	43
Central Valley of California	19
California Coast	24

Since the crust is thin and makes up only about 0.7 percent of the earth's volume and even less of its mass, the bulk of the earth's volume and mass must reside in the mantle and core. Some combination of a moderately heavy mantle and an exceptionally heavy core would meet the density requirements.

A computer study in 1968 that used a random-figure program for earthquake-wave velocities and densities in the mantle and core generated 5 million models of the earth's interior structure. Six of these models fitted known facts, but only three of them were plausible. The best fit for the density of the fluid core was found to be between 9.4 and 10.0, and the density in the solid core, between 13.3 and 13.7. Iron has a specific gravity of 7.8, but it is compressible by shock waves to 13 or more. One possible arrangement of densities is shown in Table 14.1; see also Fig. 14.3.

Magnetism   The earth is a gigantic magnet. When a compass is acted upon by the earth's magnetism, it orients itself parallel to the earth's magnetic field and points toward the earth's magnetic pole.

The earth's magnetic field is dipolar, but its magnetic poles do not coincide with the north and south poles of rotation, which are the true north and south poles (Fig. 14.6). The angle between the magnetic axis and the axis of rotation is about $11\frac{1}{2}°$. The north magnetic pole is located among the islands of northern Canada at 70°N 100°W and the southern pole is on the Antarctic Continent at 68°S 143°E. During the present century the position of the north magnetic pole has migrated, or wandered, within a range of several hundred kilometers about a mean site.

Because the magnetic and geographic poles do not coincide, a compass needle does not point to true north or south, except along a line that trends north-northwesterly across east-central United States. This is an agonic line. At points east or west of this agonic line, the compass needle makes an angle with the geographic meridian. This is the angle of declination. Repeated measurements show that the declination is gradually drifting westerly at a rate of about 2° per 100 years.

Another measurement of the magnetic field is made by using a dip needle, which is a magnetic needle mounted to swing in a graduated vertical arc. It will measure the angle of inclination, or the angle between the force lines and the earth's surface.

Both a compass needle and a dip needle are also subject to interference by a local magnetic attraction, such as that caused by a body of magnetite. Hence, a dip needle may be used to locate valuable magnetic iron ores. Reconnaissance magnetometer surveys, measuring magnetic-field strength rather than dip-needle inclination, are now made more rapidly from low-flying airplanes or from instruments towed behind a ship.

Another feature of earth magnetism is magnetic anomalies, which are local deviations of the magnetic value of the underlying rocks from the regional mean. Anomalies are also caused by a reversal in polarity. On land they are generally related to the observed rock types and structure. On the ocean floor, a pattern

FIGURE 14.6   *Diagram showing the relative positions of the magnetic and planetary poles of the earth.*

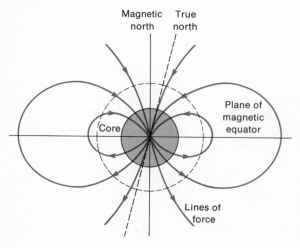

of parallel stripes with differing magnetic character-istics oriented approximately north-south is found over broad areas. Strong normal or reversed-polarity anomaly stripes also occur along mid-oceanic ridges. These are thought to be caused by the rise of magnetic field present at that time (Fig. 14.7).

These belts of magnetic anomalies beneath the various oceans extend 2,000 to 2,500 kilometers outward from the axis of the ridges (Fig. 14.7). They give way to magnetically "quiet zones" of uncertain age and origin, about 400 kilometers wide along and near the continental margins.

Research at sea by many workers in recent years has revealed a global distribution of these marine mag-netic-anomaly belts. These anomalies seem to confirm a theory called sea-floor spreading.

(*a*) *Cause of the Earth's Magnetic Field*   The earth's permanent magnetic field is hypothesized to consist of a complex system of electric currents accompanying turbulent convection in the fluid outer core of the earth. The earth thus acts as a dynamo in which the mechanical energy of this convection system gen-erates electric currents and accompanying magne-tism. The wobble of the magnetic pole, the westward

drift, and the local and temporary irregularities or vari-ations in the magnetic field can be explained by this theory. The main question to be answered is the source of heat energy to cause convection in the core, which presumably is low or lacking in radioactive ele-ments. The possibilities are that crystallization of iron is gradually taking place at the inner core—outer core boundary, thereby releasing heat, that iron is settling out of the mantle, giving up gravitational energy, or that heat is released as phase changes occur during theoretical expansion of the earth.

Some earth properties, such as magnetism, have shown variations over past geologic time. These paleomagnetic characteristics will be considered.

(*b*) *Paleomagnetic Results.   Pole Positions.* Fossil magnetism in rocks serves as a fossil compass to show previous pole positions. Pole positions during the past million years (Fig. 14.8) cluster closely about its present site and thus support the assumptions of a dipolar and geocentric magnetic field in the past. Late Tertiary rocks show a mean deviation of about 5°, and older rocks a much greater divergence. In 1964 a review was made of more than 565 magnetic-pole de-terminations from the Cenozoic to the Precambrian. It

FIGURE 14.7   *Magnetic anomalies astride the oceanic ridges. (After Wyllie, 1972.)*

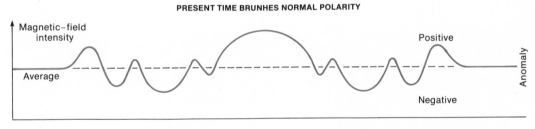

PRESENT TIME BRUNHES NORMAL POLARITY

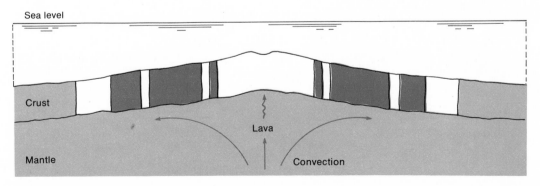

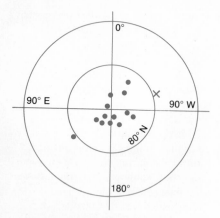

FIGURE 14.8 *Late Pleistocene and Recent pole positions. They cluster about the geographic pole (center) rather than the present magmatic pole (cross). (After Cox and Doell, 1960.)*

was found that the pole positions, though different by tens of degrees, were fairly consistent for each continent, but widely different from continent to continent (Fig. 14.9). Apparently, during geologic time, both the magnetic north pole and the earth's pole of rotation

moved from a position in the eastern Pacific Ocean in a wide arc through the equatorial Pacific to eastern Asia and thence to their present positions. The movement occurred at a rate of about 3 centimeters per year (Fig. 14.9). Results for Australia and India disagree with data from North America and Europe. Corresponding changes in climates affecting areas of glaciation, dunes, evaporites, coral growth, and life zones on land accompanied these movements of the poles or of the continents.

Various attempts have been made to explain these ancient pole positions. One theory denies any necessary connection between magnetic and geographic poles. Another would substitute for the present dipolar field a multipolar magnetic field caused by an even more complex convection system in the fluid outer core. A third, accepting polar wandering as a fact, would have the entire outer shell of the earth slide around on the plastic low-velocity layer of the upper mantle. The magnetic axis, the axis of rotation, and the bulk of the earth's mass would remain in place while segments of the earth's crust changed latitude and longitude as a unit. A fourth postulates continental drift, the differential movement of continents, such as the

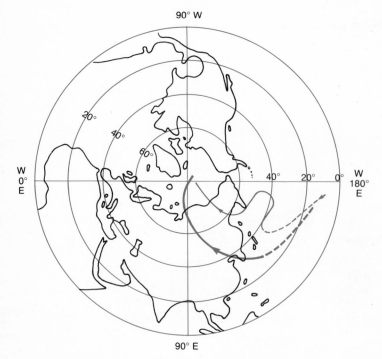

FIGURE 14.9 *Paleomagnetic-pole migration routes for North America (heavy line) and Europe (light line) from Cambrian (early Paleozoic) time onward. Later data shift the routes slightly but do not alter the general trend. [After Deutsch, 1966, in G. D. Garland (ed.), Continental Drift.]*

suggested separation of the Americas from Europe and Africa. The curves in Fig. 14.9 strongly suggest such a separation.

Polarity reversals. Paleomagnetic determinations on rocks dated by the potassium-argon method have established a time scale of polarity epochs (Fig. 14.10). This scale shows well-dated reversals of polarity during the past 4 million years. More than 170 reversals have occurred during the last 80 million years and are grouped into 33 anomalies. The reversals do not seem to follow any set pattern of regularity and the transitions take a few thousand to tens of thousands of years to complete.

The intensity of the earth's magnetic field has been weakening since 1838 at a rate of about 5 percent per century. Conceivably, it might decrease to zero in about 2,000 years, and a subsequent build-up might lead to a reversed polarity. Past changes in magnetic intensity very possibly affected the receipt of incoming cosmic rays, which may have a direct bearing on genetics. Strong bombardment by cosmic rays at times of low magnetic intensity may have modified the evolution of life on the earth.

Heat Flow   Heat flow is a measure of the quantity of heat brought to the surface from the earth's interior, measured in calories per square centimeter per second. The amount depends on its place of origin and its method of movement. Geophysicists generally agree that little if any of the present heat of the earth has been inherited from a primeval state, but that most of it is currently being produced by radioactive disintegration, mainly by uranium, thorium, and an isotope of potassium. For example, the flux of heat produced by K^{40}, U^{235}, U^{238}, and Th^{232} yields about 1.2×10^{-6} calorie per square centimeter per second.

The distribution of radioactive elements below the surface is uncertain, but uranium, thorium, and potassium all tend to be more abundant in acidic than in basic igneous rocks (Tables 14.3A and 14.3B).

Presumably, the heat output of sialic continental areas should decrease with depth in the earth, and granitic continents should have a higher heat flow than basaltic ocean basins. The first of these hypotheses cannot be tested because measurements at great depths are impossible. The second is incorrect because studies of heat flow through the rocks of the ocean floor indicate that the flow to the surface from oceanic areas is approximately equal to that through

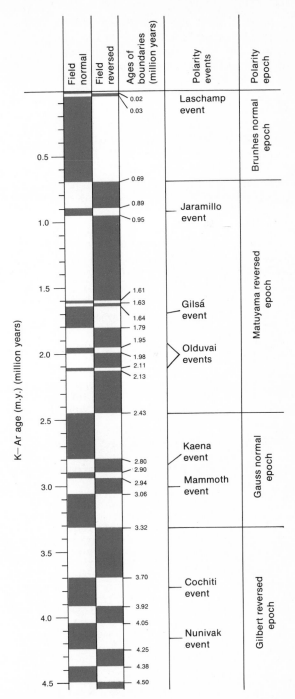

FIGURE 14.10   *Magnetic-polarity time scale for the last 4.5 million years as dated by potassium-argon ages. (After Cox, 1969.)*

continents. However, the heat flow in both areas is very uneven. If heat production is low from the thin oceanic crust, and presumably low from a dunite-peridotite upper mantle, some heat must come from deep levels in the earth.

Each determination of heat flow requires measurement of the heat gradient across a measured vertical span and the heat conductivity of the rocks within that span. Temperatures are measured by means of thermistors or thermocouples. At sea, the thermometers are combined in special probes in a coring apparatus. Oil and gas wells, mines, and tunnels provide access for instruments and heat-conductivity samples on land.

Using nearly 3,000 measurements up to 1969, Langseth surprisingly found the average heat flow to be 1.41×10^{-6} calorie per square centimeter per second for continents and 1.42×10^{-6} for ocean basins.

Local differences in heat flow may be caused by differences in the heat sources, differences in thermal conductivity of the rocks, and disturbances by

orogenic movements, igneous intrusions, or ground-water movement. Exceptionally high heat-flow values, 10 to 100 times greater than the average, occur in the geothermal areas of New Zealand, Japan, Italy, Iceland, and Yellowstone. High heat-flow readings from 2.5 to 3.0×10^{-6} to 8×10^{-6} have been obtained from crests of the mid-oceanic ridges (Fig. 14.11), the Gulf of California, and the Basin and Range province. By contrast, the Precambrian shield areas and the deep trenches of the Pacific Ocean show low heat-flow values (Fig. 14.12).

The heat generated within the earth is transmitted by conduction, radiation, and convection. Rocks are poor conductors of heat and are therefore good insulators. Thus the rate of heat movement through them is very slow. Radiation within solids increases at high temperatures, and this may be a very important process at extremely deep levels in the earth. Mass transport of materials, such as the rise of magma or by convection currents in the plastic zone of the upper mantle, would transfer heat effectively. In both the

TABLE 14.3A   Heat Production of Radioactive Elements
(cal/g· 10^6 years^{-1})

U^{238}	0.71
U^{235}	4.3
U (ordinary)	0.73
Th^{232}	0.20
K^{40}	0.21
K (ordinary)	0.000027

SOURCE: Wetherill. G. V., 1966, *Radioactive decay constants and energies,* in *Handbook of Physical Constants.* Geological Society of America Memoir 97, pp. 513–519; reproduced by permission of The Geological Society of America.

TABLE 14.3B   Radiogenic Heat Production of Various Rocks

Rock type	Concentration (ppm)			Total heat production (cal/g· 10^6 years^{-1})
	U	Th	K	
Granite	4.75	18.5	37,900	8.10
Intermediate igneous rock	2.0	—	18,000	3.40
Basalt	0.6	2.7	8,400	1.19
Peridotite	0.016	—	12	0.009
Dunite	0.001	—	10	0.002

SOURCE: MacDonald, G. J. F., 1965, *Geological deductions from observation of heat flow,* in W. H. K. Lee (ed.), *Terrestrial Heat Flow,* American Geophysical Union Monograph 8, pp. 191–210; copyright, 1965, American Geophysical Union; used by permission.

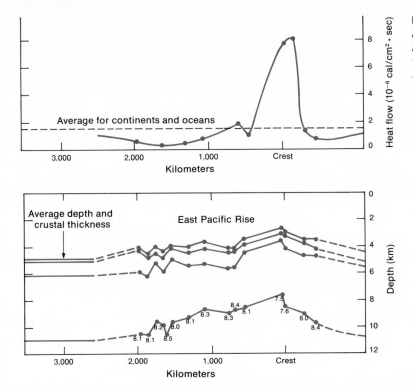

FIGURE 14.11   *Pattern of heat flow along the east Pacific rise. The series of numbers in the lower diagram are seismic velocities in kilometers per second. (From Menard, 1964.)*

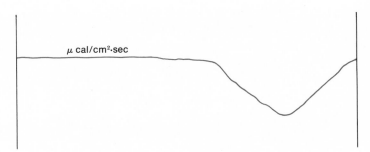

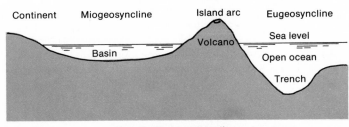

FIGURE 14.12   *Pattern of heat flow along island arcs and trenches. (After Menard, 1964.)*

solid and liquid parts of a metallic core, conduction would be important, with convection being important in the fluid outer part of the core.

It is not known whether the earth as a whole is gradually cooling, has reached a steady state, or is actually getting progressively hotter.

Temperatures in the interior of the earth can be estimated only indirectly. Higgins and Kennedy (1971) calculated that under the pressure of the weight of overlying material, iron would melt at about 3700°C at the outer core–mantle boundary, and at about 4300°C at the inner core–outer core boundary. If the earth's core is made principally of iron, molten on the outside and solid at the center, the temperature of most of the core should be roughly in the range 4000 to 5000°C.

Temperatures in the mantle presumably decrease more or less regularly upward toward the top. The mantle is solid throughout; therefore, its temperature must remain consistently below the melting point of whatever material is present at various depths. Because the exact composition is unknown, these melting points are conjectural (Fig. 14.3). Locally and temporarily, basaltic magma originates at 40- to 200-kilometer depths and is extruded as lavas at a temperature of 1000 to 1200°C.

Gravity A pendulum may be used to measure the pull of gravity at various places on the earth's surface. The period of a pendulum of a given length depends on the local pull of gravity, which varies with the mass of material directly beneath the pendulum. The stronger the pull, the faster the pendulum swings. The pendulum method is cumbersome and slow, and so portable gravimeters have been developed for use in the field. These are calibrated by frequent reference to known values at fixed stations.

Measurements of gravity are affected by local deficiencies or excesses of mass in the vicinity of the site of measurement. The residual values, after computation of the expected values and correcting for differences in elevation, are called gravity anomalies. These gravity anomalies are negative if the underlying mass has a less-than-normal density or is deficient in amount.

Gravity measurements tend to confirm the suspected distribution of mass in the earth's crust. These observations show a deficiency of mass, not only below mountains, but also in or under continents as well. It is inferred, therefore, that continents and mountains stand high because they are only the upper portions of masses of relatively lower density rocks, thus confirming mountain "roots." Although there are many local gravity anomalies, or departures from the computed values, they tend to cancel out when areas 100 to 160 kilometers in diameter are considered. The orbits of artificial satellites are appreciably affected by the mass differences of the earth. In an earlier chapter it was noted that similar anomalies from "mascons" occur on the moon.

Gravity measurements on land and at sea show that a state of balance exists between continents and ocean basins and even between smaller areas. Explaining this relationship, the theory of isostasy holds that different masses of the earth's crust stand in flotational equilibrium with each other at some depth within the earth. This depth is reached where the weights (a function of gravity) of all columns of rock of an equal area are the same, regardless of whether mountains or plains or shallow or deep seas occur at the surface (Fig. 14.13). Evidently, regions with relatively high mountains float on their foundations, like icebergs at sea, because they are light or buoyant, and the ocean basins are low because the oceanic crust is composed of relatively heavier rocks. Despite the differences in surface elevations, the masses in the underlying columns of rocks beneath unit areas are equalized by inverse differences in density (i. e., excess mass gives low density, and a deficiency of mass gives high density).

If the prevailing equilibrium is disturbed, as by long-continued erosion of a continent, the loss of mass presumably is countered by plastic flow of heavier material into the area beneath it, perhaps in the low-velocity zone. Such a process, which would cause the continent to slowly rise, is an attempt to maintain equilibrium. This adjustment is called isostatic compensation (Fig. 14.13).

How well or how promptly isostatic compensation works is debatable. Although the earth is rigid and can accumulate strains over a long time, it tends to assume a shape in accord with the theory, however tardily or imperfectly. Data from artificial satellites, for example, show that polar flattening and equatorial bulge are a little greater than they should be. Large negative gravity anomalies also exist in areas of oceanic trenches, recent crustal movements, and recent vulcanism.

A difficulty is posed by the uplift of plateaus. Old erosion surfaces have remained at a low elevation for a long time and then have been uplifted hundreds or even thousands of meters. How does this compensation work? The change is most likely due to volume rather than weight, perhaps by (1) conversion of garnet in the upper mantle to plagioclase and olivine, with more than 10 percent expansion, or (2) conversion of

FIGURE 14.13 *The upper diagram shows a design of the earth's crust according to the Airy theory of isostasy. The lower diagram illustrates how a segment of the crust undergoes isostatic compensation, as a result of erosion and sedimentaiton. (From Strahler, 1971,* The Earth Sciences, *2d ed., copyright 1963, 1971 by Arthur N. Strahler. By permission of Harper and Row Publishers, Inc.)*

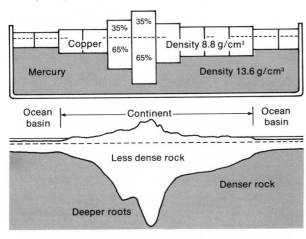

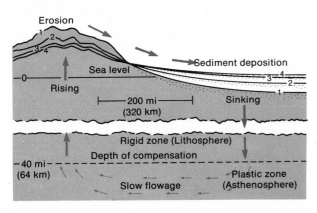

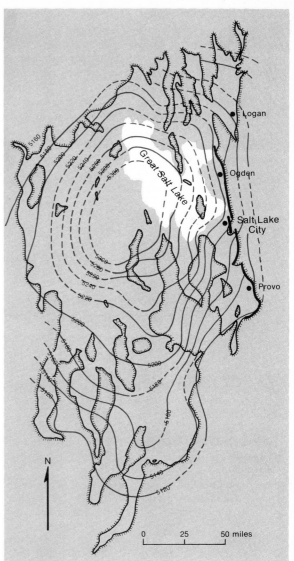

FIGURE 14.14 *Map showing isostatic doming of Lake Bonneville shorelines in Utah. The outline of the former lake is hachured. The contour interval on the deformed shorelines is 20 feet. Faults border the basin on the east. (After Crittenden, 1963.)*

olivine in the upper mantle to serpentine, with about 25 percent expansion.

It is well established that areas formerly covered with continental ice sheets were depressed hundreds of meters by the weight of the ice and that these areas have partially readjusted during the past 10,000 years. Measurements in historical time in Fennoscandia show that uplift still continues there at a rate of more than one-third of a meter per century. Similarly, shorelines on former islands near the middle of former Lake Bonneville in Utah are now at least 70 meters higher than those on the periphery of the lake basin (Fig.

14.14). This indicates isostatic rebound from formerly depressed levels. The weight of the lavas of the Hawaiian Islands has also depressed the underlying crust and caused an upward bulge in the sea floor some distance offshore. This creates a moatlike effect around the volcanic islands.

Notwithstanding some puzzling difficulties, the earth does seem to respond after a time to great load stresses by subsurface compensating adjustment, although its strength and rigidity are great enough to resist at least minor changes in isostatic equilibrium and to permit gravity anomalies to remain for a while.

REFERENCES

Anderson, D. L., Sammis, C., and Jordan, T., 1971, *Composition and evolution of the mantle and core,* Science, vol. 171, pp. 1103-1112.

Bateman, P. C., and Eaton, J. P., 1967, *Sierra Nevada batholith,* Science, vol. 158, pp. 1407-1417.

Birch, F., 1966, *Earth heat-flow measurements in the last decade,* in Hurley, P. M. (ed.), *Advances in Earth Sciences,* pp. 403-430, The M. I. T. Press, Cambridge, Mass.

Bott, M. H. P., 1971, *The Interior of the Earth,* St. Martin's Press, Inc., New York.

Cox, A., 1969, *Geomagnetic reversals,* Science, vol. 163, pp. 237-244.

———and Doell, R. R., 1967, *Reversals of the earth's magnetic field,* Scientific American, vol. 216, no. 2, pp. 44-54.

Crittenden, M.D., Jr., 1963, *New Data on the Isostatic Deformation of Lake Bonneville, Utah,* United States Geological Survey Professional Paper 454-E.

Deutsch, E. R., 1966, *The rock magnetic evidence for continental drift,* in Garland, G. D. (ed.), *Continental Drift,* pp. 28-52, Royal Society of Canada Special Publication 9.

Elsasser, W. M., 1958, *The earth as a dynamo,* Scientific American, vol. 198, no. 5, pp. 44-48.

Gaskell, T. F. (ed.), 1967, *The Earth's Mantle,* Academic Press, Inc., New York.

———, 1970, *Physics of the Earth,* Funk & Wagnalls, a division of Reader's Digest Books, Inc., New York.

Gass, I. G., Smith, P. J., and Wilson, R. C. L. (eds.), 1971, *Understanding the Earth, A Reader in the Earth Sciences,* Artemis Press, Sussex, England.

Gilbert, G. K., 1890, *Lake Bonneville,* United States Geological Survey Monograph I.

Hart, P. J. (ed.), 1969, *The Earth's Crust and Upper Mantle,* American Geophysical Union Monograph 13.

Hayford, J. F., and Bowie, W., 1912, *Effect of Topography and Isostatic Compensation Upon the Intensity of Gravity,* United States Coast and Geodetic Survey, Special Publication 10.

Jacobs, J. A., 1963, *The Earth's Core and Magnetism,* The Macmillan Company, New York.

Lee, W. H. K., and Clark, S. P., Jr., 1966, *Heat flow and volcanic temperatures,* in *Handbook of Physical Constants,* pp. 483-511, Geological Society of America Memoir 97.

———and Uyeda, S., 1965, *Review of heat flow data,* in Lee, W. H. K. (ed.), *Terrestrial Heat Flow,* pp. 87-190, American Geophysical Union Monograph 8.

Menard, H. W., 1964, *Marine Geology of the Pacific,* McGraw-Hill Book Company, New York.

Pakiser, L. C., and Zietz, I., 1965, *Transcontinental crustal and upper mantle structure,* Review of Geophysics, vol. 3, pp. 505-520.

Press, F., 1961, *The earth's crust and upper mantle,* Science, vol. 133, pp. 1455-1463.

CHAPTER 15
EARTH MOVEMENT AND ROCK DEFORMATION

Present geological observations reveal that rocks in many areas occupy different positions or have different shapes than they had originally. Such rocks are said to be deformed. Rock deformation occurs whenever rocks have been physically altered from their original state, whether this be a change of volume, shape, or position.

The scale of and type of deformation involves gentle vertical movement of large regions by vertical forces, which is called epeirogeny, and deformation by orogeny, which results in rocks that are severely bent, folded, and broken by (horizontal or vertical) forces from within the earth's crust.

MECHANICS OF DEFORMATION

Laboratory physicists and engineers have measured and classified the different ways earth materials react to deforming forces, their strengths, and they have developed a stress-strain graph (Fig. 15.1). Stress is measured as the intensity of the deforming force, shown as the vertical axis of the graph. Strain is the

FIGURE 15.1 *Contrasting behavior of brittle, plastic, and fluid deformation. Asterisk marks position of ultimate strength.*

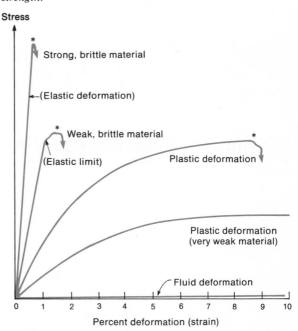

percentage of shape distortion resulting from a given stress and is indicated along the horizontal axis of the graph. Strength is defined as the limiting stress a solid can withstand without deformation failure. Ultimate strength is the greatest stress a solid can withstand under normal short-time experiments.

Elastic substances yield when force is applied but return to their original shape when the force is removed. A spring scale or a rubber band provides a familiar example. A spring scale stretches proportionally to the amount of weight applied, yet it returns to normal when the weight is removed. All elastic substances deform proportionally to applied stresses, and so their stress-strain graphs are straight lines (Fig. 15.1). All elastic substances have an elastic limit; beyond a certain point, deformation becomes permanent, not temporary. If too much weight is applied to a spring, it becomes permanently distorted and will not return to its original shape after the force is removed.

All rocks have elastic properties. On a small scale, this can be demonstrated by pebbles that bounce when dropped or thrown onto a hard surface. Forces inside the earth may produce elastic deformation,

FIGURE 15.2 *Complex folding in gneiss, a type of metamorphic rock. These rocks, in Canada, must have been very plastic when they were deformed deep beneath the present ground surface. (Geological Survey of Canada.)*

which can be observed in quarries and mines when rocks will sometimes spring loose in natural explosive rockbursts. This phenomena is caused by a release of the potential energy stored in elastic deformation.

Most rocks have a very low elastic limit and easily become permanently deformed. They bend, fold, or fracture rather than store up deformation energy. Other rocks are affected by brittle deformation or will deform plastically.

Plastic deformation is characterized by a gently curving stress-strain graph, flattening in slope toward the point of ultimate strength (Fig. 15.1). The materials bend or otherwise change shape by folding, until a rupture response occurs. Organic chemicals of the "plastic" family or moist clay have this property of plasticity.

As plastic materials become less rigid, the plastic behavior approaches fluid deformation. A fluid lacks appreciable internal strength and flows readily under its own weight. In fluid deformation the material deforms so readily in response to a force that it does not break. One might think of fluid flow as continuous deformation. With increasing internal strength or fluid rigidity, fluid deformation merges with plastic deformation. Deep within a glacier, ice moves in a manner intermediate between fluid and plastic deformation. The same is true with materials in a mud flow, which behaves almost like a fluid when very wet.

In brittle rocks, initial deformation is slight but proceeds elastically up to the elastic limit. Once the elastic limit is exceeded (marked by asterisk in Fig. 15.1), the rock quickly breaks, fractures, or crumbles so that the strength of the rock is destroyed. Limestone and sandstone are brittle rocks when subjected to deformation near the earth's surface, but have a fairly strong internal consistency. They are crushed commercially for construction materials.

The same is true for granite and metaquartzite; however, their ultimate strength at the earth's surface is so much greater than that of limestone or sandstone that it is difficult and uneconomical to crush them for industrial use. All these rocks, however, will show plastic deformation. In material such as a shale, there is a marked break in slope of the curve (elastic limit) prior to the maximum force sustained by the rock (ultimate strength). Brittle deformation is characterized by breaking and fracturing, rather than bending. When stretched, brittle rocks crack; when compressed, they

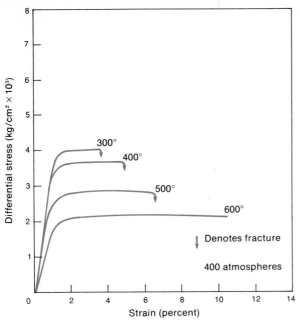

FIGURE 15.3A *Compression stress-strain curve for So-lenhofen limestone deformed at 420 kilograms per square centimeter (6,000 pounds per square inch) pressure and at various temperatures. Rock was dry during experiments. (After Heard, 1960.)*

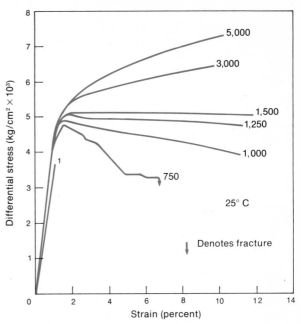

FIGURE 15.3B *Compression stress-strain curve for So-lenhofen limestone deformed at earth-surface temperature (25°C) at various confining pressures. Rock was dry during experiments. (After Heard, 1960.)*

crumble. In all rocks, the stretching strength (tensile strength) is much less than the compressive strength, and in very brittle rocks the tensile strength may be only $\frac{1}{20}$ the compressive strength.

Rock Response to Deformation

Faulted and jointed rocks are brittle, and folded rocks deform plastically. Very intricate folding (Fig. 15.2) can develop in rocks that become plastic, approaching but not attaining a fluid state (Fig. 15.1). Some rocks show evidence that they were first folded by plastic deformation and then later broken by faulting; therefore, faults may occur on or with folding.

Earlier it was noted that rocks such as sandstone, limestone, quartzite, and granite seem quite brittle when exposed at the earth's surface. A puzzling problem is presented when one considers that these rocks can be intricately folded rather than deformed by faulting. How did they deform plastically when they apparently are brittle rocks? The principal factors

involved are rock type, temperature, confining pressure, intergranular fluids, rate of application of force, and change of environment.

Rock Type Certain rocks are naturally more brittle than others, even at the surface of the earth. Unconsolidated mud and sand are soft enough to deform readily by folding. Shale is somewhat more brittle, but it will deform considerably by folding. Sandstone and limestone are brittle at the earth's surface, but at depth they are weaker than granite or the metamorphic rocks quartzite and gneiss. Various rock types differ greatly in their amount and type of deformation by a given force in a particular environment.

Temperature The higher the temperature, the weaker the rock becomes, until it eventually melts. Figures 15.3A and 15.3B show the effect of temperature on deformation of limestone. Rocks subjected to a higher temperature within the earth will generally en-

FIGURE 15.4 *Butte Valley in the Panamint Range. (Spence Air Photos.)*

dure a greater amount of plastic deformation than they do in cooler environments near the surface. Recall that the geothermal gradient follows an average increase of 20°C for every kilometer of depth. Limestone buried 5 or 6 kilometers underground can be folded more easily than limestone near the surface. Many metamorphic rocks are intensely folded, suggesting that they were heated so close to melting temperatures that the rocks became soft and plastic (Fig. 15.2).

Confining Pressure   At increasing depths, the weight of overlying rocks causes a progressively increased confining pressure. Figure 15.4 illustrates the results of a rock responding more plastically to a squeezing force when it was situated in a high-pressure confining environment. In this respect the effect of high confining pressure is similar to that of raised temperature. A cylinder of limestone, marble, or rock salt encased in a steel jacket can be changed under confining pressure to a barrel shape and still remain unbroken. Figures 15.3A and 15.3B portray the effects of temperature and confining pressure on the stress-strain behavior of dry limestone.

Intergranular Fluids   Ordinarily, limestone and other rocks within the earth contain a certain amount of groundwater or other fluids between the mineral grains. The effect of such intergranular fluids is to promote a more plastic behavior so that a rock which is normally brittle when dry will behave more plastically when wet. Its strength also becomes greatly reduced. The fluids aid internal adjustments within the rock, acting as a lubricant or a medium of recrystallization.

The effect of high-pressure intergranular fluids is similar to the effects of high temperature and high confining pressure. All three factors produce an increased plasticity as sedimentary rocks become more deeply buried before the squeezing forces of geologic deformation affect the rock. This explains why limestone and other sedimentary rocks often display intense folding when observed in some outcrops (Fig. 15.4). Such folding occurred several kilometers underground and then was exposed by extensive erosion. Deformation of the same rocks at shallow depths would result in jointing and faulting.

Rate of Application of Force   A rock will fracture (fault or joint) if force is applied suddenly, even though folding of the same rock would result from a slowly applied deforming force. An increase in the rate of deformation may cause a folding rock to break into a fault. The importance of rate upon deformation can be easily demonstrated by using children's "silly putty." Slow deformation by squeezing allows the putty to be easily molded. However, throwing the putty on the floor with a hard, quick impact will cause the putty to shatter.

Change of Environment   A change of position while deformation is in progress may cause rocks to change their style of distortion. Sedimentary rocks may subside several kilometers into the earth so that sandstone, which is brittle at the surface, becomes plastic enough at depth to fold intensely. Later, erosion may expose this folded rock, causing a reduction in confining pressure and temperature. This will cause the rock to return to a brittle state so that joints will form when renewed compressive or vertical forces are applied.

Igneous rocks were fluid when first intruded as magma. As they cooled they assumed a plastic condition, and finally they became brittle when cooled to near-surface temperatures. Forces applied after solidi-fication produce jointing in the once-fluid mass.

It remains now to consider the large-scale and visible manifestations of the various rock responses to deformation, as seen in the earth's crust and on the surface.

VERTICAL OR EPEIROGENIC MOVEMENTS

A limestone unit near the summit of Mount Everest contains fossil crinoids of late Paleozoic age. Crinoids are marine animals, and so it is apparent that the strata have been raised from beneath the sea to a high eleva-tion above sea level. Another record of vertical move-ments is provided by the ancient Roman public market now known as the Temple of Jupiter Serapis, on the coast near Naples, Italy. Columns of the temple have been bored by the marine pelecypod *Lithophaga* about 6 meters above its floor, and their shells are found in the holes. After completion of the structure by the early Romans, the land became submerged. Later the land area rose above sea level.

The dolomite bedrock near Chicago, now over 160 meters above sea level, is also marine in origin, yet it is still flat-lying the way it accumulated on the sea floor. Even if all the lakes in the world were to empty into the sea and the entire accumulation of snow and glacial ice were to melt, there would not be enough water added to the ocean to shift the present shoreline inland nearly 1,600 kilometers (1,000 miles) to inun-date Chicago. The evidence is convincing, therefore, that the Midwest was once beneath the sea, and subsequently, the region was uplifted to become land hundreds of meters above sea level. Fossils in the rocks in the Chicago area are of Silurian age, indicat-ing that the sea was there some 420 million years ago. Up to the end of Cretaceous time (70 million years) the geologic history of the Midwest was a story of suc-cessive periods of inundation and emergence. Finally, in late Tertiary time the Midwest was uplifted nearly 200 meters to approximately its present elevation above sea level.

Heavy loads covering a broad area of the earth's crust cause vertical crustal movements. The Pleis-tocene ice sheets covering large portions of the earth caused the land to subside in response to the added weight of the continental glaciers; after the ice melted and the weight was removed, the crust adjusted by

epeirogenic uplift and is continuing to do so. Such crustal upwarping is recorded by elevated beaches in regions that have had separate icecaps (Table 15.1).

Fresh-water marshes and low farm land along the shores of the Baltic Sea in Sweden and Finland con-tain many marine shells identical with those in the Bal-tic Sea today. Careful elevation surveys of rigid monuments set up at various points along the shore have demonstrated about this Fennoscandian lowland is rising at a rate of about 1 meter per century.

In the region of the Great Lakes, many old lake shorelines have been mapped that show conclusively that the lake basins have been tilted. The land north and northeast of the lakes has risen as much as several hundred meters so that ancient raised beaches now slope down to the south. The south shores of the Great Lakes, on the other hand, are marked by drowned stream valleys or estuaries, such as Keweenaw Bay and the harbors at Ashland, Wisconsin, and Mar-quette, Michigan. Such evidence of submergence by tilting of the lake basins is also conspicuous along the south shore of Lake Ontario and at the lower course of the St. Louis River where it discharges into Lake Su-perior. The elevation changes in the Great Lakes region seem to be a direct response to removal of the weight of the continental glacier at the end of the last ice age. The Straits of Mackinac are presently rising about 2 centimeters per century.

The southern tip of Louisiana provides an ex-ample of a subsiding region. Locally, near the mouth of major distributaries of the Mississippi, sedimentation is so rapid that the sea is being displaced by the growth of new land, producing a local net rise in elevation. Detailed studies of rigid survey markers show that on

TABLE 15.1   Unwarping of Regions of Com-plete or Partial Deglaciation

Area	Maximum uplift (m)
Eastern Labrador	270
Newfoundland	137
Fennoscandia	274
South Island, New Zealand	101
Iceland	119
Scotland	30

SOURCE: Daly, R.A., 1926, *Our Mobile Earth,* pp. 170–210, Charles Scribner's Sons, New York; 1934, *The Changing World of the Ice Age,* pp. 51–150, Yale University Press, New Haven, Conn.

a regional scale, however, southern Louisiana is subsiding. Portions of the region receiving additional new sediment actually are sinking beneath the surface of the Gulf of Mexico. The town of Balize was located about 1 meter above sea level in 1888. Over the years the area around the town has subsided gradually so that by 1935 an abandoned street lay 1 meter beneath the surface of the sea. Average subsidence is about 2 centimeters per year. Similar but less rapid subsidence is evident at many places along the Louisiana coast.

Plateaus

The relatively flat upland surfaces of plateaus result from occurrence of flat-lying rocks at high elevation. In the Columbia Plateau of Washington and Oregon, the high elevation has resulted from the accumulation of many lava flows until a relatively flat, high surface was built. The maximum lava thickness is over 1,600 meters and covers an area of 245,000 square kilometers. Subsequently, the area was gently folded and the Columbia River and its tributaries incised their valleys.

Plateaus more commonly result from regional uplift of flat-lying sedimentary rocks. Locally, these rocks may be broken by near-vertical faults or gently folded by unevenly distributed uplifting forces, but the more intense folding or creasing does not occur. The Colorado Plateau was uplifted over a thousand meters some 50 million years ago. The streams of this semiarid region have had little opportunity for erosion, and consequently, broad uplands remain (Fig. 15.5). A few major streams have been able to cut narrow notches deep into the Colorado Plateau, forming the spectacular scenery of Grand (Fig. 1.1), Bryce, and Zion Canyons.

Later in the cycle of stream erosion, a flat plateau upland becomes more dissected and cliff-walled valleys are many, with only small flat upland areas remaining. Examples are the Ozarks and the hill country of central Tennessee.

Extreme stream dissection of a plateau leaves almost no flat upland. The terrain is then nearly all slopes, yet the summits have about the same elevation as the original plateau. A good example of this stage of dissection is the Allegheny Plateau, which extends south from Pittsburgh across western West Virginia and into eastern Kentucky, where it is called the Cumberland Plateau. This terrain, with narrow valleys,

steep and winding roads, and immature soil, discourages industrial and agricultural development and promotes cultural isolation of the people. The plateau region is the core of the Appalachian poverty belt. The underlying structure of the plateau is vividly demonstrated by strip-mine scars that follow coal beds along the valley walls (Fig. 15.6).

Stream erosion ultimately reduces plateaus to such low hilly country that the name plateau is no longer appropriate.

The vertical forces that produce plateaus can cause uplift of hundreds or thousands of meters over a region covering tens of thousands of square kilometers. No force of such a magnitude is observed in surface phenomena. Because plateaus are so extensive and because they are deformed only by broad bending or cut by a few simple faults (some of which are sites of lava outpourings), the plateau-uplifting mechanism must be seated many kilometers inside the earth. This mechanism might be due to a chemical change deep in the earth, producing a volume expansion (such as alteration of peridotite rock in the mantle to serpentine), or it might be due to plastic or liquid materials deep in the earth that may move from one place to another to produce a bulge in the overlying rocks.

OROGENIC MOVEMENTS

Most parts of the earth's crust appear to be very stable in the short span of a person's lifetime. Many regions have been stable for thousands of years, especially most of the igneous and metamorphic shields forming the central parts of continents. Although some rocks are not now being deformed at a perceptible rate, their distorted shapes show that they have yielded to the cumulative effects of very slow deformation over millions of years in the past. On the other hand, faulting movements during earthquakes may take place in a minute or two. Major faults with displacements of thousands of meters probably develop little by little in an intermittent series of rapid episodes widely spaced in time.

The following examples will serve to illustrate orogenic deformation. (1) The anthracite coal fields of eastern Pennsylvania represent metamorphosed coal that originated from peat deposited as flat-lying strata in ancient swamps. The once-horizontal beds have been transformed to anthracite coal at depth in the

FIGURE 15.5 *Little Colorado River incised into a plateau in Arizona. The San Francisco peaks are in the distance. (Spence Air Photos.)*

FIGURE 15.6 *Coal strip mines near Charleston, West Virginia, reveal the horizontal structure of strata underlying the Allegheny Plateau. At least three coal beds have been mined along the valley walls, producing patterns resembling contour lines. (J. Dennison photo.)*

earth, and then they have been tipped up vertically or even overturned. (2) The horizontal displacement of the earth's surface during the 1906 San Francisco earthquake amounted to a maximum of about 7 meters. About a 2-meter vertical displacement was noted in the 1971 San Fernando earthquake. In recent years the main building of a winery near Hollister, California, has been splitting apart by steady horizontal creep of the ground along a fault at a rate of about 1.3 centimeters per year. (3) Finally, the movement causing an earthquake in Alaska in 1899 raised part of the shore of Yakutat Bay as much as 15 meters.

The foregoing illustrations of tilting, overturning, folding, and faulting furnish evidence of rather intense movements of the earth's crust.

Attitudes of Strata

To describe the structures produced by deformation processes and to indicate the direction and source of the forces involved, geologists determine the attitudes of the strata in the deformation structure.

The attitude of a stratum is its position as measured by an angle of inclination, or dip of the bed-

ding, which is the angle between the bedding plane and a horizontal plane, and the compass direction, or strike, of the line formed by the intersection of the stratum with a horizontal plane (Fig. 15.7).

Dip The dip of a bed is the direction of maximum slope of the bedding plane and the angle of this slope. These are called dip direction and dip amount, respectively. For example, a bed may dip into the ground 23° in a direction S60°E. The dip amount is measured down from an imaginary horizontal plane (Fig. 15.7). A standard citation for amount and direction of dip is to record that this bed dips 23° S60°E.

Strike The strike of a bed is the compass direction of a level line drawn on the inclined bed (direction *BD* in Fig. 15.7). Strike direction is always perpendicular to dip direction. The strike direction conventionally is recorded in the north half of the compass, such as N30°E. The term strike probably originated as an English folk expression among farmers and quarrymen who noted that an inclined ledge "strikes across the countryside," much as informal language today

FIGURE 15.7 *Diagram illustrating the dip and strike of a tilted stratum. Angle ABC is the dip amount. BD, a horizontal line, is the direction of strike. The dip direction is toward the right, perpendicular to the direction of strike. The arrows represent a standard strike and dip notation symbol used by geologists.*

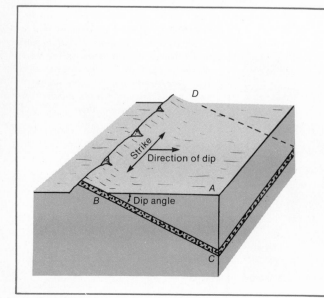

would describe a hiker striking out across an open field.

An instrument called a Brunton compass is used to measure strike and dip. It utilizes levels and a compass to determine the horizontal strike line and its direction, and it contains a clinometer for measuring the amount of dip.

A complete attitude notation for a bed would be recorded as strike N30°E, dip 23° S60°E. Horizontal strata do not have a strike direction or dip. Why is this so?

Deformation Structures

Distortion and other movements of solid rocks by forces within the earth are called diastrophism. The structures produced by diastrophic forces may be classified by origin as follows:

From plastic deformation (bending):
  Gently tilted and warped structures
  Folds
From brittle deformation (fracturing):
  Joints (cracks opening with slight movement or openings of the fracture surface)
  Faults (significant movement along the fracture surface)
From regional metamorphism induced by earth movements:
  Rock cleavage (planes of weakness caused by parallelism of mineral grains in response to pressure)

Gently Tilted and Warped Structures   Most large areas of sedimentary rocks were originally deposited as horizontal beds. If the surface on which they accumulated has not been uplifted uniformly, the strata are warped into gently dipping structures, such as irregular basins or domes, perhaps hundreds of kilometers in diameter, or broad arches over 100 kilometers across. The southern part of Michigan is a structural basin, with strata dipping inward on all sides a few meters per kilometer toward the center of the basin located near Saginaw. This large geologic structure is called the Michigan Basin. In Tennessee, the Nashville Dome is formed by beds that dip gently away from a structural summit for an average of 100 kilometers in all directions.

Very broad and gentle warps such as these represent rocks that yielded by plastic deformation to force gradually applied over millions of years. The flanks of such broad structures over areas many square kilometers in size have uniform dips.

Folds   When strata are subjected to pressures beyond their elastic limit, they may yield slowly by plastic deformation into a series of folds with alternating crests and troughs. Very weak strata may crumple into folds a few centimeters in size; thicker sequences of beds can develop folds several kilometers across. The principal forms of folded structures are monoclines, anticlines, and synclines, plus small domes and basins.

A monocline (Fig. 15.8) is a local steepening of dip in gently inclined or flat strata. Monoclines are common in the Colorado Plateau region of Arizona, Utah, Colorado, and New Mexico.

Where strata are arched up as an elongate upfold, they form an anticline (Figs. 15.8, 15.9, and 15.10). A downfold, or an elongated troughlike structure, is a syncline (Fig. 15.11). Each of these structures may have various modifications in shape. In most mountain ranges or former mountainous regions, folding, consisting of a large series of anticlines and synclines, is common.

The axial plane of a fold is the plane of symmetry of the structure, dividing the fold so that the axial plane bisects the angles formed by the two sides or limbs of the fold (Fig. 15.10). If the beds on both limbs dip about the same angle (axial plane vertical), the fold is symmetrical (Fig. 15.8). If one limb dips more steeply than the other (axial plane not vertical), the fold is asymmetrical. An overturned anticline or syncline has one limb that has been folded past the vertical so that, locally, older beds are found above younger beds (reversing the normal order of superposition of strata). A recumbent fold is an overturned fold in which the limbs are essentially horizontal. An isoclinal fold is one in which the two limbs are parallel (Fig. 15.8).

The crest of an anticline can be identified by an imaginary line connecting the highest points on the bedding surface along the axis (the straight-line summit of the fold in Fig. 15.10). On either side of the line, dips are in opposite directions. A similar line for a

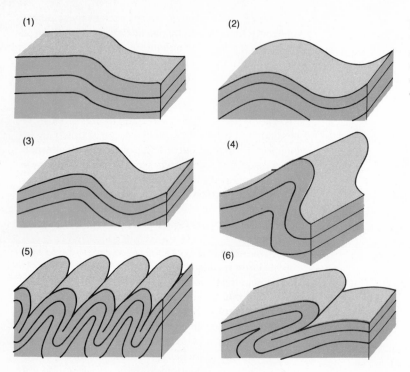

FIGURE 15.8 *Diagrams illustrating various types of elongate folds. (1) Monocline; (2) symmetrical anticline and syncline; (3) asymmetrical anticline and syncline; (4) overturned anticline; (5) isoclinal folds; and (6) recumbent fold.*

FIGURE 15.9 *Comb Ridge anticline, on San Juan River, Utah. (Spence Air Photos.)*

syncline traces the low points on a folded bed along the trough of a syncline.

All the folds shown in Fig. 15.8 extend indefinitely in the third dimension parallel to the axes. These idealized situations seldom occur in nature. Usually, the trough of a syncline is bent upward from some lowermost point so that the common shape of a bed folded synclinally is like a canoe (Fig. 15.11). When the trough is inclined, the fold is said to be plunging. The youngest beds are at the middle of a syncline (Fig. 15.11).

An anticline plunging in both directions from a high point has strata resembling the shape of an upside-down canoe (Fig. 15.12). The oldest beds are at the middle of an anticline (Fig. 15.13).

A dome is a roughly symmetrical upfold in which the beds dip in all directions outward from a central point (Figs. 15.13 and 15.14). Circular, completely symmetrical, domes are rare (Fig. 15.14), but somewhat elongated, oval-shaped domes are common. The Black Hills of South Dakota is such a dome. The high elevations of the Black Hills are caused by a core of

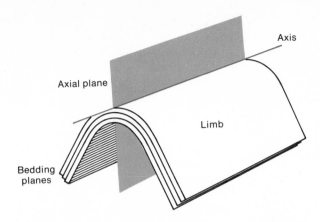

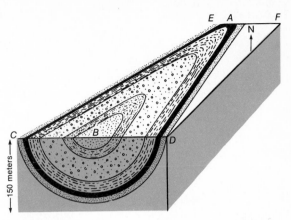

FIGURE 15.10 *Three-dimensional diagram showing the geometry of a fold. The axial plane is an imaginary plane that divides the fold into two symmetrical parts.*

FIGURE 15.11 *Block diagram of a syncline plunging toward the southwest. Plane ECDF represents the flat ground surface. At point (A) the formation shown in black is at the surface (at the troughline), but at (B) the same formation is more than 100 meters below the surface because of the plunge of the fold.*

FIGURE 15.12 *Aerial photograph of an eroded plunging fold in North Africa. The anticline at the bottom center is canoe shaped in map view and plunges toward the top of the picture. The sinuous ridge in the right part of the view is a folded bed resistant to erosion. Because of lack of vegetation, the structure can be seen easily. (U.S. Air Force.)*

FIGURE 15.13 *Aerial photograph of an eroded dome in North Africa. The beds dip outward in all directions from the crest of the dome at C. The oldest beds are at the center of the circular structure. (U.S. Air Force.)*

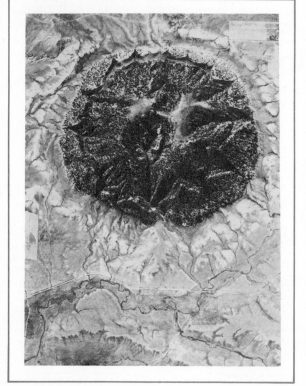

FIGURE 15.14 *Aerial view of Green Mountain, Cook County, Wyoming. This is a laccolith located in the northern Black Hills. The rocks shown are sedimentary. Erosion has not exposed the igneous core as yet. (University of Illinois Air Sterogram 172.)*

resistant metamorphic and igneous rocks protruding through the center of the uparched dome. This dome has been deeply dissected by erosion, and so its internal structure is readily observed. If a domed shape is elongated so that it is more than 3 times longer than wide in map view, the fold is called an anticline rather than a dome.

Symmetrical dome structures a few kilometers in diameter are common along the Gulf Coast from Texas to Mississippi and in the northern Gulf of Mexico. These are arched over central columns of salt 6 or 8 kilometers high and are called salt domes. The occurrence of petroleum, as well as of commercial deposits of salt and sulfur, is closely related to the salt-dome structures.

A structural basin is the opposite of a dome. It is a concave structural depression in which the strata dip toward the center from all sides, rather than away from the center. In a structural basin the strata resemble a stack of saucers, each one smaller in area than the next

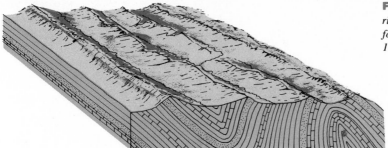

FIGURE 15.15 *Erosional mountain ridges formed by the differential erosion of folded sedimentary strata. (After Matthes, 1930.)*

one below. Small basin structures are common. Most structural basins are rather large, gentle downbends, such as the Michigan Basin.

The term basin is also used with reference to erosional depressions, topographically low areas surrounded by higher terrain. Some topographic basins coincidentally are underlain by rocks with the configuration of a structural basin, such as the Big Horn Basin of Wyoming. Other basins have contrasting structure; for example, the Nashville Basin of Tennessee is an erosional feature of the Nashville Dome. In this case the central portion of the dome consists of rocks less resistant to erosion than those around the edges of the dome.

(a) Fold Systems   Most mountainous regions are characterized by great upward folds that extend for hundreds, or even thousands, of kilometers. A large system of folds exists, including both anticlines and synclines, in which the dominant fold form is anticlinal. Similarly, large fold systems exist in which the dominant fold form is synclinal. These systems may contain individual anticlines and synclines several kilometers across and tens of kilometers long.

(b) Folds and Topography   It would seem logical to assume that high topography results from an upfolding of the ground and that low terrain is produced by downwarping. This is not always the case. High topography is generally upheld by resistant rocks, which are not so easily eroded as the bedrock underlying valleys (Figs. 15.15 and 15.16). Some mountains are indeed anticlinal in structure, but others are underlain by a syncline (Fig. 15.17). Valleys also may be anticlinal or synclinal (Figs. 15.17 and 15.18). Some valleys pursue courses completely independent of the structure of the underlying rocks, cutting across fold trends at a variety of angles, but this is uncommon.

(c) Causes of Folding   Isolated gentle anticlines or synclines may result from vertical crustal movements. Some domes or anticlines result from arching of strata above an intrusion of magma or by the upward movement of a salt core.

Intense folding commonly occurs in a fairly large region with many more or less parallel anticlines and synclines. Such fold groupings are thought to result from horizontal compressive forces in the earth's crust. Similar fold patterns can be produced in a rug by applying horizontal force to push it across the floor.

Such compressive forces shorten the crust of the earth. This is illustrated in Fig. 15.19, which shows that the bed represented by AB was originally 10 kilometers long but after folding, it measured only 6.6 kilometers in horizontal distance. Folding decreased its length by one-third. The total shortening produced by

FIGURE 15.16 *Zigzag ridge produced by the erosion of plunging folds. This is common in central Pennsylvania.*

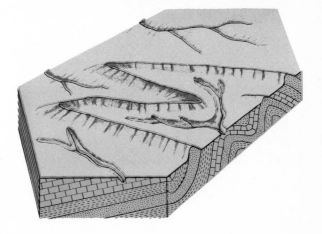

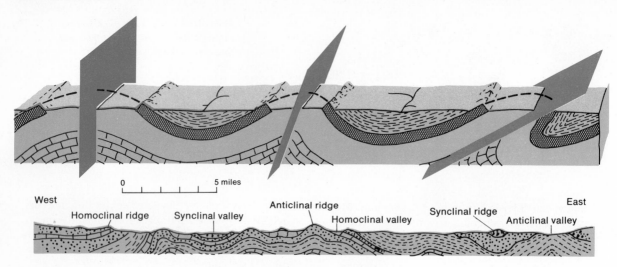

West

Homoclinal ridge   Synclinal valley   Anticlinal ridge   Homoclinal valley   Synclinal ridge   Anticlinal valley   East

FIGURE 15.17 *Mountains and valleys underlain by different types of structure. Geologic cross section of Allegany County, Maryland. Note that the ground
topography does not necessarily reflect the configuration of rock folds. (After
Maryland Geological Survey, Geologic Map of Allegany County, 1956.)*

FIGURE 15.18 *Aerial view of the Virgin
Anticline in Utah. Note that the long
valley is within the anticlinal core. (J. S.
Shelton and R. C. Frampton photo.)*

A   B

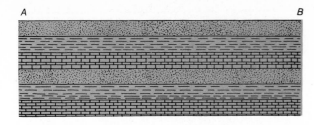

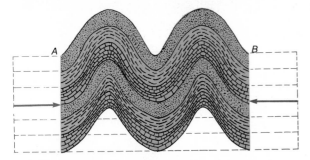

FIGURE 15.19 *The upper figure shows a system of horizontal beds as originally deposited. The lower figure shows the same strata after folding. Horizontal distance AB in the lower figure is two-thirds as long as AB in the upper figure; the shortening is one-third.*

crustal folding of the Appalachian Mountains near Harrisburg, Pennsylvania, is estimated to be 160 kilometers, and the crustal shortening in the Alps, 280 kilometers.

Deformation seems more intense at depth in the earth, and so the ultimate causes of deformation must be from inside the earth, probably some expression of the earth's internal heat energy.

Joints A joint is a fracture in which the total movement was perpendicular to the fracture surface, without any appreciable movement in a direction parallel to the fracture surface (Fig. 15.20). Slight motion of the joint produces a barely visible crack, but major jointing can open a distinct gap in the rock.

Most joints are related to tension forces pulling the rock apart. Figure 15.21 shows how joints open up along the crest of an anticline. The top and bottom of the bed originally had the same length. The top is stretched and cracked by joints, however, because the distance along the top surface of the folded bed is

greater than the folded length of the bottom surface. In some situations the rocks are plastic enough to keep joints from forming.

Slight warping of strata often will produce a pattern of intersecting joints (Figs. 15.22 and 15.23). Such joints form recognizable geometric patterns in outcrop and are important in controlling movement of groundwater. Cave systems commonly are solution enlargements of joint patterns in limestone and other soluble rocks. The more severe or more frequently repeated the brittle deformation, the more closely spaced are the joints.

As an igneous rock cools, it decreases slightly in volume. Joints are common in such rocks. These are most evident in the polygonal patterns of columnar jointing of a lava flow, sill, dike, or volcanic neck. These joint planes are oriented perpendicular to the surface of maximum cooling. Devil's Tower in Wyoming (Fig. 6.7), Devil's Postpile in California, the Palisades along the Hudson River, and the Giant's Causeway of northern Ireland are excellent examples of columnar jointing.

FIGURE 15.20 *Types of motion in rock fractures. (a) A joint moves apart perpendicular to fracture surface; (b) in a fault, the movement is parallel to the fracture surface.*

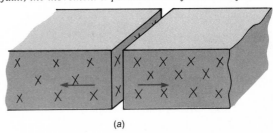

(a)

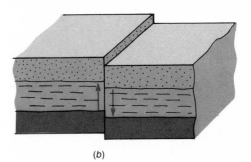

(b)

FIGURE 15.21 *Block diagram and photo of a simple anticline, showing tension joints along the crest.*

A batholith also shrinks slightly and becomes jointed as it continues to cool after solidifying. These joints will affect topographic development and also facilitate quarrying of the rock.

Some joints are formed by compressive forces, occurring as shear fractures that have not been offset by movement parallel to the fracture surface. They are not nearly so common as tension joints, however, because the rocks stretch less than they compress.

Faults   A fault is a rock fracture with detectable movement parallel to the surface of the break (Fig. 15.24). Faults occur in all types of rock, but they are best seen in sedimentary rocks because the offsetting of definite strata is readily recognized. Abrupt interruption in continuity of the border of an igneous rock body also causes a geologist to suspect a fault. The amount of fault displacement may be only a fraction of

a centimeter or several kilometers. In either case, it usually is not possible to establish whether one side of the fracture surface stood still while the other side moved, or whether both blocks of rock moved. What is certain is that one side of the fracture has moved relative to the other (Fig. 15.24).

The fracture along which displacement occurred is the fault surface, or if it seems to exist as a single surface, it may be called a fault plane. Traced along its entire length, a fault eventually dies out. The displacement commonly is greatest near the middle of the fault length and diminishes toward the fault extremities.

Faults may occur as isolated fractures, or they may form a group of distinct faults arranged in a definite pattern. The latter phenomenon constitutes a fault zone instead of a single distinct break, and the total displacement is the combined movement along a number of closely spaced, parallel, small faults.

FIGURE 15.22 *Vertical jointing in Devonian siltstone, Cayuga Lake, New York. Note that two sets of joints are parallel with and at right angles to the direction of view. (Kindle, U.S. Geological Survey.)*

Many faults break the ground surface, as well as fracturing the rocks beneath. When one side of the fault moves up relative to the other side, a cliff or fault scarp may form (Fig. 15.25). The present height or prominence of such a scarp depends not only on the amount of fault displacement but also on how recently it occurred in geologic time and the resistance of the rocks to erosion. In many places erosion has reduced the upthrown fault block to about the same level as the downfaulted block so that the fault is not topographically visible. Long after faulting movement ceases, erosion may etch out the opposing blocks so that the block containing the more resistant rock stands up in topographic relief. The resulting fault-line scarp may face in the same or in the opposite direction as the original scarp (Figs. 15.26 and 15.27).

The enormous pressures involved in faulting normally keep the surfaces of fault blocks in close contact. Friction between the blocks smooths and polishes the fault plane to slick surfaces, called slickensides. These nearly smooth slickenside surfaces are grooved or striated parallel to the last motion of the

FIGURE 15.23 *Intersecting joint systems in northern Australia as seen from the air. Joints have been enlarged by weathering and erosion to form gullies. (U.S. Air Force.)*

Normal fault

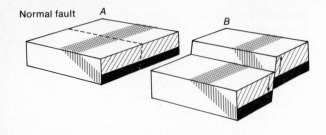

Reverse fault

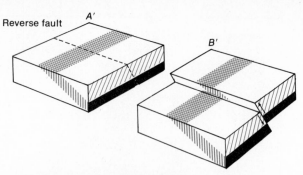

Normal fault

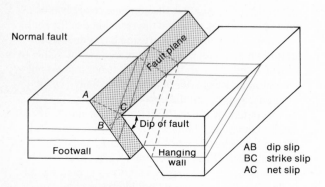

AB dip slip
BC strike slip
AC net slip

FIGURE 15.24 *Series of block diagrams showing successive stages of development of a normal and reverse fault. A and A' are before faulting, with incipient fault plane shown by dashed line. B and B' are immediately after faulting, with relative movement along fault plane indicated by arrows.*

FIGURE 15.25 *Fault scarp produced during earthquake in 1954 near Fallon, Nevada. (J. Dennison photo.)*

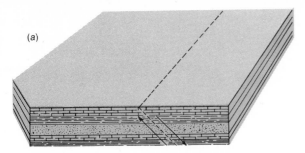

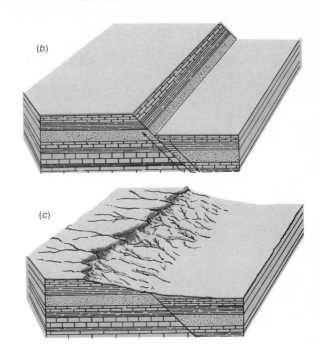

FIGURE 15.26 *Development of fault-line scarp. Unfaulted block (a) is cut by a normal fault that produced fault scarp (b). Erosion of uppermost rocks produced fault-line scarp (c) facing in same direction as original fault scarp. Further erosion leaves behind high terrain formed of resistant rock, producing a fault-line scarp.*

FIGURE 15.27 *Fault scarp forming shore of MacDonald Lake near Great Slave Lake, MacKenzie District, Canada. The faulted block at the upper left has evidently been uplifted relative to the other block. (National Air Photo Library, Canada. Print A1520-105R.)*

opposing fault blocks. These striations can be used to determine the direction of motion along the fault. When the surface is not clear-cut and definite, the opposing rocks may be more or less crushed when slippage occurs. Any resulting broken angular material is distributed along the fault surface.

Faulting movements cause displacement of rocks along fractures that may be vertical, horizontal, or at any intermediate angle. Very often, high-angle faults may be associated with epeirogeny; low-angle faults are frequently involved in orogeny.

Although a few faults are vertical, most dip in such a manner that one wall or fault block overhangs the other. Centuries ago, miners adopted the custom of calling the rocks above the fault plane the hanging wall and those below the fault plane the footwall (Fig. 15.28). These mining terms are applied to both faults and veins. The footwall is the rock on which a miner stands as he works a vein, and the hanging wall is above his head.

The block that appears to have moved upward is the upthrown side of a fault, and that which appears to have moved downward is the downthrown side.

Faults are named according to the relative motion of the opposing fault blocks (Figs. 15.24, 15.29, and 15.30). Each fault type is classified on the basis of the relative motion of the hanging versus the footwall blocks, whether the motion is vertical, horizontal, diagonal, or rotational, and the angle of the fault plane.

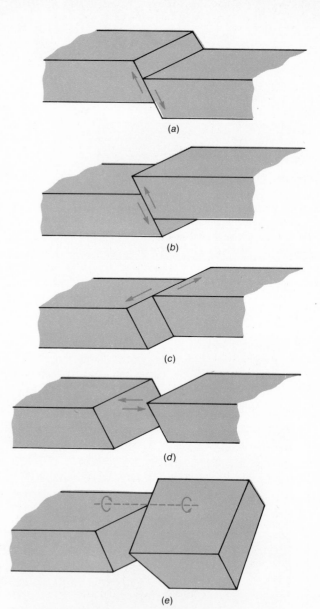

(a)

(b)

(c)

(d)

(e)

FIGURE 15.29 *Types of faults, named according to relative motion of the fault blocks. (a) Normal fault; (b) reverse fault; (c) strike-slip fault; (d) oblique-slip fault; and (e) rotational fault.*

FIGURE 15.28 *Diagram showing the relation of the hanging wall to the footwall of a fault. The miner stands on the footwall, and the hanging wall "hangs" above his head. A mineral vein has been deposited along the fault surface.*

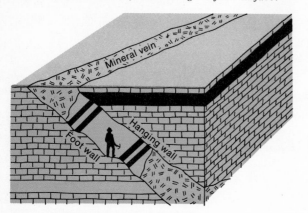

FIGURE 15.30   *Normal fault in Pleistocene gravel in Utah. (Richard A. Gilman photo.)*

Other fault structures often represent combinations of these basic types. The horst and graben fault structures are just such an example. A block depressed between two faults is a graben (Fig. 15.31), and a block raised between two faults is a horst (Fig. 15.32). These may be isolated features, or families of parallel faults may occur with both upward and downward movement so that a series of horsts and graben is developed. Horsts and graben are found in almost every complexly faulted area.

The Death Valley region of California contains many examples of block faulting. It is part of the Basin and Range geologic province in which large-scale block faulting is the dominant structure. The Basin and Range province encompasses much of Nevada, western Utah, western and southern Arizona, eastern California, and parts of New Mexico, Idaho, and Oregon. The great normal fault at the east front of the Sierra Nevada marks the western border of the Great Basin province, and the fault at the west front of the Wasatch Range near Salt Lake City is the eastern boundary. The mountain peaks in the province are uplifted blocks and the basins are downdropped blocks.

Another excellent example of a downfaulted block structure is the Great Rift Valley of eastern Africa, which consists of a series of downfaulted blocks that are now partially covered by lakes, including Lake Albert, Lake Nyasa, and Lake Tanganyika.

FIGURE 15.31   *Diagram of a graben set (A). After extensive erosion, block (A) is reduced to the level of a peneplain, and the more resistant rock is removed from the highlands. Subsequent uplift of the region will cause stream erosion of the softer limestone along the margins of the sketch, leaving the resistant rock forming high ground. This produces an inversion in topography of the original graben structure. (After Lahee, 1952.)*

FIGURE 15.32   *Diagram of a horst at (A). After extensive erosion, block (A) is reduced to the level of a peneplain, and the more resistant rock is removed from the high block. If the peneplain is uplifted and streams rejuvenated, a valley may develop as shown in block (B) in the foreground. Thus the lowlands of block (A) may become the highlands of block (B). Block (B) is still a horst in structure. (After Lahee, 1952.)*

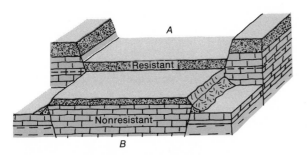

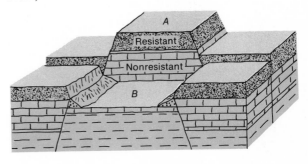

FIGURE 15.34 *Chief Mountain in northwestern Montana, an erosional remnant of the Lewis overthrust mass. Ancient Precambrian metamorphic rocks rest here on relatively young Cretaceous sediments. (Burlington Northern Railway photo.)*

Lake Tanganyika, which is about 600 kilometers long, is more than 1,300 meters deep, with its floor 360 meters below sea level. A branch of this African fault system extends northward east of the Mediterranean Sea; the Jordan Valley and Dead Sea occupy the site of a large graben. For much of its length this valley is below sea level. The elevation of the surface of the Dead Sea is 430 meters below sea level, and the shore of this evaporating lake is the lowest spot of land on earth. The Dead Sea graben is a result of the down-sinking of a large block of rock coincidental with large-scale strike-slip faulting.

The amount of movement in faulting is quite variable, depending on the forces causing the faulting and time. For example, the San Andreas fault in California (Fig. 15.33) is a strike-slip fault, and the total displacement along this fault since the late Eocene epoch, some 40 million years ago, has been more than 375 kilometers.

Some thrust faults (Fig. 15.34) associated with mountain ranges have been traced for hundreds of kilometers (Fig. 15.35), and they may have a displacement exceeding 30 kilometers. The Lewis overthrust (Chief Mountain) fault of Montana has a displacement of at least 16 kilometers (Fig. 15.34).

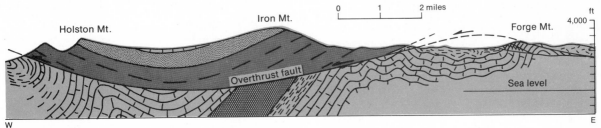

FIGURE 15.35 *Generalized cross section from Bristol, Virginia, to Mountain City, Tennessee, showing extensive overthrusting from east to west. (After Butts, 1933.)*

Not all abrupt terminations of rock masses or angular discordance of beds is produced by faulting. An unconformity may superficially resemble a fault (Fig. 11.25), but the association of slickensides with a fault and the fact that strata above an unconformity usually contain clastic fragments of the pre-unconformity rocks provide criteria for distinguishing faults from unconformities.

FIGURE 15.36 *Bad Water, a saline remnant of an ancient lake in the lowest part of Death Valley, California, 94.5 meters below sea level. The rough, flat area in middle ground consists of various salts precipitated from saturated water. Telescope Mountain is in the background. (Frasher's Inc.)*

Crustal Movements and Lake Development

It has been noted that crustal movements have created many of the largest and deepest lake basins of the world, such as Lake Baikal, Siberia, with a depth of about 1,740 meters. Downfaulting of a chain of narrow blocks of the earth's crust, the rift valley of Africa and Asia, formed the basins of many African lakes, including Tanganyika, Leopold, Nyasa, as well as the Dead Sea and the Sea of Galilee. In the United States, the basins of Upper Klamath Lake in Oregon and Lake Tahoe on the California-Nevada border are among those formed by faulting. Lake Tahoe, with a depth of 503 meters, occupies a basin formed when an earth block settled several thousand meters. Reelfoot Lake in western Tennessee was formed by depression of the valley floor during the earthquakes of 1811-1818.

Numerous ancient and modern saline lakes occur in the downfaulted blocks of the Basin and Range area. Bad Water Lake (Fig. 15.36) is a saline remnant of an ancient lake in Death Valley. The Great Salt Lake and the Bonneville salt flats are another example. Lakes of this nature are formed because the blocks have been downdropped so extensively that streams cannot drain into the sea.

Crustal movement may form a lake by raising a barrier across a river's path, as well as by downfolding or faulting. Lake Erie is an example of a lake that has been deepened 8 meters or more by the upward movements of the Niagara escarpment. This movement occurred as the earth's crust rebounded from the unloading of a kilometer or more of glacial ice that melted out of the Lake Ontario area.

REFERENCES

Badgley, P. C., 1965, *Structural and Tectonic Principles,* Harper & Row, Publishers, Incorporated, New York.

Billings, M. P., 1972, *Structural Geology,* 3d ed., Prentice-Hall, Inc., Englewood Cliffs, N. J.

Butts, C., et al., 1933, *Southern Appalachian Region,* Guidebook 3, 16th Session, International Geological Congress.

Dennison, J. M., 1968, *Analysis of Geologic Structures,* W. W. Norton & Company, Inc., New York.

Eardley, A. J., 1962, *Structural Geology of North America,* 2d ed., Harper & Row, Publishers, Incorporated, New York.

Griggs, D. T., and Handin, J. (eds.) 1960, *Rock Deformation—A symposium,* Geological Society of America Memoir 79.

Heard, H., 1960, *Transition from brittle fracture to ductile flow in Solenhofen Limestones as a function of temperature, confining pressure and interstitial fluid pressure,* in Griggs, D. T., and Handin, J. (eds.), *Rock Deformation—A symposium,* Geological Society of America Memoir 79.

Hills, S., 1963, *Elements of Structural Geology,* John Wiley & Sons, Inc., New York.

Lahee, F. H., 1961, *Field Geology,* 6th ed, McGraw-Hill Book Company, New York.

Matthes, F. E., 1930, *Geologic History of the Yosemite Valley,* pp. 94-98, United States Geological Survey Professional Paper 160.

———, 1937, *Exfoliation of massive granite in Sierra Nevada of California,* pp. 342-343, Proceedings, Geological Society of America, 1936.

Spencer, E. W., 1969, *Introduction to the Structure of the Earth,* McGraw-Hill Book Company, New York.

Thornbury, W. D., 1965, *Regional Geomorphology of the United States,* John Wiley & Sons, Inc., New York.

CHAPTER 16

MOUNTAIN BUILDING

Mountains represent a combination of regional uplifting to a high elevation followed by erosion of valleys in the softer rocks to leave the resistant materials standing as mountains peaks.

A mountain is an area that is substantially higher in elevation than the surrounding terrain and that has steep slopes. If the difference in elevation (relief) exceeds 300 meters, most geographers consider the name mountain to be appropriate. Hills are irregular terrain of less abrupt proportions.

In the United States, the Sierra Nevada, the Coast Ranges, the Rockies, the volcanic peaks of Hawaii, the Brooks Range in Alaska, and the Appalachians are mountains. The rolling terrain of upstate New York or parts of Wisconsin and Alabama is considered hill country.

Mountainous areas are of two types: constructional and erosional. The constructional areas are built up from a rather flat surface, chiefly as volcanic peaks. Erosional mountain areas are remnant landscapes developed by the selective erosion by streams or glaciers. Forces within the earth cause uplifting of rock masses to a high elevation and subsequent deepening of valleys isolates mountain peaks. Such deeply eroded terrain commonly occupies thousands of square kilometers with scores of summits, forming a mountain range.

VOLCANIC MOUNTAINS

Some majestic volcanic peaks tower alone above their surroundings. Examples include Mt. Vesuvius in Italy (Fig. 6.2), Kilimanjaro in Africa, Mayon in the Philippines (Fig. 6.14), and Fujiyama in Japan. All are composite volcanoes and are representative of some of the greatest mountains on earth.

More often, the fractures that allow lava outpourings have a linear trend so that an elongate cluster of volcanoes results. The Cascade Range extending from northern California to southern British Columbia consists of hundreds of volcanoes, including several of extremely large dimensions (Fig. 16.1): Mt. Lassen, Mt. Shasta, Mt. Hood, Mt. St. Helens, Mt. Adams, Mt. Rainier, Glacier Peak, and Mt. Baker, listed from south to north. Of these, only Mt. Lassen has erupted in historical time. The Aleutian Range of Alaska is another example of an elongate chain of volcanic peaks.

FIGURE 16.1 *Map showing distribution of volcanic mountains in Washington, Oregon, and California. They are lined up atop the volcanic eastern portion of the High Cascade Range. (After H. Williams, 1962.)*

A crustal weakness in a broadly domed area of Arizona south of the Grand Canyon was the site of numerous eruptions shortly before the beginning of local recorded human history. Most eruptions in this area, called the San Francisco Volcanic Field, formed cinder-cone hills, but some attained mountainous heights, notably the ski resort area of San Francisco Mountain.

Most mountainous oceanic islands are volcanoes. The state of Hawaii consists of many volcanoes rising from the ocean floor (Fig. 6.1). Mauna Kea is the tallest mountain on the earth's surface, extending 5.4 kilometers above the floor of the Pacific and reaching 3.6 kilometers above sea level, for a total height of approximately 9 kilometers.

Many island chains rise as volcanic peaks along arcuate fractures near the continental edges (Fig. 6.16). These include the Kuril, Mariana, and Solomon Islands in the western Pacific, the Leeward Islands of the Caribbean, and the Sandwich Islands of the South Atlantic. These islands represent zones of sharp folds, with deep ocean trenches occurring seaward of the island arcs (Figs. 17.7 and 17.8). The islands are high points, often active volcanoes, along the crest of great anticlinal arches. Many other volcanic peaks of mountainous proportions are found within the ocean basins but are not tall enough to break surface above the average 3,800-meter depth of the oceans.

One of the most remarkable volcanic mountain chains is the Mid-Atlantic Ridge (Fig. 16.2). This broad mountainous zone extends north-south nearly the entire length of the Atlantic Ocean. It represents in part the accumulations of volcanic eruptions that poured out of a great fracture or rift on the ocean floor. Volcanic peaks projecting above sea level include the mountains of Iceland, Surtsey, the islands of Flores and Corvo in the western Azores, Ascension Island, the Rocks of St. Paul, and Tristan de Cunha. The tallest is Mount Pico of the Azores, towering over 2,500 meters above sea level. The base of Mount Pico is over 6,600 meters below sea level.

DOMED MOUNTAINS

The upward arching of a dome can form a mountainous area, especially if the center of the arch under erosion contains unusually resistant rock. Such a dome may be less than 1 kilometer to over 165 kilometers across. Domal mountains can originate in three ways (Fig. 16.3).

1. A batholith or stock is intruded into country rock, frequently less resistant material. The overlying rock may be arched upward by a concordant intrusion (Figs. 16.3*b, c* and 16.4), or more commonly it may cut across discordantly (Fig. 16.3*d*). Differential erosion leaves the igneous core standing in mountainous relief.

2. Laccoliths are arched concordant intrusions. Sundance Peak and Missouri Buttes in South Dakota are laccolithic mountains. The Henry Mountains in southern Utah are domed structures formed by a complex of laccolithic bodies extending outward from a central, trunk-shaped stock (Fig. 16.5). They stand singly or in clusters upon a desert plain that has an altitude of more than 1,600 meters.

There are five individual mountains, the highest of which is Mount Ellen, more than 3,600 meters high. The domes show considerable diversity in amount of erosion. Some are still partially or completely capped by overlying strata. In others, the intruded igneous rock is exposed at the crest of the structure, and steeply dipping strata encircle the uplift.

3. If strata accumulate unconformably above igneous and metamorphic rocks, and subsequent arching forms a domal structure, the resistant core may become exposed in raised relief after erosion breaches the dome (Fig. 16.3*d*). This is the origin of the central mountainous region of the Black Hills of South Dakota and the higher portions of the Big Horn Mountains of Wyoming.

BLOCK-FAULTED MOUNTAINS

Nearly parallel, steeply dipping faults can isolate rock masses so that they stand up as large mountainous blocks. The most notable example in North America is the series of fault-block mountains in the Great Basin

FIGURE 16.2   *A profile across the South Atlantic ocean, showing the broad mountainous zone of the Mid-Atlantic Ridge. The lower cross-sectional profile across the United States uses a similar spacing of reference points to allow a comparison to the Mid-Atlantic Ridge. (After Shepard, 1963.)*

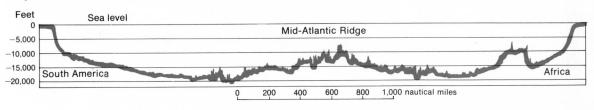

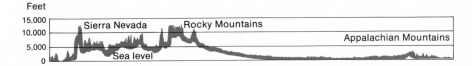

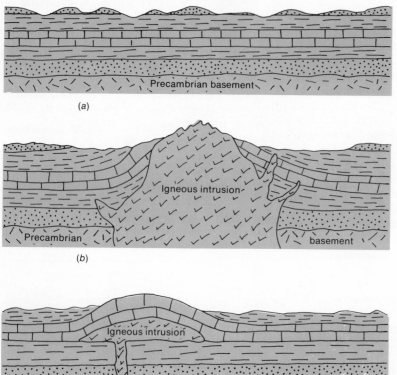

(a)

(b)

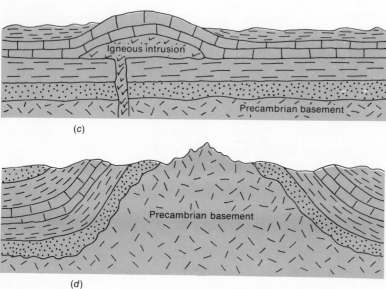

(c)

(d)

FIGURE 16.3 *Three types of domal mountains formed in originally horizontal strata. (a) Original strata overlying basement; (b) dome over stock (example: Packsaddle Mountain, Idaho, and Mole Hill, Virginia); (c) laccolithic dome (example: Henry Mountains, Utah); (d) dome formed by uparching of older igneous rocks buried uncomfortably beneath strata (example: Black Hills, South Dakota).*

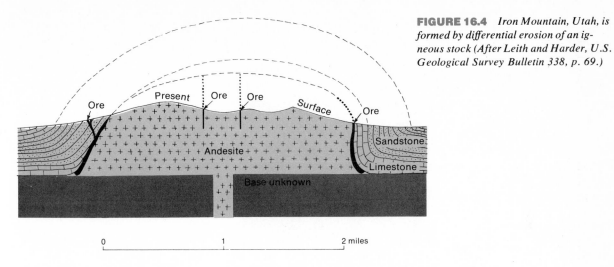

FIGURE 16.4 *Iron Mountain, Utah, is formed by differential erosion of an igneous stock (After Leith and Harder, U.S. Geological Survey Bulletin 338, p. 69.)*

FIGURE 16.5 *Cross sections of laccolith complex on north side of Mount Pennell in the Henry Mountains, Utah. (After Hunt, Averitt, and Miller, U.S. Geological Survey Professional Paper 228, 1953, p. 120.)*

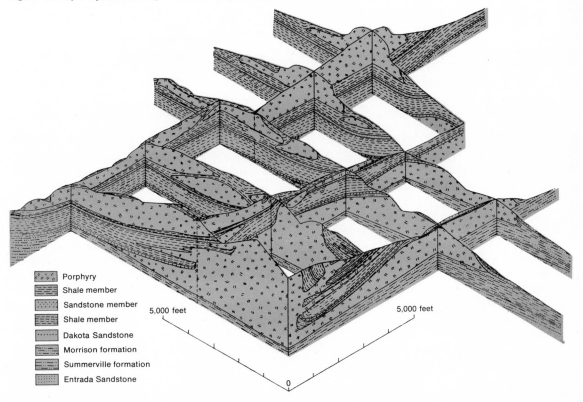

FIGURE 16.6   *Basin and range structure results in angular mountains standing in high relief surrounded by talus and alluvial fans. (Spence Air Photos.)*

region of Nevada and parts of adjacent states (Fig. 16.6). In most cases, block mountains represent segments of the earth that probably have maintained their positions while the adjoining blocks subsided after faulting. The Great Basin extends for nearly 1,800 kilometers north and south and for 830 kilometers east to west. The basin area is transversed by numerous mountain ridges generally extending north and south. Many of the ranges rise 1,000 to 1,600 meters above their bases and are outlined by the slightly eroded escarpments of normal faults. The normal faulting apparently resulted from tension in the earth's crust and subsequent collapse of large rock masses, rather than from compressive forces.

The abrupt change from valley floor to mountain slope, together with the uniform slopes of the mountain sides, is a striking characteristic of the basin and range topography (Fig. 16.6). In the more arid portions of the region, the slopes range from 20° to near vertical. In the southern part of the basin, many of the fault blocks have been so deeply eroded that their original angular blocklike form has been nearly destroyed. Many of the intermontane depressions there are so filled with rock waste that the mountains are partly buried by talus slopes derived from the weathering of

the fault escarpments. Sediments from alluvial fans deposited from intermittent streams also clog the depressions.

Faulting and crustal uplift of great magnitude produced the Sierra Nevada on the eastern border of California (Fig. 6.21). These mountains form a continuous mountain range about 125 kilometers wide and nearly 660 kilometers long. They represent a huge block of resistant, mostly granitic rock uplifted by faulting and tilted toward the west. The crest of the range is near its eastern margin, and the eastern fault slope is steep and clifflike. Where the crest is highest, the average eastward slope of the Sierra exceeds 206 meters per kilometer. In contrast to the steep eastern front, the western slope of the Sierra descends gradually about 46 meters per kilometer to the east margin of the Central Valley of California some 83 kilometers to the west. Mount Whitney rises to approximately 4,830 meters, the highest elevation in the United States, excluding Alaska. All structural features of the east front of the Sierra Nevada indicate that it is a huge fault scarp (Fig. 16.7). The displacement did not take place along a single fracture but, rather, along a fault zone with numerous compound faults. This type of displacement left the spurs and offsets of the present topography. The abundance of recent earthquakes in this area and the recorded elevation changes accompanying some of them indicate that uplift along the Sierra faults is still an active process.

OROGENIC SYSTEMS

Most of the world's great mountain ranges occur in narrow belts, several hundred kilometers wide and hundreds to thousands of kilometers long (Fig. 16.8). These are all characterized by intensely folded anticlines and synclines, thrust faults, and commonly, regional metamorphism and igneous intrusion of grani-

tic batholiths. Such a complex structure constitutes an orogenic system in which individual mountains tend to be long ridges of upturned resistant strata, metamorphic rocks, or granite etched out in erosional relief.

Orogenic systems occur on every continent in varied stages of erosional destruction. In the United States, the mountainous areas of the Appalachians, Rockies, and Ouachitas are orogenic systems that were formed at different times in the geologic past. Other orogenic systems include the Andes, the Himalayas, the Alps, the Atlas Mountains of northern Africa, the Great Divide Range of eastern Australia, and the mountains of the Antarctic Peninsula.

FIGURE 16.7 *Structure sections across the Sierra Nevada, California. Top, Granite Peak to the Great Valley; middle, Carson Range, Nevada, to the Great Valley, California; Mono Lake, California, to the Great Valley. (Redrawn after fig. 82, p. 146, in Phillip B. King,* The Evolution of North America, *copyright © by Princeton University Press. Reprinted by permission of Princeton University Press.)*

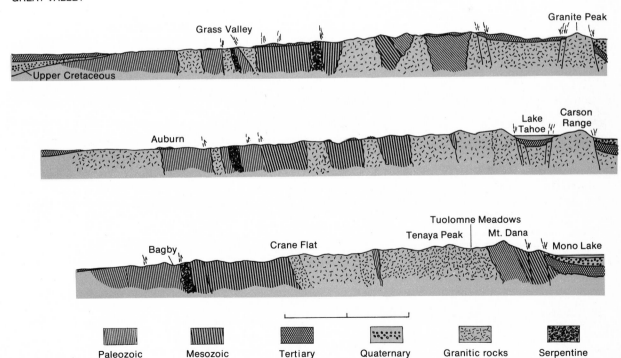

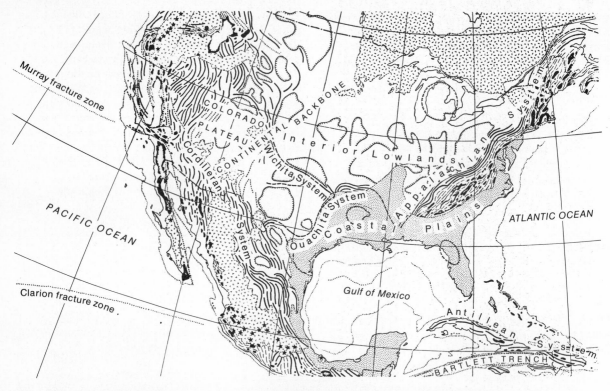

Murray fracture zone

COLORADO
PLATEAU
CONTINENTAL BACKBONE
Cordilleran
Interior Lowlands
Wichita System
Appalachian System
System
Ouachita System
Coastal Plains

PACIFIC OCEAN

ATLANTIC OCEAN

Clarion fracture zone

Gulf of Mexico

Antillean System

BARTLETT TRENCH

CENTRAL STABLE REGION

 Shields

Exposed ancient rocks that were long ago stabilized and worn down.

 Interior lowlands

Sedimentary rocks, gently tilted or folded. Principal domes and basins outlined by hachured lines.

OROGENIC BELTS

Marginal parts

Sedimentary rocks, largely of miogeosynclinal origin, folded and thrust-faulted. Lines indicate generalized trends of structures; dashed where concealed by later deposits, or submerged.

Interior parts

Sedimentary and volcanic rocks, largely of eugeosynclinal origin, partly metamorphosed. Lines indicate generalized trends of structures.

 Acidic plutonic rocks

Forming small to large bodies in interior parts of orogenic belts.

 High-angle faults

Through-going faults of great length. Includes normal faults, transcurrent faults, and high-angle thrust faults.

POST-OROGENIC FEATURES

 Volcanic rocks

Mainly laid over deformed rocks of orogenic belts. Of Tertiary and Quaternary age.

 Volcanoes

Active or recently active.

 Coastal plains

Areas of young sediments along coasts, with low seaward dips. Outer edges continue beneath continental shelves.

OCEANIC FEATURES

Areas mainly floored by sialic crust

Includes epicontinental seas, continental shelves, and continental slopes.

Areas mainly floored by simatic crust

Ocean basins, mainly lying at depths of more than 10,000 feet.

Deep-sea trenches

Line indicates axis of trench.

Fault or fracture zone

Mainly inferred from submarine topgraphy.

FIGURE 16.8 *Tectonic map of the United States. (Redrawn after plate 1 in Phillip B. King,* The Evolution of North America, *copyright © 1959 by Princeton University Press. Reprinted by permission of Princeton University Press.)*

Characteristics of the Appalachian orogenic system are shown in map view in Fig. 16.9 and in cross section in Fig. 16.10. A central core showing the most intense metamorphism was the site of the most active igneous intrusion. West of this is a belt of Paleozoic sediments that have been intensely folded and faulted (Figs. 16.9, 16.10, and 16.11). Information is generally lacking east of the central metamorphic core because that portion of the Appalachians was buried unconformably beneath the Coastal Plain sediments. The sediments were deposited during the Cretaceous and Tertiary periods when the western Atlantic Ocean inundated the edge of the continent (Fig. 16.10).

In 1857, James Hall noted a peculiar aspect of the Appalachians. The sedimentary accumulation in the folded belt is 4 or 5 times thicker than the sedimentary layers of equivalent age in central United States. The rocks deformed by the Appalachian orogeny seem to occupy a geologic area representing an unusually thickened accumulation of strata. Such an elongate area of thickened strata that later becomes a mountain system is called a geosyncline. Nearly all the world's orogenic systems with their associated intense folding and thrust faulting occur at sites of former geosynclines in which thousands of meters of strata were deposited (Fig. 16.12a and b). Interestingly, the geosynclinal sediments generally show evidence of a shallow-water origin, even though the layers are thousands of meters thick. Shallow-water marine fossils and limestones are common, and mudcracks frequently are found in the geosynclinal rocks. Apparently, the slowly downsinking geosynclinal trough was kept filled nearly to sea level by sedimentation. Some geosynclines do show evidence of local deep-water environments, which are indicated by turbidity-current deposits (with characteristic graded bedding and current markings) and radiolarian cherts.

Orogenic Cycle

There seems to be a definite sequence or cycle in the development of an orogenic system (Fig. 16.13). First, an area 1,000 or more kilometers long and 200 or 300 kilometers wide subsides beneath the sea, and sediments accumulate to 6,000 to 20,000 meters in thickness (Figs. 16.13a and b and 16.12b).

The geosynclinal sediments may include limestone, chert, or other chemical accumulations and great thicknesses of clastic sediments derived from nearby land areas. These clastic sediments may come from volcanic islands. The origin of these sediments is similar to those accumulating today in linear, narrow, and deep oceanic trenches adjacent to volcanic island arcs. In other cases, the sediments come from weathered masses of continental rocks. Today, similar sediments are being deposited along the Atlantic and Gulf coastal shores of the United States (Fig. 16.12b). The sediment accumulation phase in the geosynclinal lasts hundreds of millions of years.

The principal deformation of a geosyncline, called an orogeny, takes place over a period of a few million years. The site of thickest sediment accumulation is deeply buried. It is subjected to increased temperatures and compressive forces of dynamic metamorphism. This part of the geosyncline is characterized by intensely contorted slate, schist, gneiss, quartzite, and marble. Granitic batholiths and stocks are intruded as magma from kilometers beneath the surface or, perhaps, represent the extreme product of metamorphism. Away from the central part of the geosyncline, the temperatures are not high enough to produce metamorphism, but the strata are folded into a series of anticlines and synclines (Fig. 15.17). If the forces exceed the folding strength of the rocks, thrust faults may form (Figs. 15.34 and 15.35). The Wills Mountain anticlinal structure in Maryland can be traced for a total of 260 kilometers north and south. The Saltville thrust fault can be traced for about 600 kilometers from Virginia to Alabama. This thrust fault has several kilometers of horizontal displacement.

The Coast Ranges of California appear to be in an active orogenic deformation phase. Pleistocene sediments in these mountains have been folded in the past 1 to 1.5 million years, and they now exhibit steep dips (Fig. 16.14).

After orogeny has folded the mountains and uplifted them, erosion continues to attack the raised terrain, carrying sediment toward lower elevation. The earth's surface responds to the resulting weight loss by uplift through isostatic adjustment. Ultimately, enough material becomes eroded so that the mountains no longer continue to respond by regional uplift,

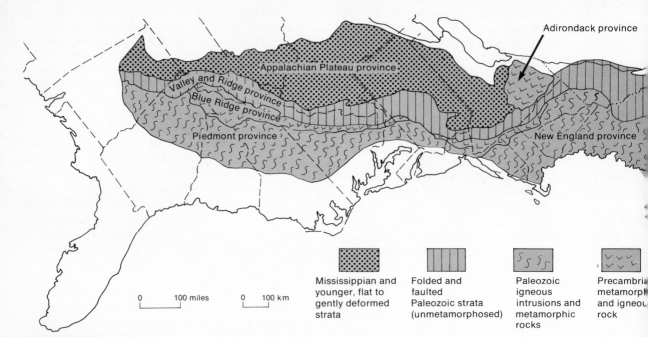

FIGURE 16.9   *Map of the Appalachians showing major geologic features and provinces.*

FIGURE 16.10   *Geologic section from Cincinnati to the Atlantic Ocean, showing the structure of the Appalachian Mountains.*

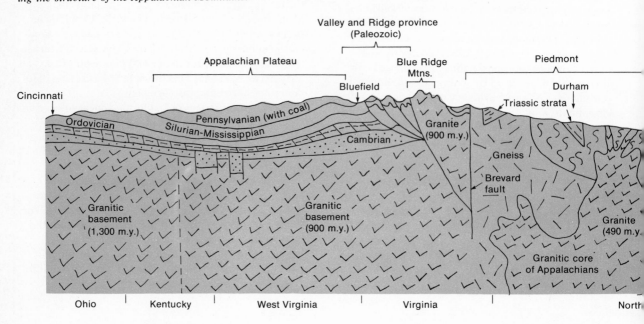

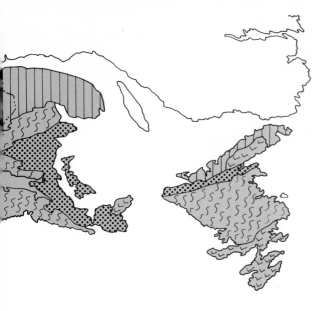

and the topography of the mountain range is obliterated by erosion. Convincing evidence of the site of a former mountain range remains in the folded and faulted strata, dynamic metamorphism of the rocks, and batholithic intrusions. The erosional destruction of a mountain range may require 100 to 200 million years.

Ultimately, the mountains may be reduced to such a low elevation that the sea can inundate the site of former mountains, burying the old mountain system. The metamorphic and igneous rocks that make up the roots of an ancient mountain range are exposed beneath a nonconformity in the bottom of the Grand Canyon (Fig. 16.15).

FIGURE 16.11   *Folded Ordovician limestone in Virginia.*

0   100   200 miles

Outer banks

Coastal plain sediments

Sea level

Jur.-Pleist.

Granite gneiss

Mohorovičić discontinuity

Basaltic crust

Mantle

Carolina   Atlantic Ocean

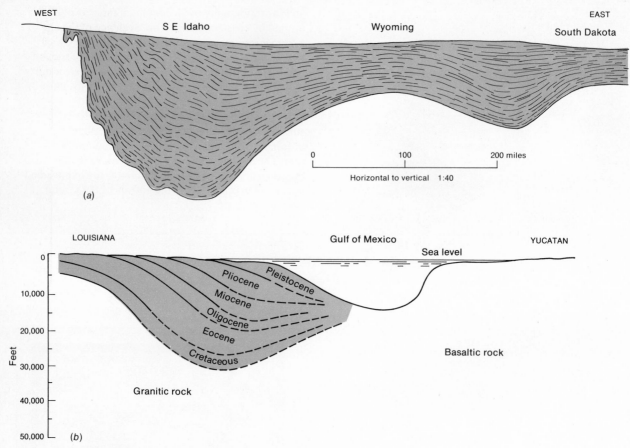

WEST

S E Idaho

Wyoming

EAST

South Dakota

0   100   200 miles

Horizontal to vertical   1:40

(a)

LOUISIANA

Gulf of Mexico

Sea level

YUCATAN

0

10,000

20,000

Feet

30,000

40,000

50,000

(b)

Pleistocene

Pliocene

Miocene

Oligocene

Eocene

Cretaceous

Granitic rock

Basaltic rock

FIGURE 16.12   *Geosynclines. (a) Cordilleran geosyncline, showing the
downwarp in southeastern Idaho that filled with sediments prior to the folding of
the Rocky Mountains (after Kay, 1942). (b) Diagrammatic section of the
geosyncline now in the process of formation along the north coast of the Gulf of
Mexico (after Barton, 1933).*

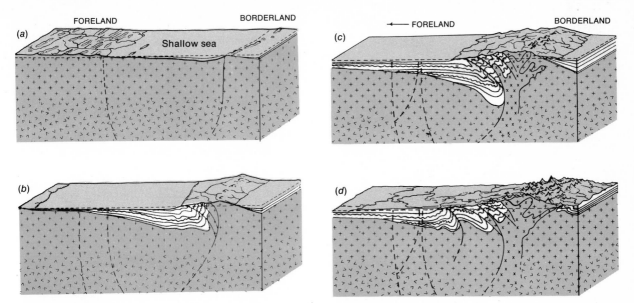

FIGURE 16.13 *Stages in development of an orogenic mountain system. (a) Initial downwarping of region to below sea level; (b) geosynclinal trough developed, with borderland rising along faults that extend deep into the earth; (c) deep subsidence causes dynamic metamorphism and melting of magma that intrudes folded and faulted sediments and metamorphic rocks (orogenic phase); (d) vertical uplift and withdrawal of sea, followed by erosion. (After Weeks, 1952.)*

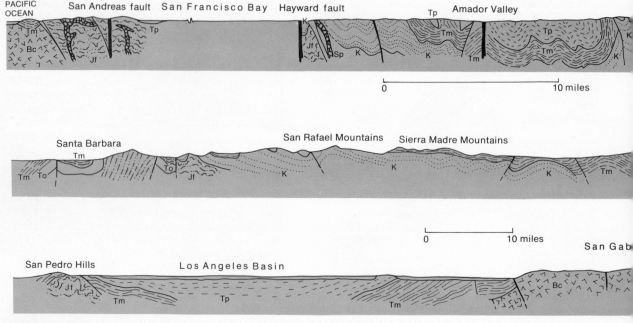

FIGURE 16.14 *Structure sections across the California Coast Ranges. Bc, granitic basement complex; Jf, Franciscan group; K, Cretaceous; Te, Eocene; To, Oligocene; Tm, Miocene, including Miocene volcanics; Tp, Pliocene and Pleistocene. (Sections from Reed and Hollister, 1936. By permission of the American Association of Petroleum Geologists.)*

FIGURE 16.15 *Root zone of an ancient mountain range exposed in the bottom of the Grand Canyon, Arizona. An orogeny 1.3 billion years ago intruded the Zoroaster Granite and deformed the Brahma Schist and the Vishnu Schist before nonconformable burial by the Late Precambrian Unkar group. The Unkar group was later faulted and tilted to form an angular unconformity at the base of the Cambrian strata (540 million years old). (Cross section based on a geologic map by J. H. Maxson.)*

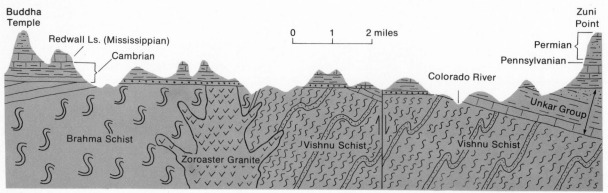

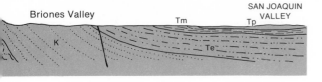

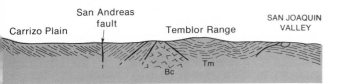

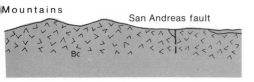

FIGURE 16.16 *Inferred section across part of Japan (modified after Matsuda and Uyeda, 1971). For simplicity, numerical values of heat flow and gravity anomalies are omitted.*

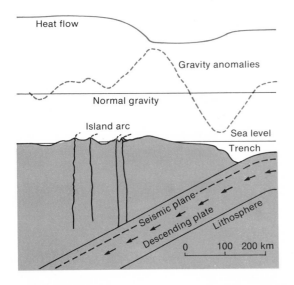

Another but larger mountain range was created some 1.6 billion years ago during the Precambrian Hudsonian orogeny, forming a metamorphosed fold belt extending eastward from Minnesota across Michigan into Ontario. By the beginning of the Cambrian period, gradation had reduced this mountain system to a low elevation. Paleozoic seas flooded the area and deposited marine fossiliferous strata. This ancient mountain-root zone has remained quiescent and at a low elevation ever since.

The geosynclines that are receiving thick accumulations of sediment at the present time will probably be the sites of mountain ranges millions of years hence. A geosyncline receiving sediments off the Gulf Coast from Texas to Mississippi now has some 13,000 meters of sediment in it, and this region could well become an orogenic belt, given millions of years (Fig. 16.12*b*).

Cause of Orogeny

The cause of orogeny is not clear. The sequence of events discussed begins with downsinking of a geosyncline. The sequence culminates in an orogeny, with tremendous compressive forces, regional metamorphism, and igneous activity some hundreds of millions of years later. The obviously powerful source mechanism is located at depth within the earth.

One possible mechanism is that of convection cells in the upper mantle, which move lithosphere plates of the earth. When this plate descends into the mantle at a continent–ocean basin boundary, it will form a large trench (geosyncline); given time, the forces of this moving plate may create orogenic deformation (Figs. 16.16 and 16.13*c* and *d*). This mechanism will be examined further in Chap. 17.

REFERENCES

Anderson, D. L., 1971, *The San Andreas fault,* Scientific American, vol. 226, no. 3, pp. 143-157.

Badgley, P. C., 1965, *Structural and Tectonic Principles,* Harper & Row, Publishers, Incorporated, New York.

Barton, D. C., 1933, *Gulf Coast geosyncline,* American Association of Petroleum Geologists Bulletin, vol. 17, pp. 1446-1458.

Billings, M. P., 1972, *Structural Geology,* 3d ed., Prentice-Hall, Inc., Englewood Cliffs, N. J.

Briggs, P., 1971, *200,000,000 Years Beneath the Sea,* Holt, Rinehart and Winston, Inc., New York.

Clark, T. H., and Stearn, C. W., 1968, *Geological Evolution of North America,* 2d ed., The Ronald Press Company, New York.

Cox, A., and Doell, R. R., 1960, *Review of paleomagnetism,* Geological Society of America Bulletin, vol. 71, pp. 645-768.

Dewey, J. F., and Bird, J. M., 1970, *Plate tectonics and geosynclines,* Tectonophysics, vol. 10, pp. 625-638.

———— and ————, 1970, *Mountain belts and the new global tectonics,* Journal of Geophysical Research, vol. 75, pp. 2625-2647.

Eardley, A. J., 1962, *Structural Geology of North America,* 2d ed., Harper & Row, Publishers, Incorporated, New York.

Frazier, K., 1970, *Building mountain ranges—A plate tectonics model,* Science News, vol. 98, pp. 143-145.

Hammond, A. L., 1971, *Plate tectonics: I. The geophysics of the earth's surface; II. Mountain building and continental geology,* Science, vol. 173, pp. 40-41 and 133-134.

Heath, M., 1962, *Great American Mountains,* Pacific Coast Publishers, Menlo Park, Cal.

Heezen, B. C., and Hollister, C. D., 1971, *The Face of the Deep,* Oxford University Press, New York.

Hunt, C. B., 1953, *Geology and Geography of the Henry Mountains Region, Utah,* United States Geological Survey Professional Paper 228.

Huxley, A. (ed.), 1962, *Standard Encyclopedia of the World's Mountains,* G. P. Putnam's Sons, New York.

Isacks, B., Oliver, J., and Sykes, L. R., 1968, *Seismology and the new global tectonics,* Journal of Geophysical Research, vol. 73, pp. 5855-5899.

Keen, M. J., 1968, *An Introduction to Marine Geology,* Pergamon Press, New York.

King, P. B., 1959, *The Evolution of North America,* Princeton University Press, Princeton, N. J.

Le Pichon, X., 1968, *Sea-floor spreading and continental drift,* Journal of Geophysical Research, vol. 73, pp. 3661-3699.

McKee, B., 1972, *Cascadia: The Geologic Evolution of the Pacific Northwest,* McGraw-Hill Book Company, New York.

McKenzie, D. P., and Parker, R. L., 1967, *The North Pacific—An example of tectonics on a sphere,* Nature, vol. 216, pp. 1276-1280.

Morgan, W. J., 1968, *Rises, trenches, great faults, and crustal blocks,* Journal of Geophysical Research, vol. 73, pp. 1959-1982.

Oakeshott, G. B., 1971, *California's Changing Landscape,* McGraw-Hill Book Company, New York.

Oliver, J., Sykes, L. R., and Isacks, B., 1969, *Seismology and the new global tectonics,* Tectonophysics, vol. 7, pp. 527-541.

Reed, R. D., and Hollister, J. S., 1936, *Structural Evolution of Southern California,* American Association of Petroleum Geologists, Tulsa, Oklahoma.

Shepard, F. P., 1967, *The Earth Beneath the Sea,* 2d ed., The Johns Hopkins Press, Baltimore.

Spencer, E. W., 1969, *Introduction to the Structure of the Earth,* McGraw-Hill Book Company, New York.

Strangway, D. W., 1970, *History of the Earth's Magnetic Field,* McGraw-Hill Book Company, New York.

Takeuchi, H., Uyeda, S., and Kanamori, H., 1967, *Debate About the Earth,* Freeman, Cooper and Co., San Francisco.

CHAPTER 17
GEOTECTONICS

What is geotectonics? The very word itself indicates a study of deformation on a scale that involves the entire earth, rather than, say, mountain building that occurs in relatively narrow belts. However, saying what the word seems to mean is not good enough. Very often a question such as, What is geotectonics? can best be answered by asking further questions and attempting to answer them. For example, appropriate questions we might ask are:

Do the physical nature and internal structure of the earth below the lithosphere have any bearing on the nature of the lithosphere itself?

Have outlines of the continents and ocean basins always been as they are now? Have continents and ocean basins always occupied the same relative positions on the earth's surface?

Do earthquakes, earth magnetism, or fossils provide any clues as to the physical type and the chemical nature of the earth's lithosphere? Do they provide evidence for internal earth processes that may have affected the earth's lithosphere?

Is it necessary to seek separate answers for these questions, or is there perhaps a unifying answer?

Seeking the answers to these and possibly many other questions is the study called geotectonics. Prior to pursuing this study, it will be advisable to review and re-emphasize some significant characteristics of the earth, namely, the nature of the lithosphere and the distribution of earthquakes.

THE LITHOSPHERE

The lithosphere of the earth was defined as that strong outer layer of the earth consisting of the continental crust, the underlying oceanic crust, and the upper mantle. The lithosphere represents a solid layer 60 to 100 kilometers in thickness and overlies a low-velocity seismic layer called the "plastic layer" (weak sphere), or asthenosphere (Fig. 17.1). This plastic layer is about 100 kilometers thick, extending from 100 to 200 kilometers below the earth's surface.

Continental Crust

The continental crust consists of a central shield area,

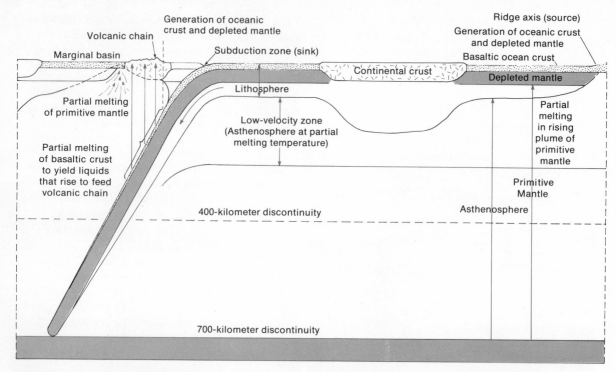

FIGURE 17.1 *Design of a lithosphere plate upon low-velocity layer. Note that new crust is generated at the ridge axis and subsequently is destroyed at the subduction zone or sink. (From Dewey, 1972,* Plate Tectonics, *Scientific American, vol. 226, no. 5, pp. 56-68. Copyright 1972 by Scientific American, Inc. All rights reserved.)*

of metamorphic and igneous rocks, and is bordered by progressively younger belts of deformed rocks, most of which once formed mountains (Fig. 17.2).

The margins of the continents extend below sea level to depths of 1,500 to 3,500 meters. This continental margin may be subdivided into three topographic, or physiographic, divisions (Figs. 17.3 and 17.4).

Continental Shelves   The continental shelves are gently sloping edges of the continents beneath shallow water. On the average, they slope about 1.8 meters per kilometer to a depth of about 200 meters, where the floor begins to descend more abruptly (Figs. 17.3 and 17.5). The shelves range in width from less than 1 kilometer to 1,100 kilometers, averaging about 66 kilometers. The continental shelves resemble the continents in both composition and structure. For example, the

deep crustal structure for the Atlantic shelf off the United States shows a trough, or trench, parallel to the coast that contains as much as 6,000 meters of sediments (Fig. 17.5).

Continental Slopes   At the outer edge of the continental shelves, the sea floor drops off at an average of 66 meters per kilometer drop over a distance of 16 to 32 kilometers (Figs. 17.3 and 17.5). Then in many places it gradually merges with the main deep-sea floor through a width of several tens of kilometers in what is distinguished as the continental rise. The continental slopes are found on the seaward limits of the continental crust, which gives way seaward to heavier (simatic) rocks that underlie the ocean basins proper.

Continental Rise   The continental rise is a gently

sloping (0.5 to 25 meters per kilometer) surface commonly present at the foot of the continental slope (Figs. 17.3, 17.4, and 17.5), which consists of sediments deposited on the lower margin of the continental slope.

Oceanic Crust

The oceanic crust is thin compared to the continental crust. It consists of about 3 to 4 kilometers of simatic, basaltic rock that is overlain by 0.5 to 1.0 kilometer of unconsolidated sediment and an average of 3,800 meters of water (Fig. 17.1).

The deep-sea basin beyond the continental rise

consists of a highly diversified series of physiographic, or topographic, features (Figs. 17.3, 17.4, 17.6, 17.7, 17.8, and 17.9).

In this chapter, only three major physiographic features will be discussed: the abyssal floor, the oceanic rises or ridges, with their fracture zones, and the island arc–trench systems.

The Abyssal Floor or Plains   The plains are the deep, flat, generally sediment-covered bottom of the ocean floor (Figs. 17.3 and 17.4). They vary in dimensions and shape from ocean to ocean, as well as within an oceanic area. The landward margin of the abyssal

FIGURE 17.2   *Continental framework. Note that some geosynclinal belts correspond to modern-day island arc systems. (From Weeks, 1952. By permission of the American Association of Petroleum Geologists.)*

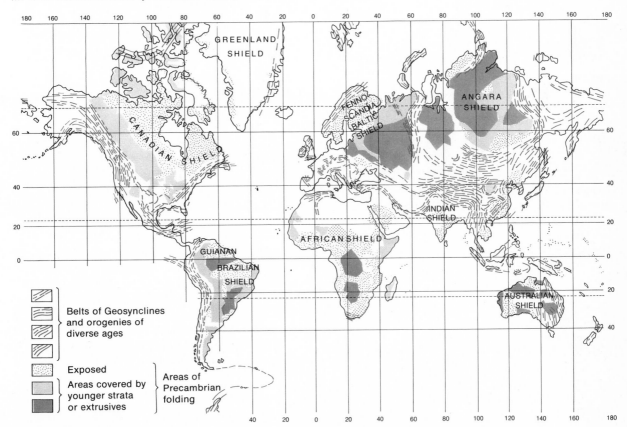

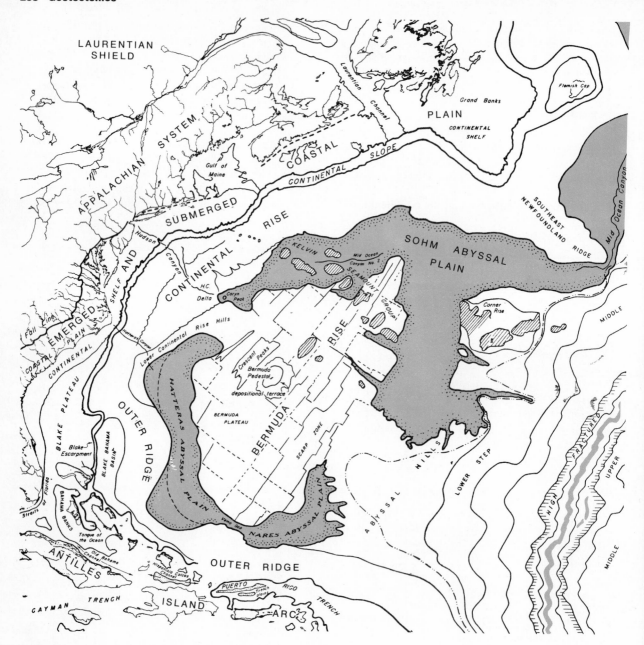

FIGURE 17.3 *Physiographic map of the floor of the North Atlantic Ocean. (From Heezen et al., 1959, Geological Society of America Special Paper 65. Courtesy of the Geological Society of America.)*

PHYSIOGRAPHIC PROVINCES

ATLANTIC OCEAN

Abyssal Plains   Seamounts

Axis of Maximum Depth

Intermontane Basins

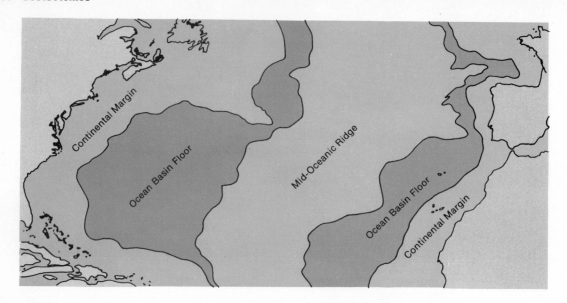

FIGURE 17.4 *Outline map of the major divisions of the North Atlantic Ocean basin. Below is a representative profile from New England to the Sahara coast of Africa. The vertical exaggeration of this profile is about 40 times. (From Heezen et al., 1959, Geological Society America Special Paper 65. Courtesy of the Geological Society of America.)*

FIGURE 17.5 *Cross sections of the deep crustal design of the continental margins with the sediment removed. (After Drake et al., 1959.)*

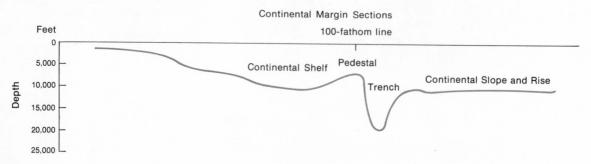

plains generally merges with continental rise, while the seaward margin merges with abyssal hills or gentle irregularities in the basaltic oceanic crust.

The Oceanic Rises or Ridges   These represent a nearly continuous undersea mountain system found in all oceans (Figs. 17.3, 17.4, 17.6, and 17.10). The length of this system exceeds 64,000 kilometers. The ridges or rises (the names indicate a varying topography) extend lengthwise and across oceanic areas, occupying both central and marginal positions on ocean floors and locally intersecting continental areas (Fig.

17.10). On many ridges, the central crest has an extensive rift or grabenlike feature associated with it. The ridges and rises are also displaced along cross-trending fracture zones known as transform faults (Figs. 17.4, 17.6, and 17.10). These faults are situated approximately at right angles to the trend of the rise crests.

The Island Arcs, or Island Arc – Trench Systems   These are features common to the margins of ocean basins. Such systems are found in all oceans, but they are best developed around the circum-Pacific rim (Figs. 17.7 and 17.8). The system

FIGURE 17.6   *The surface of the solid earth. The stable continental platforms and the stable ocean basin floor are traversed by active mountain belts and submarine ridges. The rifted crest of the submarine ridge (thick dark lines) is displaced into segments by fault zones (thin dark lines). The heavy dotted lines show the deep ocean trenches, adjacent to volcanic island arcs or to continental margins. Earthquakes are associated with the mountain belts, the submarine ridges, the ocean trenches, and island arcs. (From Wyllie, 1972, University of Chicago Magazine.)*

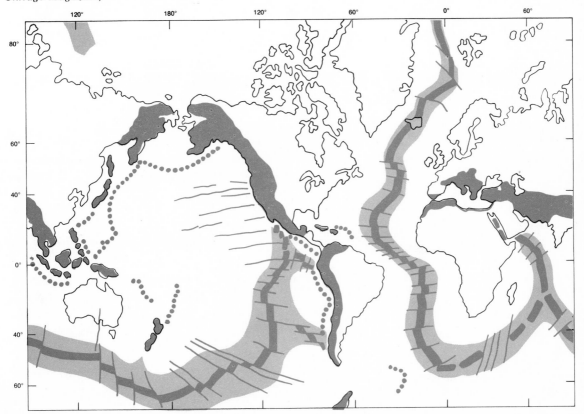

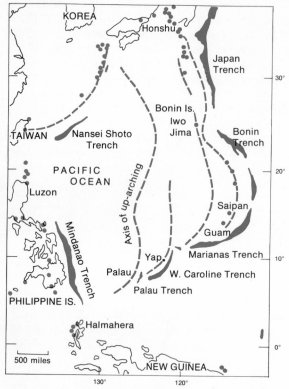

FIGURE 17.7 *Map of the western Pacific, showing island arcs (dashed lines), deep-sea trenches (solid color), and active volcanoes (circles). (After Hess, 1962.)*

in its simplest form consists of an arcuate (convex toward the ocean) series of volcanic islands with a seaward-deep trench and a landward-shallow epicontinental sea (Fig. 17.8).

DISTRIBUTION OF EARTHQUAKES

The distribution of earthquakes will be reviewed in light of the design and topography of the continental and oceanic crust. A general summary of distribution is as follows (Fig. 13.6).

All earthquakes with a foci depth greater than 200 kilometers (intermediate to deep earthquakes) are generally restricted to active island arc–trench systems. These areas also contain the highest percentage of shallow (less than 60 kilometers) earthquakes (Figs. 17.7 and 17.8).

Other earthquake areas of shallow- and intermediate-foci depth include the oceanic ridge system, the mountain system from the Mediterranean to the Himalayas (Fig. 17.2), the continental rift zones, such as the East African Rift Valley, the intersection of the oceanic ridge or rise system with a continental area (Fig. 17.10), and along the fracture zones (transform fault zones) of the ridge-rise system (Figs. 17.6 and 17.10). Areas such as the continental shields, central oceanic basins, and older mountain systems show minimal earthquake activity (Fig. 13.6).

Let us now ask whether it is possible to relate the lithosphere to the distribution of earthquakes by means of some unifying concept. Yes it is, and the idea behind this concept is similar to a spherical jigsaw puzzle, with the puzzle pieces being lithosphere plates whose boundaries are delineated by the earthquake zones. Over geologic time, the completed puzzle takes

FIGURE 17.8 *Cross section of Kuril Islands, Japan Trench. (Data from B. Gutenberg and C.F. Richter, 1949, Seismicity of the Earth,* Princeton University Press, Princeton N.J. From Strahler, 1971, *The Earth Sciences, 2d ed., copyright 1963, 1971 by Arthur N. Strahler. By permission of Harper and Row Publishers, Inc.)*

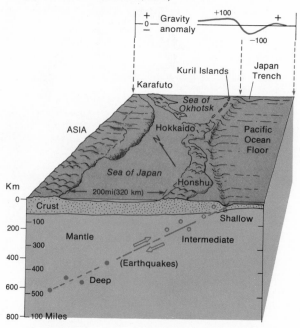

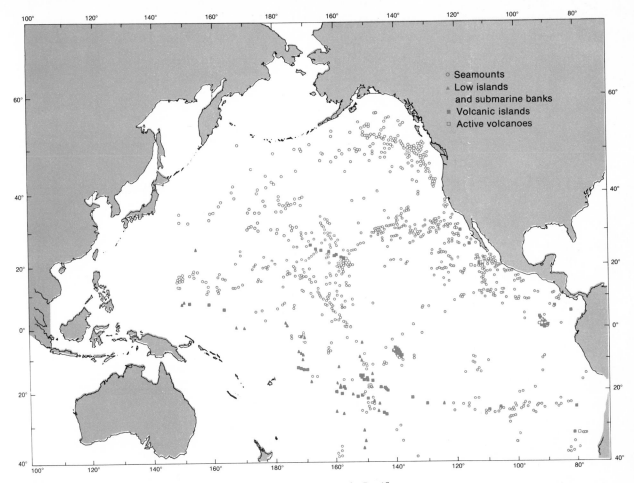

FIGURE 17.9 *Modern volcanic islands and submarine volcanoes in the Pacific Basin. (From Menard, 1964.)*

on different views as the pieces are shifted and moved around.

LITHOSPHERE PLATES

The distribution of earthquakes in belts and a consideration of the earth's crustal features allow us to visualize the lithosphere as being composed of plates (Fig. 17.10). The lower boundary, or base, of the plates is generally considered to be the low-velocity zone (Fig. 17.1). These plates, which number between 6 and 22, have a variety of marginal boundaries. (1) Tensional, where lithosphere plates appear to be pulled apart, as in the oceanic-ridge system. (2) Compressional, where one plate appears to be either pushing over or, more commonly, sliding beneath another plate. Such a boundary would typically occur at an island arc–trench system. (3) A fracture-zone boundary, where plates are apparently sliding past one another. This boundary is the major transform fault zones that transect the oceanic rise-ridge system (Figs. 17.6 and 17.10).

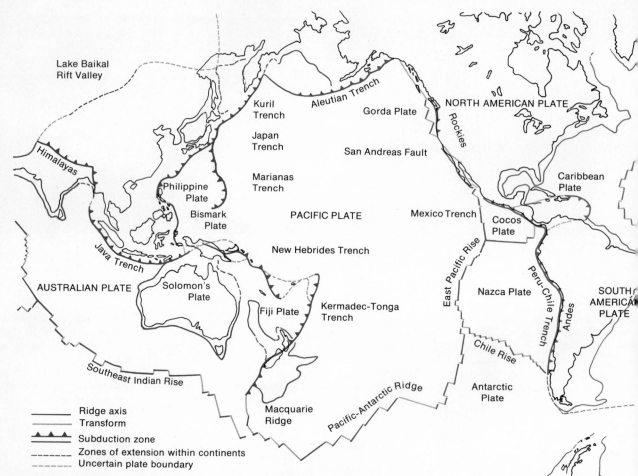

FIGURE 17.10 *Mosaic of plates forms the earth's lithosphere, or outer shell. According to the recently developed theory of plate tectonics, the plates are not only rigid but also in constant relative motion. The boundaries of plates are of three types: ridge axes, where plates are diverging and new oceanic floor is generated; transforms, where plates slide past each other; and subduction zones, where plates converge and one plate dives under the leading edge of its neighbor. Triangles indicate the leading edge of a plate. (From Dewey, 1972,* Plate tectonics, *Scientific American, vol. 226, no. 5, pp. 56-68. Copyright 1972 by Scientific American, Inc. All rights reserved.)*

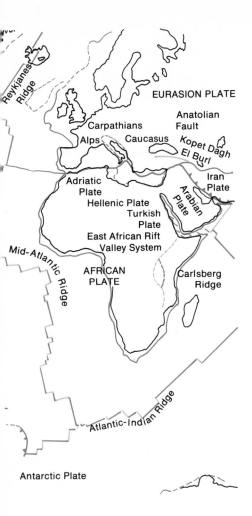

EURASION PLATE

Reykjanes Ridge

Carpathians

Alps

Caucasus

Anatolian Fault

Kopet Dagh
El Burl

Adriatic Plate

Iran Plate

Arabian Plate

Hellenic Plate

Turkish Plate

East African Rift Valley System

Mid-Atlantic Ridge

AFRICAN PLATE

Carlsberg Ridge

Atlantic-Indian Ridge

Antarctic Plate

The various types of lithospheric-plate boundaries imply that plate motion creates tensional, compressional, or fracture-zone conditions. The source of the motion is thought to be centered in the plastic zone of the mantle. The cause of this motion can possibly be explained in terms of convection currents in the plastic zone (Fig. 17.11).

Convection Currents in the Mantle

Convection currents were originally proposed as a mechanism for downbending of the earth's crust, with the resultant accumulation of great thickness of sediments as geosynclines and the ultimate crumpling of these geosynclinal sediments into folded mountain ranges. At first the convective overturning was thought to involve the entire mantle. More recent seismic data have, however, revealed a concentric density zonation of the mantle rather than a homogeneous uniform mixture from top to bottom. Thus, many geologists have envisioned convection as operating only in the upper mantle (Figs. 17.1 and 17.11). Others, however, still consider the possibility of the convective rise of minor quantities of material from the lower mantle.

Convection is known to operate only in fluids, such as air and water. Many geophysicists and geologists consider convection in a solid mantle to be impossible, or at least improbable. Other geologists say that the usual standards of rigidity, strength, and viscosity of solids do not apply to mantle materials and that the rocks there flow by high-temperature creep, as hot metals do.

Opponents of convection currents in the mantle doubt the existence of an adequate source of heat, inasmuch as a mantle of ultramafic rocks can be expected to contain little uranium or other radiogenic elements.

Possible sources of energy for the convective cells are (1) radioactivity, (2) heat-releasing phase changes of basalt to eclogite in the mantle, (3) gravitational settling of certain components, such as the sinking of iron from the mantle into the core, (4) release of the heat of crystallization as iron crystallizes at the inner core–outer core boundary, and (5) tidal effects within the earth, or earth tides. Each of these possible theories for energy sources has its supporters and its opponents. If one accepts any mechanism that produces lithospheric-plate motions (Fig. 17.11), some interesting theories result.

CONTINENTAL DRIFT

The idea that continents may have changed their positions during geologic time dates mainly from the ideas of Taylor in 1910, Wegener in 1912, Van der Gracht in 1928, and Du Toit in 1937. It was advocated by European and South African geologists and generally rejected by Americans. Taylor envisioned equatorial sliding of continents as a primary mechanism for creating mountain ranges. Wegener's scheme postulated: one original land mass, Pangaea; its fragmentation; isostatic drift of continental-sialic crust across a "sea"

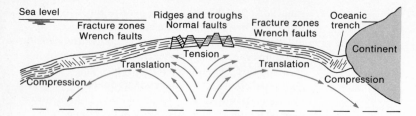

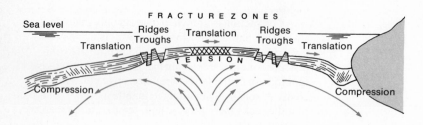

FIGURE 17.11 *Conceptual presentation of convection currents. (From Menard, 1964.)*

of sima; and the plowing up of mountain ranges along the leading edges of the continental sialic crust, especially of the westward-moving Americas. According to Du Toit, there were two initial land masses, Laurasia in the Northern Hemisphere and Gondwanaland in the Southern. They began to break up in Mesozoic time and still continue to drift (Fig. 17.12).

The accumulation of new data about the oceanic crust, heat flow from the earth, mid-oceanic ridges, and paleomagnetism plus a renewed interest in the possibility of continental drift has stimulated a flood of recent publications on this controversial subject. In a recent article in *Science* (1973), continental drift was called upon as an explanation (although implausible) for the orientation of the Great Pyramids of Giza. The east and west sides of the pyramids are 4 minutes west of true north, which apparently cannot be blamed on a builder's error. The measurements of the pyramids have been made over a 50-year period of time, which is difficult to reconcile to measurements of sea-floor

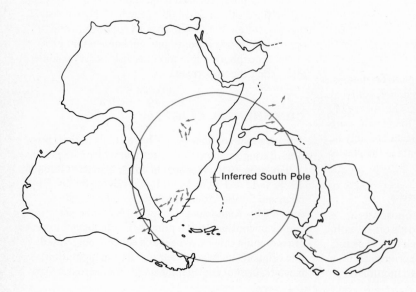

FIGURE 17.12 *Arrangement of South America, Africa, India, Australia, and Antarctica about the South Pole as part of the hypothetical Gondwanaland in late Paleozoic time, according to Du Toit (1937). The arrows show the directions of movement of former ice sheets. Note that in some places the ice seems to have come from areas now offshore. The suggested fit of the continents has been redesigned.*

spreading and continental drift over millions of years.

Generally regarded before 1960 as an incredible hypothesis, continental drift has found an increasing acceptance, relying on the new hypothesis of plate motion as the explanation of continental drift.

SEA-FLOOR SPREADING

The theory of sea-floor spreading postulates that the ocean floors split along the axes of the worldwide oceanic ridge-rise system and that the separated crustal segments slowly moved outward from the axes on both sides of the rifts (Fig. 17.11). Heezen (1958) noted that many shallow earthquakes originate along the rifts associated with mid-oceanic ridges (Fig. 13.6) and suggested that these rifts were sites of sea-floor spreading. Thus, by the hypothesis of sea-floor spreading, the Mid-Atlantic Ridge is the central area from which the Americas have been separated from Europe and Africa. The worldwide mid-oceanic ridge-rise system (Fig. 17.6) is the locus of continuing separation on a global scale. Other movements would include that of India, which has moved 70° of latitude, about 8,000 kilometers (Fig. 17.13), northward since Jurassic time.

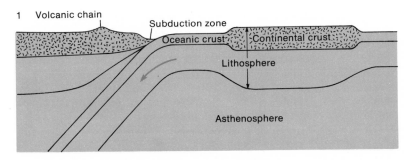

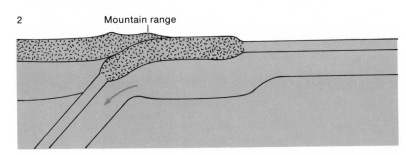

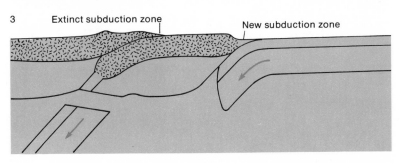

FIGURE 17.13 *(a) Collision of continents occurs when a plate carrying a continent is subducted at the leading edge of a plate carrying another continent. (b) Since the continental crust is too buoyant to be carried down into the asthenosphere, the collision produces mountain ranges. (c) The Himalayas evidently formed when a plate carrying India collided with the ancient Asian plate some 40 million years ago. The descending plate may break off, sink into asthenosphere, and a new subduction zone started elsewhere. (From Dewey, 1972,* Plate tectonics, *Scientific American, vol. 226, no. 5, p. 56-68. Copyright 1972, Scientific American Inc. All rights reserved.)*

and that of Iceland, which is calculated to have been widened by 200 to 400 kilometers. This widening occurred by emplacement of dikes and the formation of rifts along the extension of the Mid-Atlantic Ridge, which occurs underneath the island itself (see cover photo). The Gulf of California, the Red Sea, the Gulf of Aden, the Gulf of Aqaba, and the rift valleys of eastern Africa and of the Dead Sea-Jordan Valley represent early stages of crustal separation where the mid-oceanic ridges impinge upon or intersect the continents.

Descriptions of the sea-floor-spreading hypothesis differ in detail, but all envision the mid-oceanic ridges as passive sources of rising new oceanic crust (Figs. 17.1 and 17.11) and the deep-sea trenches or certain continental margins as areas (Figs. 17.1 and 17.13) in which the crust is removed. In one version, upwelling of a convection current by high-temperature creep in the upper mantle develops a surface ridge that splits in two along its crest. The two flanks, or branches, are then forced laterally away from the ridge at a rate of a few centimeters per year.

The upwelling material, forming basaltic lava flows, dikes, and possibly masses of serpentine, creates new oceanic crust in the central rift area of the ridges (Fig. 17.1) and thereby adds new material to form the oceanic crust. As this material cools below the Curie temperature (578°C), it becomes magnetized with normal or reversed magnetic polarity, according to the earth's general magnetic field at that time (Fig. 14.7).

Descending currents carry the top of the mantle, the free-riding crust, and overlying sediments inward toward the mantle beneath the island arc–trench systems. If the basal layer of the oceanic crust did include serpentine, its dehydration at the descending downturn would release large quantities of water, which would facilitate mountain building and metamorphism at continental margins.

Spreading rates from the central ridge-rise areas are calculated to be 1 centimeter per year in the Arctic Ocean, 1 to 2 in the North Atlantic, 1 to 3 in the Indian Ocean, and 2 to 6 centimeters per year in the Pacific. Observations of abrupt changes in sediment thickness on the ocean floor show that these rates have varied in the past.

Supporting evidence for the concept of sea-floor spreading includes these facts about the mid-oceanic ridges: they are nearly barren of sediment, they are sites of shallow earthquakes (Fig. 13.6), they show higher than average heat flow (Fig. 14.11), and they have a peculiar crustal structure, possibly indicative of altered mantle material (Fig. 17.14). Moreover they are offset repeatedly by transform faults (Figs. 17.4 and 17.10), which has resulted in shifting of the ridge axes during spreading, and the rocks at the ridge axes are fresh or altered basalt, with locally ultramafic plutonic rocks and in some places serpentine.

Other factors upholding the theory are the negative gravity anomalies of the deep-sea trenches, the low heat flow at the trenches (Fig. 14.12) and their faulted structure (Fig. 17.8), the occurrence of shallow earthquakes on the ridges, and the dominance of nearly all the intermediate and deep earthquakes at the

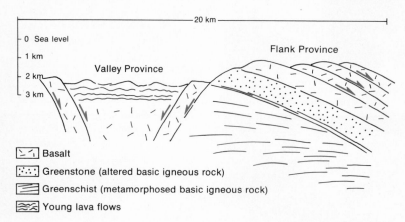

FIGURE 17.14 *Schematic cross section of the Mid-Atlantic Ridge. (After van Andel and Bowin, 1968.)*

20 km

- 0 Sea level
- 1 km
- 2 km
- 3 km

Flank Province

Valley Province

Basalt

Greenstone (altered basic igneous rock)

Greenschist (metamorphosed basic igneous rock)

Young lava flows

island arc systems (Figs. 13.6, 17.7, and 17.8). The foci of the deep earthquakes are along a shear plane slanting under the volcanic arc or under a continental margin (Fig. 17.8).

Deep-sea drilling by the research ship *Glomar Challenger* in well-scattered areas of the ocean basins since 1968 has also shown that the sediments of the sea floor are thin and are progressively older away from the axes of mid-oceanic ridges; in fact, no sediments older than Jurassic have been found anywhere. The apparent spreading rates indicated by the sediments closely match those inferred from the patterns of the magnetic anomalies, and cores of sediments show similar alternations of magnetic polarity. Siliceous sediments of a cold-water marine environment may be succeeded by warm-water calcareous ooze, or vice versa.

Because these findings are readily explainable by the sea-floor-spreading hypothesis, it has rapidly gained a wide acceptance and has revolutionized geologic thought on tectonics and the stability of continents and ocean basins. However, this hypothesis led to another, even broader in its scope.

PLATE TECTONICS

The discovery of high heat flow along mid-oceanic ridges, the localization of many earthquakes along continental borders and oceanic fracture zones (Fig. 13.6), and the plausibility of sea-floor spreading led to a new hypothesis of geotectonics—plate tectonics. It was first proposed in 1967, and since that time it has gained the support of a large number of geologists.

According to this hypothesis, the earth's crust and upper mantle are divided into plates that grow at mid-oceanic ridges and slowly move horizontally by sea-floor spreading (including horizontal sliding along fracture zones). They disappear by sinking into slanting subduction zones at ocean trenches beside a continent or island arc (Figs. 16.16, 17.1, 17.8, 17.11, 17.12, and 17.13). In their descent, the motion of the cold plates generates numerous shallow, intermediate, and deep earthquakes. These occur in a narrow zone about 20 kilometers thick within the upper portion of the downsinking plates. In areas where these descending plates carry a considerable thickness of sediment and volcanic rocks, they deform these sediments and vol-

canics into folded and thrust-faulted mountain structures (Fig. 16.13). A sinking lithosphere plate, thus freed from its load of buoyant light-weight earth material, may possibly remain intact to a depth of 400 to 500 kilometers before it is remelted in the mantle. At depths of 50 to 100 kilometers, partial melting of the oceanic crustal portion of the descending plate results in the generation of magmas that rise to the surface (Fig. 17.1). The products are basalt, andesite, or rhyolite, each of which would form according to the degree to which the original composition of the magma has been modified. More complete melting at the base of the orogenic zone will form batholiths of granodiorite or granite surrounded by characteristic metamorphic zones.

The number of plates would apparently range from 6 to 22 or possibly more and would consist of oceanic or continental crust, or both (Fig. 17.10). North America and South America, for example, are considered to be parts of the same Atlantic plate as the western half of the North and South Atlantic Oceans, while virtually all the Pacific plate is oceanic crust.

The Andes of South America, the Himalayas, and the Cascade and Coast Ranges of the western United States are typical results of plate tectonics where an oceanic plate collides with a continental plate (Fig. 17.15). The Pacific coasts of South America and southern Mexico–Central America still have offshore trenches due to a downsinking plate, but from California northward such a trench is absent. The North America–North Atlantic plate, by drifting westward, overrode not only a trench area but also part of the East Pacific Rise (Fig. 17.15). This interpretation has been used to clarify many puzzling features of the geology of the Coast Ranges of California (Fig. 17.15). The Himalayas are the product of the movement of the Indian plate from a former position near Antarctica across the Indian Ocean against the southern edge of Asia (Fig. 17.13). By interpreting past geosynclinal accumulations as the equivalent of those now forming at active island arc systems or on continental shelves around the world, the plate-tectonics hypothesis possibly can be applied to the Appalachians, the Alps, and other complex mountain ranges of the Paleozoic or Precambrian.

The base of the moving plates is generally considered to be within the low-velocity zone, although plate

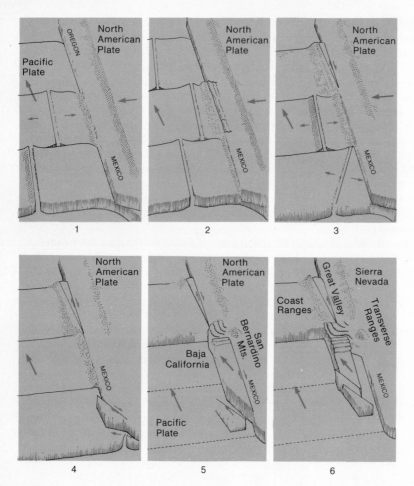

thicknesses up to 400 to 700 kilometers have also been proposed (Fig. 17.1). The difference in these estimates is due to a lack of knowledge on the details of the convection mechanism supposedly responsible for seafloor spreading and plate drift.

The plate-tectonics hypothesis explains (1) the relatively young geologic age of the ocean floors, (2) the association of elongate trenches and island arcs, (3) the negative gravity anomalies and low heat flow at ocean trenches, and the positive gravity anomalies and higher heat flow in island arcs, (4) the high-angle faulting on the inner margins of trenches, (5) the basaltic and andesitic volcanoes in the island arcs, (6) the gen-

eration of intermediate and deep earthquakes along a slanted high-angle-fault zone and the focal mechanisms of these earthquakes due to the shattering within the top of the descending plate, (7) a possible source of local heat in the shearing of the plate beneath the underridden part of the mantle, (8) known underthrusting (as occurred in the 1964 Alaskan earthquake), (9) the addition of sedimentary material to the front of an island arc, (10) the conversion of unsinkable material into folded mountains, (11) the apparent lithologic and spatial changes in island arc–trench systems to mountain belts with time, (12) the permanence of continents, (13) the relatively low heat-flow gradients beneath shield areas, and (14) the spatial distribution of intrusions and metamorphic zones in relation to folded mountains.

Thus, some answers to the questions posed in the beginning of this chapter have been given, but it is necessary to critically examine this hypothesis.

PALEOMAGNETISM

Lava flows, dikes, and sills that contain magnetically susceptible minerals become magnetized during their cooling. Thus, a newly crystallized basalt containing 1 or 2 percent magnetite, upon cooling below a critical temperature (called the Curie temperature, 578°C for magnetite), becomes weakly magnetized in accordance with the general magnetic field of the earth. This magnetism persists as "frozen-in" natural remanent magnetism; in this instance, thermo-remanent magnetism.

If a specimen of naturally magnetized basalt is placed in a laboratory at an angle to its former natural position and then heated to a temperature above the Curie point, it loses its magnetism. On cooling it is remagnetized in a new direction corresponding to the present direction of the magnetic field.

Remanent magnetism is subject to change by weathering of the rocks, by lightning, by induced magnetism from the earth's general field (which may or may not coincide with the original remanent magnetism), by heating in a changed position by an igneous intrusion, or by metamorphism at depth. In some instances the effects of weathering and the secondary induced magnetism can be erased by chemical treatment, by heating in incremental stages below the Curie temperature, or by partial electrical demagnetization.

Because of the possibility of the alterations mentioned above, sampling for remanent magnetism is generally restricted to fresh strongly magnetized rocks, in stable nontectonic areas. The remanent magnetism of an oriented sample is measured in the laboratory by a very sensitive magnetometer, or in the field by a portable instrument, to find the declination and the inclination at the time of the original magnetization. Under the assumptions that the global magnetic field has always been dipolar and that the magnetic and planetary axes, despite their present difference of about $11\frac{1}{2}°$, have generally coincided in the geologic

past, the measured inclination and declination serve to indicate the former position of the poles (Figs. 14.8 and 14.9). Many geologists, however, disagree with these postulations.

Supporting Evidence

Paleomagnetism supports continental drift, and continental drift is a possible explanation of the apparent polar wandering in the geologic past.

The pole-migration path for North America during Paleozoic and Mesozoic time is displaced about 20 to 30° from that of Europe. This could result if the two continents had drifted apart relative to the North Pole, which did not move (Figs. 14.8 and 14.9).

A set of symmetrically paired, alternately polarized, magnetic-anomaly stripes is situated astride a strong central anomaly along Reykjanes Ridge on the sea floor southwest of Iceland (Fig. 17.16). The pattern is interpreted as evidence of the slow rise of material from the upper mantle, its magnetization during cooling near the surface and during a series of periodic reversals of polarity (Fig. 17.17), followed by a slow spread away from the axis as new oceanic crust. Assuming these facts, the magnetic-anomaly stripes are a fossilized history of magnetic reversals and sea-floor spreading.

The Reykjanes Ridge magnetic-anomaly pattern, when fitted to the known polarities of the geomagnetic-polarity epochs of the past 4 million years (Fig. 17.17), indicates a spreading rate of 1 centimeter per year. Similar linear magnetic-anomaly stripes bordering the East Pacific Rise (Fig. 17.18) and the Pacific-Antarctic Ridge yield magnetic-polarity time scales that are in agreement with the record over Reykjanes Ridge, if adjusted for different spreading rates. These rates generally range from 1 to 6 centimeters per year. A magnetic-anomaly pattern off the coast of Oregon, Washington, and British Columbia also correlates with the Reykjanes and East Pacific Rise patterns and the known polarity time scale (Fig. 17.19). Scientists have succeeded in matching and correlating the magnetic-anomaly patterns in the Atlantic, Pacific, and Indian Oceans. Correlations extending back 79 million years, to Late Cretaceous time, have been made (Fig. 17.19). Reversals in earlier geologic periods seem to have been less frequent, or they may have gone undetected because of insufficient data.

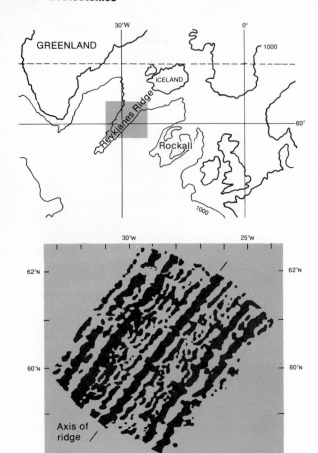

FIGURE 17.16 *Magnetic anomalies at Reykjanes Ridge southwest of Iceland, symmetrically paired beside the central axis. The central positive anomaly is correlated with the present polarity epoch and the others with progressively older polarity epochs outward from the axis. (After Vine, 1966.)*

Negative Aspects

The pole positions indicated by paleomagnetism, according to some scientists, permit but do not require continental drift. Others propose that the number of magnetic determinations is not large enough and the scatter of their results is not precise enough to warrant any far-reaching conclusions. Critics say that the assumptions of a dipolar and geocentric magnetic field in

the past may not be warranted and that polar wandering may have occurred without continental drift.

A proposed alternative is that anomalies on the sea floor are not old, but instead, represent induced magnetism in susceptible rocks of the crust by the earth's present magnetic field. Finally, others argue that the remanent magnetism in the rocks of the sea floor is not alternately reversed and normal in polarity but only correlated with rock types. In any event, the paleomagnetic data are not unequivocal.

PALEOCLIMATES

Supporting Evidence

In Permian time, continental ice sheets occupied parts of Brazil, southern Africa, India, and Australia. Deposits left by these ice sheets occur in latitudes of 10 to 30°N and S. Moreover, the ice in some areas moved away from the present equator. A simple explanation is that these areas were clustered about the South Pole during the Permian period and later drifted to their present positions (Fig. 17.12).

Most arid climates today lie along the subtropical high-pressure belts near latitudes 30°N and 30°S, yet Ordovician salt deposits occur in Siberia, Silurian salt deposits in New York, Devonian salt in Saskatchewan, Mississippian salt in Michigan, and Permian salt in Kansas, Oklahoma, and Texas. Also, certain thick sedimentary rock layers of wind-blown desert sand present in middle latitudes seem out of place, considering present world climate. Either the atmospheric circulation during Paleozoic time was vastly different from that of today, or these areas of evaporite deposits and desert sands have changed latitudes by means of continental drift.

Reef-forming corals are now restricted to clear, warm water, generally within 30° of the equator; but remnants of fossil coral reefs occur not only in central United States, they also occur beyond the Arctic Circle in northern Canada and Greenland. Early in 1973, Russian geologists reported finding coral-reef deposits in the Kara Kum desert in the Soviet province of Turkmenia. The deposits included Lower Cretaceous (120 million years ago) corals, sea lilies, and sponges. This fossil coral reef was found at a depth of 2 kilometers below the surface desert deposits. The

Kara Kum desert lies between 40° and 50° north latitude. Poleward movement of the continents would help to explain this anomalous distribution.

Negative Aspects

The high-latitude coral reefs, evaporites, and desert deposits may not indicate continental drift but may point to a much warmer global climate with different latitudinal wind belts than exist now. However, Permian glaciation at low elevations within the present tropic climatic zones is not so easily disregarded. Fossil plants of Cenozoic age indicate climatic zones parallel to those of today, gradually shifting equatorward as the general cooling occurred that foreshadowed the Pleistocene Ice Age. It is assumed, however, that the greater part of continental drift had already taken place by the Cenozoic.

FAULTING

The San Andreas fault of California shows, over geologic time, horizontal, lengthwise movement of at least 290 and possible 580 kilometers. The Great Glen fault of Scotland, the Alpine fault of New Zealand, the Philippine fault, and a fault in Japan have similar horizontal displacement. On the sea floor, the north-south-trending magnetic belts in the northeast Pacific are offset about 1,200 kilometers at the Mendocino fracture zone and 775 kilometers at the Murray fracture zone (Fig. 17.18). In the Atlantic, the mid-oceanic ridge is offset by transform faults in many places. Locally, the displacements are as much as 1,100 kilometers (Fig. 17.10). The large, measured horizontal displacements along these faults lend credibility to the movement of the earth's crust on the scale of continental drift. Thus, the inferred movement of large plates of the earth's

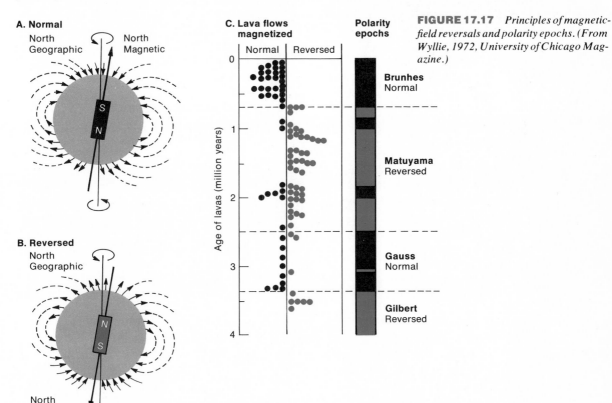

FIGURE 17.17 *Principles of magnetic-field reversals and polarity epochs. (From Wyllie, 1972, University of Chicago Magazine.)*

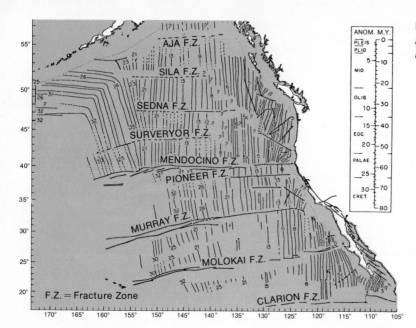

FIGURE 17.18 *Correlation of magnetic-anomaly belts in the northeast Pacific. (After Atwater, 1970.)*

crust, which are bounded in part by faults, becomes part of the basis of plate tectonics.

On the negative side, some believe that the African rifts, which are also bounded by faults, are the result of vertical rather than horizontal spreading movements, and conceivably the apparently horizontal displacements of magnetic anomalies on the sea floor may also be the effects of vertical movements.

PALEONTOLOGICAL EVIDENCE

Supporting Evidence

It can be argued that the occurrence of similar marine faunas of Paleozoic age in Africa and South America can be explained by shallow-water migration routes. However, identical coal-forming land plants, associated amphibians, and fresh-water fishes seem to require direct land connections. A peculiar tongue fern, *Glossopteris,* was widely distributed in the Southern Hemisphere from India to Antarctica during the Permian period.

Many early Mesozoic reptiles (*Mesosaurus,* a fresh-water swimmer) in Africa also had counterparts in South America. Fossils of fresh-water amphibians and land reptiles of Triassic age (about 200 million years ago) on Antarctica and South America furnish strong support for continental drift. A fossil reptile found in 1969, 660 kilometers from the South Pole, that is a kin to South American and South African forms is convincing.

Negative Aspects

It is argued that the spread shown by fossil organisms does not require any such postulated land masses as Laurasia and Gondwanaland. Instead, land bridges, isthmian links, island arcs, and ocean currents may be adequate to explain their presently disjunct distribution.

The taxonomic diversity and geographic distribution of marine organisms of the past presumably were strongly affected by temperature. The variety and range of planktonic foraminifera are similarly temperature-controlled today. Accordingly, a study of the diversity and geographic ranges of all known Permian brachiopods was to show that the geographic pole in Permian time was about at its present position. This location would be highly discordant with the pole position inferred from paleomagnetism. A natural inference is that the rotational and magnetic poles may not have been coincident. In that case, only the magnetic pole may have wandered. But many earth scientists do not advocate polar wandering, only moving of the continents.

A recently published analysis of Mariner 9 data and photomaps of the surface of Mars has presented some intriguing speculations concerning polar wandering, climatic changes, and continental drift. Three distinct features were noted, each having an interesting interpretation. First, in the north and south polar regions, quasi-circular topographic features were noted. They are described as plates with outward sloping edges and are composed of light and dark laminae about 20 to 80 meters thick. The laminae are interpreted to be a mixture of volatiles (mostly carbon dioxide) and eolian-atmospheric dust. The roughly circular plates and their concentric pattern are related to past location of the Martian planet spin axis. The wandering of the axis is about 15° and covered a time span of 100 million years. A second feature is the presence of large constructional volcanic (large shield volcanoes) forms near the Martian equator. The forms, four in all, have a diameter of about 400 to 500 meters and a height of 10 kilometers. If Mars could be assumed not to be a rigid planet, then the mass of these volcanoes might have caused a sufficient imbalance to cause a nutation of the spin axis. The nutation can be easily visualized by the wobbling or swaying of the spin axis of a child's toy top. Still a second interpretation of the volcanic forms is that they are indicative of surface expressions of mantle convection. This in itself would have a role in polar wandering.

The third feature noted, from other data, is the presence of a gravitational field that is "rougher" (more variable) than that of the earth or moon. While many interpretations are possible, the correlation of gravity harmonics with the constructive volcanic features can be interpreted to indicate large-scale mantle convection. Polar wandering on the earth (slow drift of the spin axis) is advocated by many geologists, and they believe this wandering can explain the radical climatic changes that have occurred on the earth over geologic time. These same proponents of polar wandering do agree that the process may create instabilities that aided continental drift. The recent interpretation of Mars seems to support their theories.

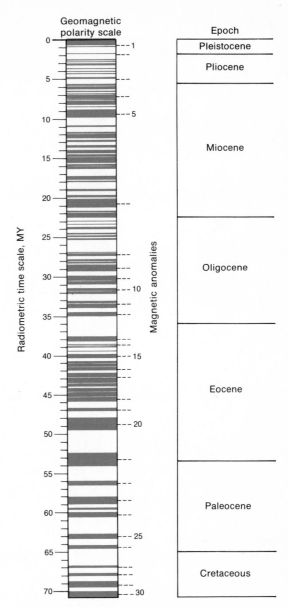

FIGURE 17.19 *The geomagnetic scale and time scale for the last 70 million years. Modifications arising from Deep Sea Drilling Project studies show that this scale may be extended to anomaly 33 at 76 to 79 million years.*

FIT OF CONTINENTS

Supporting Evidence

South America and Africa fit well-selected isobathic depths below sea level. The offshore boundaries of North America correspond to those of Greenland and northwest Europe (Fig. 17.20). If suitably rotated, South America, Africa, Antarctica, India, and Australia can all be fitted together to form ancient Gondwanaland (Fig. 17.21).

Paleozoic stratigraphy and structure of eastern North America are very similar to those of northwestern Europe. The fold system of the Appalachians, which terminates abruptly in the northeast, is matched by a European orogenic belt similarly interrupted at the Atlantic coast line. Striking similarities are found between South America, Africa, and Antarctica in types of rock, their ages, and their structural trends. These characteristics are indicative of former interconnections of these continents (Fig. 17.22).

Negative Aspects

The fitting of the continents across the Atlantic leaves some incongruent areas, notably the Caribbean region. These areas make full acceptance of the hypothesis difficult for some scientists. Resemblances in stra-

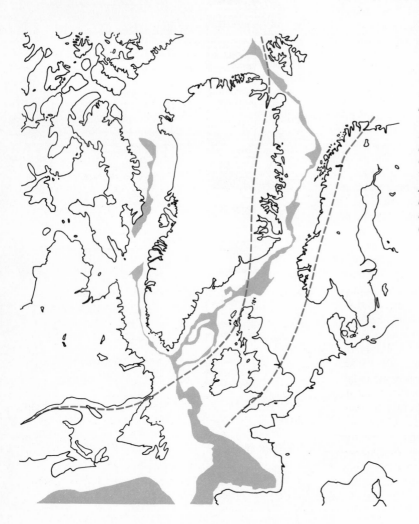

FIGURE 17.20 *Fit of North America to Greenland and Greenland to northwest Europe at the 500-fathom level. Ellesmere Island at the upper left has not been shifted to fit. The belt (between dashed lines) extending from Gaspe Peninsula, Nova Scotia, and Newfoundland via the British Isles to Norway, eastern Greenland, and Spitzbergen is the site of folded early and middle Paleozoic sediments, which continue southwesterly through the Appalachian Mountains. Their alignment in this reconstruction strongly supports the former junction of the continents. (After Bullard, Everett, and Smith, 1965.)*

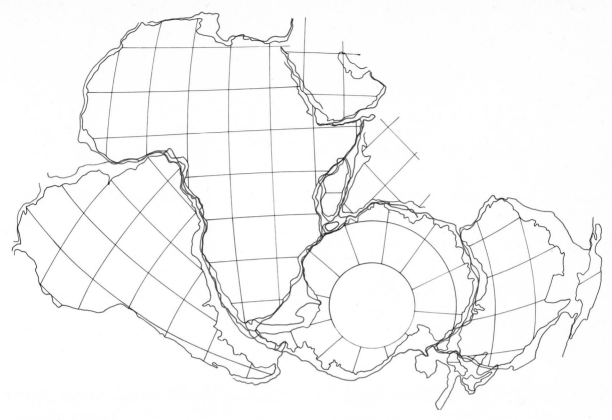

FIGURE 17.21 *Fit of South America, Africa, India, Antarctica, and Australia at the 500-fathom contour. (After Smith and Hallam, 1970.)*

tigraphy, structure, and fossil faunas and floras can possibly be explained in other ways without involving continental drift.

OTHER LINES OF EVIDENCE: PRO AND CON

Sea-Floor Sediments

Sediments and sedimentary rocks in the deep sea floor have been explored only to depths of 1,300 meters below the sea floor and have been dated as being no older than Jurassic. Older rocks may exist at greater depths. If any Paleozoic or Precambrian rocks should be found in place, the clearing of the ocean floor by a supposed convectional conveyor belt would be negated. Rocks of Lower Miocene age, adjacent to the

Mid-Atlantic Ridge, imply that the Atlantic spreading system has been inactive locally for a considerable time. On the other hand, the sedimentary rocks in the Pacific do increase in thickness and become progressively older further away from the axis of the East Pacific Rise.

Heat Flow

The higher than average heat flow observed along the mid-oceanic ridges (Fig. 14.11) is an expected consequence of a rising convection current that would make continental drift and sea-floor spreading or plate tectonics possible. A low heat flow is found, as expected, along belts of supposedly descending currents at deep-sea trenches (Fig. 14.12).

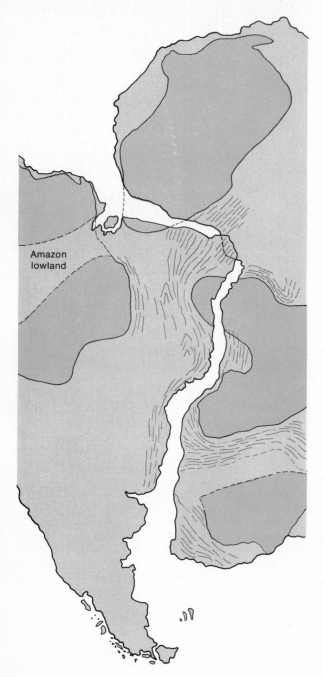

FIGURE 17.22 *Juxtaposed portions of South America and Africa, showing the close correspondence of the age and structure of their rocks. The shaded areas are underlain by very ancient rocks (more than 2 billion years old). The light lines follow the structural trend in somewhat younger rocks elsewhere. The ages are known from radiometric determinations. The disjunct segments in Brazil and the alignment of structural grain are impressive evidence for the former continuity of both sets of rocks, now separated by the Atlantic. (After Hurley, 1968.)*

Character of the Ocean Floor

The type of sediments and sedimentary rocks, their thicknesses, ages, and magnetic polarities, provide convincing evidence that sea-floor spreading is a very acceptable mechanism for moving continents. The nature of the oceanic crust also supports the concept.

The Atlantic and Indian Ocean floors bear possible "microcontinents." An example is the Seychelles Bank (off east Africa), where the crust is 32 kilometers thick.

Mantle Characteristics

The possibility of convection in the mantle requires evaluation of uncertain factors of viscosity, creep, temperature, and an energy source, over which opinions are divided. The existence of differences in the mantle from region to region and at different depths would seem to imply an absence of mixing by convection currents.

Mechanics

Some scientists disagree with the mechanics of sea-floor spreading. Some insist that the deep structure of the earth is incompatible with continental drift by mantle convection and that thin continental blocks cannot be pushed along by near-surface currents on a viscous substrate. Others are critical of the flow pattern proposed for sea-floor spreading. Many accept it as a working hypothesis, however, because it seems to account best for many geological and geophysical features of the ocean.

One great continuous convection cell around the earth has been proposed to account for sea-floor spreading. However, such a pattern is difficult to fit to the central and northern Pacific Ocean basin. There, the current design would have to be one-sided rather than symmetrical. Likewise, the observed displacements on certain large faults, such as the San Andreas fault, do not fit such a drift mechanism. A bothersome question is whether the convection currents are steady and spatially persistent or are discontinuous in both time and space. Situated between the East Pacific Rise and the Mid-Atlantic Ridge, South America should be the "scene of a gigantic struggle" between opposed east-west forces. Only the Andean side appears to be active, however.

Rejecting convection currents in the mantle, some geologists attribute the separation of paired magnetic belts astride the mid-oceanic ridges to a wedging effect caused by repeated intrusions of basalt dikes. Still another theory accounts for the movement of crustal plates by using the pull of gravity along deep-sea trenches, where the plates turn downward. According to this theory, the plate would be subject to tension in its upper part and to compression at a depth of 500 to 700 kilometers.

SUMMARY

The evidences for continental drift, sea-floor spreading, or plate tectonics taken singly are inconclusive. Considered together from unrelated fields of inquiry, they become extremely potent arguments. To make the hypotheses even more acceptable, additional information, about the mantle, the oceanic crust, sediments, and rocks of the deep sea floor, as well as the radiometric ages of the rocks, must be obtained. The present-day findings from gravity studies, heat flow, geomorphology, seismology, structural geology, historical geology, and petrology seem reconciled in unifying concepts. Only the unseen pattern of mantle convection is still mysterious.

Perhaps King and King, in their letter to *Nature* (1971), can provide the best summary of geotectonics.

THE WORLD RIFT SYSTEM AND PLATE TECTONICS OR 1971 AND ALL THAT

"I'll put a girdle round about the earth in forty minutes"—*Puck*[1]

They put a girdle round the Earth
  And named it the Worldwide Rift;
It helps explain the ocean floor
  And Continental Drift.

Vine and Matthews sailed away
  Exploring the ocean bed;
It took much longer getting them back;
  They said it was seafloor spread.

It appears that the oceans were mostly
    formed
  By Caenozoic streams
Of mantle flooding up the cracks
  And gumming up the seams.

Our Earth can scarce make up its mind;
  It flips its magnetic poles
And the magnetized basalts so produced
  Give temporal controls.

When all believed the Earth was flat
  It was but a single plate,
But even then the edges were
  Hot subjects for debate.

They can't curl down; they must curl up
  To form a kind of dish
To stop the oceans spilling out
  And losing all the fish.

We now all know the Earth is round
  And moves about the Sun[2],
So what the ocean spreaders said
  Could not be simply done.

Unless the Earth itself enlarged
  And kept on getting bigger,
But gravity opposes this:
  It is a constant figure.

And so the plate has been revived
  In present day tectonics,
Though sial and sima still remain
  As crustal term mnemonics.

Both kinds of crust now constitute
  The grander types of plates
And as they move upon the Earth
  They suffer subtle fates.

The edges which are growing still
  Are hid beneath the oceans,
While those around the island arcs
  Show self-consuming motions.

Yet others seem to hit or slide
  Performing curious functions,
And where they can't make up their minds
  You there have triple junctions.

On continents the crustal plates
  Are edged by earthquake foci,
And little plates proliferate
  By joining up the loci.

As alchemists once sought the stone
  For magic transmutations,
The motions of the plates are shown
  By seismic computations.

Geologists naively thought
  That rifts were due to faulting,
By subsidence of crustal strips
  Along a pre-rift faulting.

Their evidence was solely based
  On visual observations
Of structure and stratigraphy
  And such out-moded notions.

But others now hold better views
  And think that each "mañana"
Brings Africa a step close
  To the fate of old Gondwana.

McKenzie sees his moving plates
  Wedging the rift asunder,
And one day ships will sail the rift
  To maritime Uganda.

Girdler and Khan from the gravity highs
  That they have in the Gregory rift
See the mantle rearing its ugly head
  And East Africa going adrift.

But the OAU[3] no doubt will vote
  At a suitable early date
To stop that drift and plug the rift
  Before it is too late.

B. C. King
Department of Geology,
Bedford College,
London

G. C. P. King
Department of Geodesy and Geophysics,
Cambridge

SOURCE: *Nature,* vol. 232, July 2, 1971. Reproduced with permission of the authors and *Nature.*

[1]Shakespeare, W., *A Midsummer Night's Dream* (1599).
[2]Copernicus, N., *On the Revolution of the Celestial Orbs* (1543).
[3]Organization of African Unity; Addendum Headline in *The Times,* June 18, 1971, page 6: "Emperor Tries to Close African Rift."

REFERENCES

Atwater, T., 1970, *Implications of plate tectonics for the Cenozoic tectonic evolution of western North America,* Geological Society of America Bulletin, vol. 51, pp. 3513-3536.

Bullard, E. C., Everett, J. E., and Smith, A. G., 1965, *The fit of the continents around the Atlantic,* Philosophical Transactions of the Royal Society of London, Series A: vol. 258, pp. 41-51.

Cox, A., 1969, *Geomagnetic reversals,* Science, vol. 163, pp. 237-244.

Dewey, J. F., 1972, *Plate tectonics,* Scientific American, vol. 226, no. 5, pp. 56-72.

Dietz, R. S., 1961, *Continent and ocean basin evolu-tion by spreading of the sea floor,* Nature, vol. 190, pp. 854-857.

———— and Holden, J. C., 1970, *Reconstruction of Pangaea—Breakup and dispersion of continents, Permian to present,* Journal of Geophysical Research, vol. 75, pp. 4939-4956.

Drake, C. L., Ewing, M., and Sutton, G.H., 1959, *Continental margins and geosynclines. The east coast of North America, north of Cape Hatteras,* in Ahrens, L. H. (ed.), *Physics and Chemistry of the Earth,* vol. 3, pp. 110-198, Pergamon Press, London.

Ewing, M., Ewing, J., and Talwani, M., 1963, *Sediment distribution in the ocean—*The Mid-Atlantic Ridge,* Geological Society of America Bulletin, vol. 74, pp. 17-36.

Gilluly, J., 1966, *Continental drift—A reconsideration,* Science, vol. 152, pp. 946-950.

Heezen, B.C., Tharp, M., and Ewing, M., 1959, *The Floor of the Oceans: I. The North Atlantic,* Geological Society of America Special Paper 65.

———, ———, and ———, 1960, *The rift in the ocean floor,* Scientific American, vol. 203, no. 4, pp. 98-110.

Heirtzler, J. R., 1968, *Sea-floor spreading,* Scientific American, vol. 219, no. 6, pp. 60-70.

Hess, H. H., 1962, *History of ocean basins,* in Engel, A. E., et al. (eds.), *Petrologic Studies: A Volume in Honor of A. F. Buddington,* pp. 599-620, Geological Society of America.

Hurley, P. M., 1968, *The confirmation of continental drift,* Scientific American, vol. 218, no. 4, pp. 52-64.

——— et al., 1967, *Test of continental drift by comparison of radiometric ages,* Science, vol. 157, pp. 495-500.

Irving, E., 1964, *Paleomagnetism and Its Application to Geological and Geophysical Problems,* John Wiley & Sons, Inc., New York.

Isacks, B., Oliver, J., and Sykes, L. R., 1968, *Seismology and the new global tectonics,* Journal of Geophysical Research, vol. 73, pp. 5855-5899.

Johnson, H., and Smith, B. L., 1970, *The Megatectonics of Continents and Oceans,* Rutgers University Press, New Brunswick, N. J.

Le Pichon, X., 1968, *Sea-floor spreading and continental drift,* Journal of Geophysical Research, vol. 73, pp. 3661-3698.

Menard, H. W., 1967, *Sea-floor spreading, topography, and the second layer,* Science, vol. 157, pp. 923-924.

Orowan, E., 1966, *Age of the ocean floor,* Science, vol. 154, pp. 413-416.

Pittman, W.C., III, and Heirtzler, J.R., 1966, *Magnetic anomalies over the Pacific-Antarctic Ridge,* Science, vol. 154, pp. 1164-1171.

Runcorn, S. K. (ed.), 1962, *Continental Drift,* Academic Press, Inc., New York.

Strahler, A. N., 1971, *The Earth Sciences,* 2d ed., Harper & Row Publishers, Incorporated, New York.

Talwani, M., Le Pichon, X., and Heirtzler, J. R., 1965, *East Pacific Rise; The magnetic pattern and the fracture zones,* Science, vol. 150, pp. 1109-1115.

Umbgrove, J. H. F., 1947, *The Pulse of the Earth,* Martinus Nijhoff Press, The Hague.

Vine, F. J., 1966, *Spreading of the ocean floor: New evidence,* Science, vol. 154, pp. 1405-1415.

——— and Matthews, D. H., 1963, *Magnetic anomalies over oceanic ridges,* Nature, vol. 199, pp. 947-949.

Wegener, A. 1924, *Origin of Continents and Oceans,* E. P. Dutton & Co., Inc., New York.

———, 1968, *Static or mobile earth; The current scientific revolution,* Proceedings of the American Philosophical Society, vol. 112, no. 5, pp. 309-320.

——— (ed.), 1970, *Continents Adrift* (Readings from Scientific American), W. H. Freeman and Company, San Francisco.

——— and Hess, H. H., 1970, *Sea-floor spreading,* in Maxwell, A. E. (ed.), *The Sea,* vol. 4. pp. 587-622, Interscience Publishers, a division of John Wiley & Sons, Inc., New York.

GRADATION

Geologic Surface Events in 1972, as Reported by the Smithsonian Institution Center for Short-lived Phenomena

JANUARY — Landslides near Ostrava, Czechoslovakia

MARCH — Jökulhlaups (glacial outburst floods) from the Vatnajökull icecap on Iceland

JUNE — Landslides on Victoria Peak in Hong Kong

JULY — Floods in the Drava, Mura, Raba, and Lajta Rivers in Hungary

SEPTEMBER — Earthflow between Bad Schandau and Krippen, near Dresden, German Democratic Republic

OCTOBER — Formation of a storm ridge on the eastern reef of Funafuti Atoll, Ellice Islands, South Pacific Ocean

Cyclone at Somalia, Aden, and the French Territory of Afars and Issas, leaving up to 198 mm of rainfall in some areas

Mudslides at Big Sur, California

Flooding of the Nerang River and adjacent Gold Coast streams in the States of Queensland and New South Wales, Australia

NOVEMBER — Flooding in eastern Michigan and northern Ohio shore areas

Huge (6.1 meters) swells and waves along the coasts of Washington, Oregon, and California

CHAPTER 18
CHANGE UPON THE EARTH'S SURFACE

In the chapters to follow, we turn our attention to the earth's surface. The concern will be with the more familiar, visible characteristics of geologic change, as exemplified by the ever-changing landscape (Fig. 18.1). Each chapter will present a gradational process, and each will consider in some detail the process itself and the features the process forms to develop a landscape.

LANDSCAPE DEVELOPMENT PROCESS

Very early in Chap. 1, the term gradation was introduced. Gradation is the summation of all processes that decay and erode rock materials, transport and move or shift these rock materials, and deposit the transported materials. These are the processes of weathering and erosion. As a result, the higher topographical levels of the earth are lowered, and the lower topographic levels are built up. Thus, given the earth's irregular surface at the time of its conception, one would suppose that over geologic time the ultimate end result would be a level surface. Between these two end points, the transition stages are revealed to man as landscapes (Fig. 18.2).

It should be noted that the ultimate leveling has never been reached, nor is there a likelihood that it ever will. The study in previous chapters has shown that igneous activity, mountain building, epeirogeny, or even sea-floor spreading all act to oppose the leveling process. However, gradational processes do work to achieve a level called base level. This base level is in reality a gravity-potential level. All gradational processes above sea level work to lower the land surface to that level. Artificial levels similar to sea level, caused by lakes or dams, act temporarily in the same manner as sea level on the gradational processes.

It might be asked, What about gradation below sea level? Is it not true that the ocean basins have their own landscape, much of which is due to gradational processes? If so, what is their base level? To answer these questions, consider what drives all gradational processes. What causes a stream to flow downhill? What causes a landslide to move downhill? The gradational processes operate downslope under the influence of gravity. Thus, ideally, the base level for processes below sea level would have to be another temporary gravity level, the floors of the ocean basins. One still might ask, Is there an ultimate base level?

FIGURE 18.1   *The starkly beautiful landscape of Glacier National Park, Montana. Note the glacially shaped skyline, as well as the U-shaped glacial valley in the foreground. Streams, lakes, and gravity transfer continually modify the results of the glacial process. (Spence Air Photos.)*

Divisions of Gradation

Degradation   The initial breakdown of earth materials by weathering and the removal and transportation of these materials by the various erosional processes is called degradation.

Aggradation   Following the removal and transportation of materials, there is deposition, or aggradation, usually by the same agents that cause degradation. Remember, the erosional process that removes and transports earth material may not be the same process that subsequently deposits the material in its final resting place. Consider the slumping of a stream bank into a stream due to gravity. The initial degradation process is gravity transfer. It provides the materials for the stream, which in turn transports them and subsequently deposits them.

Gradational Processes

The processes of gradation that bring about gradational changes are categorized as follows.

Gravity Transfer   This is the downslope movement of earth materials under the direct influence of gravity. Water or ice may aid the process, but they are usually subordinate to gravity. Gravity transfer is not limited in its areas of operation, as are the other processes. Gravity as a force exists all over the earth; the value of the force is affected by height or relief, and the results of gravity transfer are determined by slope, rainfall, vegetation, or type of material (Fig. 18.2).

Stream Erosion   The extensive development of streams on a new land surface begins with rainfall. Rainfall in itself is an erosional process whereby a type of spatter erosion can occur. The unguided flow of rain down a slope can be termed overland flow, and this causes sheetwash or sheet erosion. Given time and continued rainfall, the overland flow begins to concentrate in rills or gullies, eventually enlarging into streams and river systems (Fig. 18.2).

Streams degrade by the abrasive wear of water, water and sediment, the hydraulic force of the water, and solution. These same forces acting to create a valley also act on the materials carried by the stream itself. The transportation capabilities are very dependent on the discharge characteristics and velocity of the stream. Aggradation results when the transporting capabilities or factors affecting these capabilities are changed or checked (Fig. 18.2).

Solution by Groundwater   This is an effective erosional process below the water table. Groundwater erosion can trigger or cause additional erosion by gravity transfer or assist in stream erosion.

To be effective as a degradation or aggradation process, groundwater is dependent on the water-holding and transporting capabilities of the earth material. The effectiveness of groundwater lies primarily in its solution characteristics; therefore, its resultant land

FIGURE 18.2   *The Columbia River at Rock Island Dam near Wenatchee, Washington. Damming of the river has created a lake behind the dam. Notice that the steep cliffs of basalt flows to the left of the river are being modified by talus slopes due to gravity transfer. The gentle landscape on the right bank of the river reflects in some degree the landscape development by wind. (Spence Air Photos.)*

forms are best exemplified in areas with soluble rocks (Fig. 18.3). Aggradational land forms are generally a result of chemical precipitation from groundwater solutions.

Wind Erosion This is the effect of horizontal air transfer caused by unequal heating of the earth's surface. Associated air movements, however, are often vertical or have a large rotary motion, as in a tornado. Wind erosion is most effective in areas of low rainfall, but gradation capabilities are not confined there.

Degradation is accomplished by the abrasive wear of wind and sediment, as well as by the simple removal of fine sedimentary materials from a land surface. As a moving, transporting energy system, aggradation occurs when the energy factors are changed or checked (Fig. 18.4).

Erosion by Ice Glaciers are masses of ice, usually with lateral limits, that move under the influence of gravity. Quite obviously at the present time, the erosional gradation by glaciers covers limited areas, but during times in the geologic past, glaciers were more extensive, and the resulting landscapes still remain in areas far removed from ice today.

Glacial wear is due to the abrasive force of the ice mass and is even more effective if the ice is embedded with earth materials. Glaciers are also capable of caus-

FIGURE 18.3 *Aerial view of a karst landscape developed by groundwater erosion. The area is near Park City, Kentucky. The flat areas are underlain by limestone, upon which small and large craterlike sinkholes have developed. (U.S. Geological Survey air photo.)*

FIGURE 18.4   *View of Death Valley, looking south from Stove Pipe Wells. The landscape to the right of the highway reflects wind aggradation in the form of dunes. To the left of the highway, and in the upper right portion, alluvial fans have been formed by streams, leaving the higher elevations and reaching the flatter valley floor. (Spence Air Photos.)*

ing degradation by plucking, or dislodging earth material by the force of ice wedging (Fig. 18.1). Deposition of glacially transported materials occurs as a result of ice melt or frictional drag between the moving ice and the land surface.

Marine Processes   The oceans cover about 70 percent of the earth's surface area. The edges of the ocean basins, the shores or coasts (Fig. 18.5), and the basins themselves represent earth-ocean landscapes. Some of the deep-ocean relief features previously discussed (Chaps. 14 and 17) owe their initial existence to internal processes, but their subsequent modification represents gradation below sea level.

The gradational processes in the oceans involve waves and tides, the surface wind-driven currents and deeper currents driven by water density or sediment density changes, the more irregular (time and origin dependent) processes resulting from tsunamis or internal waves, and finally, submarine gravity transfer. Also note that similar processes operate in lakes on the terrestrial landscape.

CONCEPTS OF LANDSCAPE DEVELOPMENT

Each of the gradational processes mentioned has its own characteristics for gradation and its own characteristic landforms. There are similar characteristics

FIGURE 18.5 *Aerial view of the coast near Point Arguello, California. Notice the present-day wave attack and the sea-cliff development. The flat terrace inland represents landscape development by wave process at a previously higher sea level. This terrace is also being eroded by streams leaving the highland areas. (Spence Air Photos.)*

for all, such as the role of gravity in gravity transfer, streams, glaciers, groundwater, and even wind. There are also contrasting characteristics, such as the state of the water medium, be it liquid, ice, or gas.

They all, however, develop the landscape, and a consideration of ten concepts will aid in the interpretation of the geomorphic processes and their landscapes (Table 18.1).

Concept 1 Simply, this concept indicates the principle of uniformitarianism, but the emphasis is placed on intensity varying through geologic time. Perhaps in the geologic past a canyon the scope of Grand Canyon (Fig. 1.1) was created in half the geologic time it took

Grand Canyon to be developed. It may be just as valid to assume it took twice as long. Compare the example illustrated in Fig. 18.2, the present Columbia River, with that in Fig. 18.6, which shows the path and canyon of the ancestral Columbia River.

Concept 2 The term structure implies gross structure, such as a volcanic dome or an anticline, as well as smaller-scale structures, such as joints, permeability, bedding, foliation, or mineral hardness.

Concept 3 Earlier it was noted that the earth's surface has not been and will not be graded level, and that landscapes exhibit highs and lows, or relief, on the surface. Given two areas undergoing stream erosion, would not the relief of the landscape differ if the precipitation amounts yielding stream water differed or the rocks of one area were easy to erode, whereas those in another area were harder to erode? The concept says the processes (or the same process) operate at different rates, depending on the variables involved in the process or the area the process is acting upon.

Concept 4   The concept says that the erosional processes leave distinctive fingerprints known as landforms. Knowing the fingerprint design, one can genetically relate the landforms to the process.

Concept 5   This is an extension of concept 4. Not only do landform fingerprints develop, but each landform may develop as an evolutionary sequence.

Concept 6   This concept emphasizes that, in open systems with many variables, geomorphic evolution is not apt to be simple.

Concept 7   The concept relates to the Tertiary period, beginning 70 million years ago, and to the Pleistocene epoch, beginning perhaps 1 to 1.5 million years ago. Since the earth may have had its beginning 4.5 billion years ago, many landscapes have been formed and continually destroyed since then. The life expectancy of a landscape is not long, geologically speaking. Think about the reasons why such a statement can be made!

Concept 8   The Pleistocene epoch, or ice ages, began 1 to 1.5 million years ago. During the ice ages, over 26 million square kilometers of the earth's surface were covered by glacial ice. Characteristic landscapes were developed over a vast area. Consider this, however: What happened to the streams and rivers on continents partially or wholly covered by glacial ice? What happened to the streams when the ice melted or when streams became dammed by glacial ice (Fig. 18.6)?

Since about 85 percent of our water is in the

TABLE 18.1   Some Fundamental Concepts of Geomorphology

Concept 1
  The same physical processes and laws that operate today operated throughout geologic time, although not necessarily all with the same intensity as now.
Concept 2
  Geologic structure is a dominant control factor in the evolution of landforms and is reflected in them.
Concept 3
  To a large degree the earth's surface possesses relief because the geomorphic processes operate at different rates.
Concept 4
  Geomorphic processes leave their distinctive imprint upon landforms, and each geomorphic process develops its own characteristic assemblage of landforms.
Concept 5
  As the different erosional agencies act upon the earth's surface, there is produced an orderly sequence of landforms.
Concept 6
  Complexity of geomorphic evolution is more common than simplicity.
Concept 7
  Little of the earth's topography is older than Tertiary, and most of it is no older than Pleistocene.
Concept 8
  Proper interpretation of present-day landscapes is impossible without a full appreciation of the manifold influences of the geologic and climatic changes during the Pleistocene.
Concept 9
  An appreciation of world climates is necessary to a proper understanding of the varying importance of the different geomorphic processes.
Concept 10
  Geomorphology, although concerned primarily with present-day landscapes, attains its maximum usefulness by historical extension.

SOURCE: Thornbury, W. D., 1969, *Principles of Geomorphology*, 2d ed., pp. 16-34, John Wiley & Sons, Inc., New York.

FIGURE 18.6 *Ancestral path of the Columbia River south of Grand Coulee, Washington. The river flowed toward the viewer, creating this immense canyon, or coulee. An ancestral waterfall (Dry Falls) can be seen in center of the photo along with ancestral plunge pools. The Columbia River's ancestral path was formed when the original channel was blocked by a glacial ice dam. (Spence Air Photos.)*

oceans, if this water helped to form glacial ice, did sea level lower? If sea level lowered, did waves and tides affect different areas than they do now? What about wave and tide erosion as the ice melted? Also, with so much water tied up in ice, did arid or desert areas cover different or larger areas?

Concept 9   This concept points out direct and indirect effects that climate must have on the processes that develop landscapes. For a direct factor, consider the amount of precipitation, while indirect effects could be the amount and coverage of vegetation.

Concept 10   This concept explains the necessity of concern that geologists have for recognizing and interpreting landforms produced by processes no longer active, and it reaffirms the fact that geology is a historical study of change (Fig. 18.1).

REFERENCES

Spencer, E. W., 1972, *The Dynamics of the Earth,* Thomas Y. Crowell Company, New York.

Thornbury, W. D., 1969, *Principles of Geomorphology,* 2d ed., John Wiley & Sons, Inc., New York.

On October 15, 1972, newspapers in California reported that "...mudslides created by blinding rains poured down the scenic hills of Big Sur... burying parts of Coast Highway No. 1 by over a yard of mud." A picture caption in the March 1972 issue of *National Geographic* tells of the "flowing earth" of the tundra, where moisture-laden topsoil flows slowly (during a summer season) down the foothills of the Alaska Range (Fig. 19.1). In 1971, newspapers all over the world reported the total destruction of the Quebec, Canada, town of St. Jean-Vianney. Areas within the town started to subside vertically and then proceeded to collapse and "flow as a river of mud." Mudslides, "flowing earth," and subsidence are three examples of the downslope movement of earth materials by gravity transfer.

All rocks, soil, and organic matter are under the influence of the inexorable force of gravity. Without support, they move directly, if possible, or obliquely, if controlled by slope or another force, toward the center of the earth.

Narrow streams and slope wash move earth materials in limited areas, but gravity transfer is a far more extensive mechanism for leveling the land. The development of hill slopes involves a complex interplay of many erosional processes, but this chapter stresses gravity. In polar regions, gravity transfer is more important in moving material than any other process.

Slow vertical subsidence of the land in some places follows withdrawal of groundwater and oil from pore spaces in the rock. Catastrophic collapse occurs in caves and mine workings. However, the most familiar examples of gravity transfer are slides and mudflows (Fig. 19.2). These movements are common on all steep slopes, both above and below sea level. Less obvious is the almost imperceptible creep of earth material on slopes (Fig. 19.1). Quantitatively over geologic time, creep can be more important than landslides in moving material to lower levels.

THE GRAVITY-TRANSFER PROCESS

Running water, glaciers, and wind all move because of gravity. This chapter is concerned with the movements of unconsolidated earth materials and broken rock over a period of years, or even minutes or seconds, entirely or largely due to the gravitational force alone. This process is gravity transfer.

FIGURE 19.1 *Solifluction waves with vegetation nets from frost processes in soil in the tundra of central Seward Peninsula, Alaska. (R.F. Black photo.)*

While movements by gravity transfer are producing distinctive deposits and landforms under a variety of conditions today, relic forms representing similar occurrences in the past have been found in many areas. The assumption is that gravity transfer has been operating as an erosional process as long as gravitational attraction has been acting on objects at the earth's surface.

To adequately consider gravity transfer as an erosional process, it is necessary to briefly discuss those factors that contribute to the effectiveness of the process. Actually, it is nearly impossible to specifi-

FIGURE 19.2 *Madison landslide, Montana, and a newly formed lake resulting from damming of the Madison River. (J.R. Stacy, U.S. Geological Survey.)*

cally isolate one contributing factor that plays the major role in the effectiveness of gravity transfer.

While certain factors have been separated for the following discussion, the descriptions of the various gravity-transfer mechanisms later in the chapter will show the interdependence of contributing factors.

Materials The various mechanisms of gravity transfer will move coarse-rock waste called debris, fine-rock waste called soil, and fine-grained loose silts and clays or mud. The particle size of the material transferred (competence) and the amount of material transferred (capacity) will be a function of the gravity-transfer mechanism and its characteristics.

Slope The slope of the land surface has many characteristics that are important in gravity transfer. These characteristics include the steepness of the slope (some slopes are oversteepened due to erosion or excavation by man), the characteristics and composition of the material forming the slope (i. e., jointed, foliated, bedding directions), and overloading. Some slopes are overloaded due to the natural addition of earth material or additions by man.

Geologic Aspects While the type of material is a factor, gravity transfer is also dependent upon the compositional, textural, and structural characteristics of the material. Consider a few examples: a large amount of the glacial and postglacial clays in Canada and Alaska are quickclays, whose compositions are such that they spontaneously liquefy under physical disturbance. Often, as was illustrated during the 1964 Alaska earthquake, little slope is necessary to precipitate movement if quickclay is found in the subsurface. Many rock outcrops on slopes may be highly jointed or fractured, and this helps to promote the simple breakaway and downslope movement of material. Many sedimentary rocks exhibit a wide variety of bedding characteristics, from thin laminae to massive beds. Others may have alternate thick and thin bedding, and still others have competent and incompetent beds. Metamorphic rocks exhibit varying degrees of foliation, and igneous rocks have a wide variety of joint and/or fracture patterns, as well as having bedding, as illustrated by lava flows.

Climates The characteristics of climate are also important in gravity transfer and help to define the specific type of transfer. Forest and brush fires during the dry seasons of coastal California destroy slope cover and eliminate the protection from the often torrential rains that follow.

In very cold climates a permafrost layer is common. This layer, a short distance beneath the surface, prevents infiltration of surface moisture. As a result, the topsoil remains saturated. Then during the summer months when it thaws, the soil moves slowly downslope.

Triggering Mechanisms The triggering mechanisms for gravity transfer are varied. They can include vibrations from earthquakes, oversteepening or overloading of slopes, changes in water pressure within the earth material, intense rain infiltration, freezing and thawing or the presence of springs, solution of rock cements, or sudden compaction of water-saturated materials.

CLASSIFICATION OF GRAVITY-TRANSFER MECHANISMS

As surmised from the discussion of those factors of gravity transfer, there are several different mechanisms of gravity transfer. Any attempt to classify these various mechanisms must consider the following:

1. The type of material being transferred
2. The state of lubrication of the material, whether it be dry or wet with the presence of water, ice, or snow (Fig. 19.3)
3. The inclination and conditions of the slope
4. The speed of the moving material
5. The "triggering" mechanism
6. The condition of the material during and after transport
7. The method of movement, such as, rolling, bouncing, sliding
8. The presence, or absence, of a shearing or slip surface

It almost goes without saying that a classification system which would attempt to interrelate all these factors would be unwieldy just in terms of the extensive

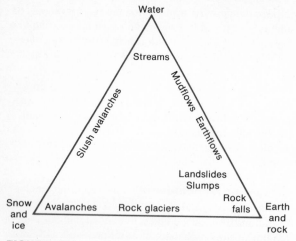

FIGURE 19.3 *Diagram illustrating the interrelationships of movements involving water, snow, and ice with earth and rock.*

terminology introduced. Observations of gravity transfer, and the results, generally indicate that a clear-cut separation or delineation of gravity-transfer mechanisms does not exist.

As a working model for further discussion, four mechanisms of gravity transfer will be defined and utilized (Table 19.1).

Falling

Free fall is a common and characteristic movement of fragments of rock from cliffs (Fig. 19.4). Rocks falling in air commonly break by impact when they strike the surface below them. From that point they may bounce, roll, or slide if not embedded.

The rock fragments are freed from cliffs by frost and organic wedging (such as root growth), wetting and drying, chemical weathering, undercutting, and expansion cracking. This action may take place piece by piece or involve large masses of rock. For example, in 1941, a huge sandstone block, 45 meters long, 30 meters wide, and about 10 meters thick, fell and damaged a portion of the ruins of Pueblo Bonito, Chaco Canyon, New Mexico.

The free fall of earth material is the chief process supplying fragmental material, known as talus, found at the foot of cliffs in rugged areas. Talus accumulations develop as wedge-shaped or cone-shaped deposits (Fig. 19.4). Cliffs rise above the talus, and cliff retreat is a discontinuous process. Sudden rock falls are separated by long periods of stability during which the talus is gradually broken down and removed.

Sliding

Material transferred by sliding or slumping can be rock masses, rock fragments, or soil. In this type of movement, gravity must exceed the friction of one body against another on a sliding surface. The surface of any solid, if magnified enough, appears rough and causes it to oppose sliding across another surface; thus, the nature and slope of the slide plane are extremely important.

Water confined under pressure along the sliding surface reduces friction by "floating" the material above, reduces the force holding the objects together, increases the weight of the material, and decreases its shearing resistance, or strength.

Slides   When large masses of earth and rock slip bodily down steep slopes and break into discrete fragments, the movement is called a slide.

Most slides are greatly deformed masses in which flowage also occurs. Movements range from extremely slow (less than 1 meter in 5 years) to extremely fast (more than 3 meters per second). Different terms are used for slides to indicate the nature of the material, such as debris slides and rockslides.

Slides leave arcuate headwall scarps and open

TABLE 19.1   Gravity-Transfer Mechanisms

Falling	This is best and simply characterized as "free fall" of earth material
Sliding	Movement on a shearing slide plane. Particles of the earth material tend to retain their relative orientation after movement
Flowage	Movement that generally does *not* involve a distinct basal shearing. Particles of earth material tend to become totally disoriented in rapid flowage
Subsidence	Movement is primarily a settling, or vertical collapse, and may involve components of, or lead into, additional sliding or flowage

FIGURE 19.4   *Talus-slope and rock-slide materials at the Lake of Jade, Mount Revelstoke National Park, British Columbia. (Canadian Government Travel Bureau.)*

FIGURE 19.5   *Toe of landslide near the Glenn Highway, south-central Alaska. R. F. Black photo.)*

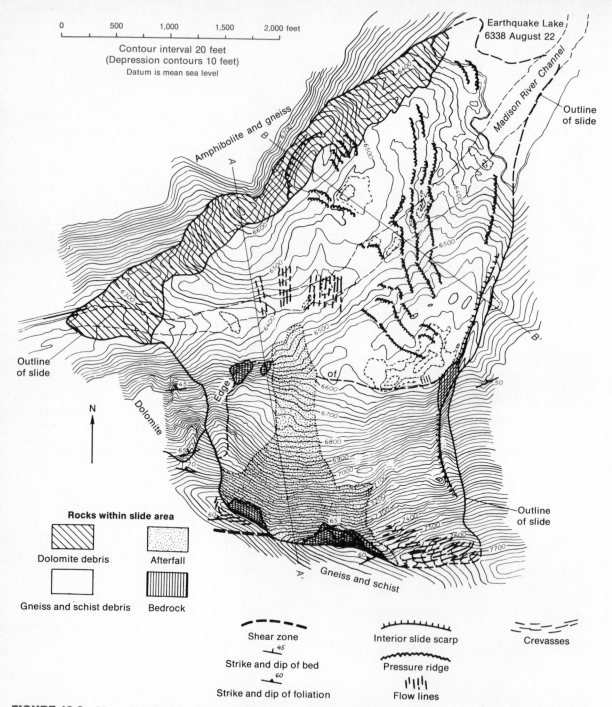

0   500   1,000   1,500   2,000 feet

Contour interval 20 feet
(Depression contours 10 feet)
Datum is mean sea level

Earthquake Lake
6338 August 22

Madison River Channel

Outline
of slide

Amphibolite and gneiss

Outline
of slide

N

Dolomite

Edge   of

Outline
of slide

Rocks within slide area

Dolomite debris

Afterfall

Gneiss and schist debris

Bedrock

Gneiss and schist

Shear zone

Strike and dip of bed ⎯⎯ 45

Strike and dip of foliation ⎯⎯ 60

Interior slide scarp

Pressure ridge

Flow lines

Crevasses

FIGURE 19.6 *Map of Madison landslide, Montana. Sections A-A' and B-B' are shown in Fig. 19.7. (From U.S. Geological Survey Professional Paper 435, fig. 59.)*

cracks near their origins and tilt and jumble their blocks (Fig. 19.5). They may create hollows for ponds and lakes on their irregularly hummocky surfaces and dam up valleys temporarily to make lakes. These geographic forms are distinctive of slides.

(a) Madison River Slide, Montana Thousands of destructive slides have taken place in recent centuries, many initiated by earthquakes. On August 17, 1959, along the Madison River, Montana, a slide involving 50 million metric tons of material was triggered by an earthquake. The Madison slide (Figs. 19.2, 19.6, and 19.7) carried 27 million cubic meters of broken rock into the canyon. This slide buried 1.6 kilometers of the river and highway, interrupted the flow of the river (until the slide was deliberately breached by bulldozers), and formed a lake nearly 10 kilometers long and 57 meters deep at its deepest point. The slide moved perhaps 160 kilometers per hour and owed its size and destructiveness to local structural and geomorphic conditions. These conditions had produced a dynamically unstable slope ready to be set into motion. Deep weathering and fracturing of gneiss and schist and a resistant mass of dolomite determined the depth of the surface of rupture (Figs. 19.6 and 19.7).

The center of mass of the slide (in section *A-A'*, Fig. 19.7) traveled only 300 meters on a 30° slope, but the slide carried dolomite blocks as large as 10 meters across.

(b) Anchorage, Alaska, Slides The susceptibility of many geologic areas to sliding is well known; most slides are predictable. For example, consider the massive destruction accompanying the Good Friday earthquake, March 27, 1964, in Alaska. A U.S. Geological Survey report had warned that an earthquake might trigger slides in Anchorage and vicinity. Slides did more damage than the earthquake; the damage reported for the Anchorage area alone amounted to $200 million.

The slides moved downslope without much rotation, and despite wide variations in size, appearance, and complexity, slid on an underlying layer of quick-clay. Some slide failures were assisted by the build-up of high pore-water pressures in sand layers.

Slumps Another category of the sliding mechanism is slumps. Slumps are blocks of material that tend to retain their integrity. The blocks move downward and

outward or rotate along one or more curved shear planes (Figs. 19.8 and 19.9).

Slumping is quite commonly seen on stream banks. Given a bank that is composed of unconsolidated material, silts, sands, and clay, lateral stream erosion can quickly cause the bank to become oversteepened. The unstable bank will then tend to break away in more or less coherent blocks, resembling large terraces or steps.

The Palouse Hills in eastern Washington and some slopes of the Coast Ranges of California bear minutely terraced surfaces. A great number of these small terraces are small-scale slump blocks. The appearance is a series of long zigzag interwinding paths from the bottom of the hill upward. In areas of the country where these slump terracettes appear, extensive cattle grazing is common. It is difficult to assign these terracettes a geologic origin or to attribute them solely to roaming cattle. The two causes may be interrelated. It can be shown, however, that on many slopes which exhibit this effect, there is little or no animal grazing.

On a larger scale, slumping is quite common near coastal cliffs. Near Point Fermin, California, in 1941, a large slump completely displaced the coast paralleling a highway (Fig. 19.9). The large slump block also had a tendency for its leading edge (cliff edge) to show further breakup by smaller slumps. The cause for the slumping was lateral wave erosion at the foot of the cliff, creating conditions similar to the stream-bank slumps cited earlier.

Flowage

This mechanism of gravity transfer involves movement where a distinct slip plane generally does not exist. During flowage, the material involved is disoriented during and after transport. In this category belong such events as earth, mud, or debris flows and rock or soil creep. No sharp line of demarcation is possible between slow and rapid flowage or between rapid flowage and another gravity-transfer mechanism, slippage.

Earth, Debris, and Mudflows Earth flows are made up of smaller fragments of earth materials, and debris flows are composed of the coarser detritus. Mudflows refer to those flows made up largely of the

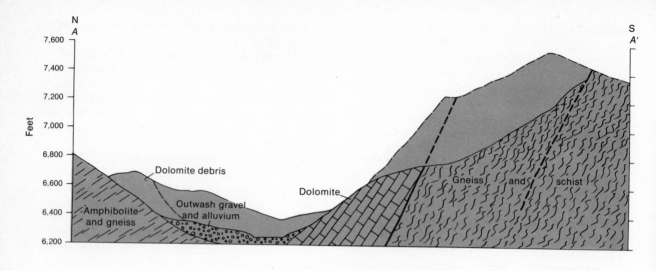

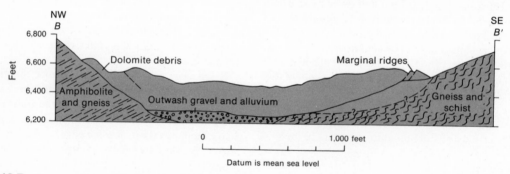

FIGURE 19.7 *Cross sections of Madison Landslide, Montana. (U.S. Geological Survey Professional Paper 435, fig. 60.)*

finest mud (Fig. 19.10). Earth flows, mudflows, and debris flows thus constitute a family subdivided mainly on the basis of material they contain. Generally, rapid flowage involves movements slower than the sliding and falling movements previously described. Arbitrarily, these flows are distinguished from soil flow. The latter term is usually applied to thin flows taking place slowly in summer in cold climates.

Mudflows are most common in alpine areas of the temperate zones of the globe, but a flow along the Saguenay River, Quebec, moved 20.5 million cubic meters in an area of 21 square kilometers on slopes as low as 2°. Mudflows also are common in desert regions drenched by sudden cloudbursts (Fig. 19.10). They are especially likely to occur in areas where deeply

weathered material contains clay or silt. Mudflows are less common in humid regions than in deserts because of the protective effects of vegetation, but especially favorable conditions exist in rain forests, as on Oahu, Hawaii. There, heavy rains periodically set off debris flows in narrow belts down steep mountain slopes. In semiarid regions, fires following abnormally dry seasons can destroy the vegetation, thus providing another potential cause of mudflows.

Active volcanoes with thick accumulations of unstable volcanic ash and cinders are subject to flows and creep. Important contributing factors are steep slopes, unstable material, and steam that condenses to form heavy rains during eruptions.

Big Sur Mudflow, California, 1972   An example of a

FIGURE 19.8 *Forward tilting and sliding and backward rotation and slumping of blocks of conglomerate, New York. (R. F. Black photo.)*

FIGURE 19.9 *Slump at Point Firmin on the coast of California. (Spence Air Photos.)*

FIGURE 19.10 *Margin of a recent mudflow (now dry and sun cracked) on an alluvial fan along the east side of the Stillwater Range, Churchill County, Nevada. (Eliot Blackwelder.)*

FIGURE 19.11 *Mud and debris around and nearly burying the store and post office at Big Sur Village, California. The mud- and rock flow occurred in Pheneger Creek Canyon. (California Division of Highways.)*

mudflow occurred October 14-18, 1972, in the Big Sur area of California. During the previous summer, a series of intense fires had burned 4,700 acres in the area of the Pfeiffer-Big Sur State Park. The fires eliminated the cover of trees, brush, and grass on the slopes in the area. The denuded slopes could not absorb the heavy, intense rain that followed in October.

The results were described as "hillsides melting like ice cream in the sun." Mudflows pouring down the damaged slopes were described in such terms as "a wall of mud 2 meters high and 12 meters wide." Observers likened the flows to "a brown troop train" moving for 35 minutes with trees, rocks, and debris floating by. At Big Sur Village, debris stood up to 2 meters deep in shops and buildings (Fig. 19.11). Coast Highway 1 was blocked by tons of mud, in some areas 2 meters deep. The area was finally declared a local disaster area as officials noted that 72,000 metric tons of mud and debris were still precariously remaining on slopes above the village 7 days after the initial rains, with more forecast. This sequence of events was repeated in February 1973, again after extremely heavy rains.

Creep and Solifluction   Creep is the slow, often intermittent but persistent, downward migration of soil (Fig. 19.1) and loose rock (Figs. 19.12, 19.13, and 19.14) under the force of gravity. Freezing-thawing and wetting-drying produce alternating expansion and contraction that locally are dominant mechanisms at the surface.

Solifluction was defined initially as surface soil flowage in northern areas where the ground commonly remains perennially frozen. It is now widely used everywhere for any soil flow, although some geologists still restrict the term to soil flow on perennially frozen ground. In such areas the uppermost zone that thaws in summer, the active layer, creeps and flows downhill over the underlying frozen ground (Fig. 19.1). The top of the perennially frozen zone prevents penetration of soil meltwater, thus augmenting the conditions favorable for soil flow. Viscous and plastic movements are common on slopes of more than 2° and may occur on slopes of only $\frac{1}{2}$°.

Rock creep is the slow downhill movement of rock blocks occurring where well-jointed rocks are favorably situated on a slope.

The role of creep versus runoff from rain on slopes also is very difficult to distinguish, except under very favorable circumstances. Both occur simultaneously or alternately on most slopes.

Soil Creep   On a soil-covered slope, soil creep may

FIGURE 19.12   *Part of schist block weighing many tons is separating by creep processes from the hillside above the Alaska Railroad, endangering future passengers. (Black, U.S. Geological Survey.)*

FIGURE 19.13 *Creep of soil and warping of railroad tracks north of Nome, Alaska. (R.F. Black photo.)*

FIGURE 19.14 *Creep of shale, northern British Columbia. (R.F. Black photo.)*

be responsible for tilted fenceposts, telephone poles, or gravestones and for broken or displaced retaining walls or tilted tree trunks. It can move highways and railroads out of alignment (Fig. 19.13).

Detailed studies of slopes in northern Alaska and northeast Greenland show that downslope movements are most pronounced during fall freeze-up and spring breakup and are erratically distributed over a particular slope. Slope orientation, water content, material, gradient, and vegetation are important factors.

Creep may go on beneath a continuous covering of sod or trees, where the lower part of the soil mantle on bedrock may move faster than the surface, or depending on the situation, it may move slower than the surface. Where boulders are present in the soil, turf or soil ridges are pushed up downslope from the creeping boulders.

Subsidence

A large number of natural or man-caused instances of gravity transfer occur in which subsidence (settling or vertical collapse) is the prime mechanism during the initial transfer of material. For example, there are numerous accounts that relate the subsidence and collapse of the land surface above old mine workings.

In areas underlain by limestone and limestone caves, subsidence, settling, and collapse are common occurrences. The solution erosion of the soluble rock by groundwater creates an unstable land surface. The collapse of cave or cavern roofs has been reported many times and is observable in areas of Virginia or Puerto Rico. Where the subsidence is not complete enough to expose the subsurface caverns, surface depressions, known as sinks, result. Groundwater collects in the sinks and creates ponds. Thus in an area of soluble rocks and groundwater erosion, the distinctive topography formed owes its existence to subsidence, as well as to groundwater-solution erosion.

A more spectacular example of initial subsidence occurred in the town of St. Jean-Vianney in Quebec, Canada, in 1971. The event illustrates the near-simultaneous occurrence of several mechanisms of gravity transfer. St. Jean-Vianney, with a population of 1,308, is situated about 62 kilometers north of Quebec city. It is located near the Saguenay River, called the "river of broken lands." The town was actually built on the de-

posit of a huge slide that had occurred 500 years earlier. This earlier slide developed on relatively flat land, and the prime cause appeared to be the geologic setting of the area. Below the topsoil are extensive marine clay layers with pockets of sand. The clay is about 30 meters thick. The sand pockets, when they become oversaturated, exert pressure on the clay. The pressure or disturbance on the clay causes the clay to lose its cohesiveness and become fluid. Because of its geologic setting, the St. Jean-Vianney area is considered to be one of six landslide danger zones in Quebec Province.

The 1971 subsidence, sliding, and flow in St. Jean-Vianney apparently were due to the same geologic problems. Triggering the event was a month of heavy rain and warm temperatures that were producing a fast thaw. An earth tremor on April 20 may also have contributed. Starting on April 23, 1971, the following events occurred:

1. Half of a 12-meter hill (Blackburn Hill) outside of the town disappeared. In its place was a V-shaped hole, 24 meters deep, 61 meters wide, and 151 meters long.
2. Driveways, streets, and house foundations started settling; subsidence up to 10 to 20 centimeters occurred. Residents started hearing a "flowing of water" beneath the ground surface.
3. On May 4, the remainder of Blackburn Hill completely disappeared, and large gaping holes developed all over town. The subsidence apparently began at Blackburn Hill and moved west. As the land subsided, it appeared to collapse and flow, creating a canyon nearly 30 meters deep. In the canyon, liquefied clay, 20 meters thick, was flowing westward toward the Saguenay River at 7 meters per second (16 miles per hour). The canyon continued to widen, reaching 0.8 kilometer across, 30 meters deep, and 1.6 kilometers in length.
4. The subsidence, sliding, and flowing continued for 2 hours, virtually destroying the community. All that remained was the canyonlike hole. Thirty-one residents were killed.

Submarine Gravity Transfer

All the previous examples of gravity transfer have been descriptions of this process above sea level. Gravity transfer also occurs below sea level. It most commonly occurs on the continental shelves, slopes, and rises of the sea floor; however, evidence of gravity transfer can also be found in the deeper ocean basins, or abyssal plains.

The mechanisms of submarine gravity transfer can be placed, for the most part, into the land-surface classification system. Even though different terminology is used, the principles of flowage, or sliding, can be applied.

Submarine Slumps, Slides, and Turbidity

Currents   Slow deposition of sediments in the oceans through long periods of time produces thick deposits, not only on the continental shelves but also on the continental slopes. These deposits are susceptible to subaqueous gravity transfer, easily triggered by earthquakes, unusual tides and currents, or additional sediment weight.

Sediment deposited rapidly has excessive pore space, that is, open packing of the grains that are not fitted tightly together. If water fills the pores of such a sediment and then is subjected to vibration, the individual particles shake down to fill the gaps. Earthquake vibrations commonly provide the energy needed to pack the material more tightly, expel the excess water, and initiate submarine slides.

Turbidity currents are currents that have a high density due to sediment in turbulent suspension. Once an unstable sediment is induced to slide down a subaqueous slope, it mixes with water and picks up speed. Turbidity currents may be triggered by earthquakes, storm waves, floods, unusually heavy tides, or an over-steepened depositional slope. The amount of sediment and triggering effects control the size and frequency of such currents. Off the mouths of major rivers, such as the Congo, Magdalena, and Mississippi, continuous sedimentation builds up thick unstable delta deposits that periodically develop turbidity currents from slides off their steep fronts. Turbidity currents caused in this way are probably more important in terms of total material transported than the more infrequent but larger currents induced by earthquakes from the continental slopes. Some excellent examples of submarine slides have been recorded. Two of these are described here.

(a) Seward and Valdez, Alaska   In 1964, submarine slides and large waves accompanying and following the Good Friday earthquake did much damage to the towns of Seward, Whittier, and Valdez, Alaska. At Seward, 1,500 meters of the waterfront slid into Resurrection Bay, taking with it associated waterfront installations and industries. The ground that slid was water-saturated mud with a submarine slope near shore of 30 to 35°. The water near the former dock, which was about 10 meters deep, is now 50 to 65 meters deep. Large waves were generated as a result of the slide.

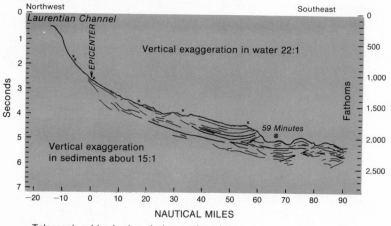

Telegraph cables broken during earthquake ×
Telegraph cable broken 59 minutes later ⊗

FIGURE 19.15   *Cross section of the Grand Banks slump area. (From Heezen and Drake, 1964, Copyright American Association of Petroleum Geologists.)*

At Valdez, at the northeastern corner of Prince William Sound, the dock area disappeared in a sudden submarine slump generated by the Good Friday earthquake. Valdez is on an alluvial delta of unconsolidated sediments built out into the head of the deep, steep-sided fjord of Port Valdez. The submarine slump was 1,200 meters long and 200 meters wide. Water depth at the former dock increased from 12 to 36 meters, and the small boat harbor, which formerly was exposed at low tide, now is covered by about 20 meters of water.

(b) Grand Banks   A seismic-reflection profile (Fig. 19.15) and a map (Fig. 19.16) across the area affected by the Grand Banks earthquake of 1929 in the North Atlantic reveal what appears to be a large gravitational slide over 400 meters thick, 100 kilometers long, and possibly of greater width. Another area of slumplike features lies south of Georges Bank. Apparently, the 1929 earthquake generated several slides that transformed into turbidity currents which flowed around and in part over the gravitational slides (Fig. 19.16). The slumps broke many transatlantic communication cables in the main earthquake area, while the subsequent turbidity currents then caused a systematic sequence of delayed breaks from north to south over a period of 13 hours. Some cables were up to 700 kilometers away from the epicenter. Maximum velocity achieved by the turbidity current has been estimated at 96 kilometers per hour. At the time of the last cable break, the current was probably moving about 22 kilometers per hour.

Lakes and Gravity Transfer

The gravity transfer of rock debris, such as a mudflow or slide, may dam a valley and produce a lake basin. A dam generated by a slide is generally temporary. This occurs unless there are enough boulders present to allow the outlet stream to form a natural spillway.

A few examples may be cited. Lake San Cristobal in Colorado was formed by the Slumgullion mudflow. Another lake produced by the Gros Ventre landslide in northwestern Wyoming overtopped and washed out the slide dam formed in 1927, wiping out half of the small town of Kelly. In Madison Canyon, Montana, west of Yellowstone Park, a landslide caused by the Hebgen earthquake formed Quake Lake (Fig. 19.2) in 1959 by blocking the Madison River. In order to

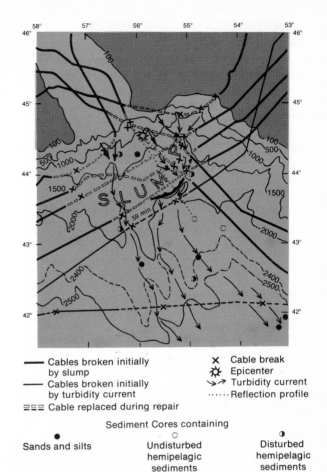

FIGURE 19.16   *Sketch interpreting the Grand Banks slump area, where an earthquake apparently generated several slumps, slides, and turbidity currents. (From Heezen and Drake, 1964, copyright American Association of Petroleum Geologists.)*

prevent a comparable repetition of the events at the Gros Ventre slide, the Corps of Engineers hurriedly cut a spillway, lowering the level of Quake Lake before it overtopped the Madison slide. The spillway appears to have become armored or protected by large blocks within the slide and seems to be stable.

Other lake basins are found in the scars created by slides and in depressions on irregular slide surfaces.

MEASUREMENT AND PREDICTION

Quantitative measurement is not particularly difficult in small local movements, but it becomes a complicated task over large areas and long periods of time. One of the first major studies was done in the 1950s in the Swiss Alps. Quantitative effects of gravity movements were compared with stream processes, glaciation, wind work, solution, and others. Although the reliability of the information varied, the data showed that rock slides involved the greatest mass and volume of material, but they affected only 6.5 percent of the area.

A more detailed investigation was attempted in the Karkevagge valley of northern Sweden. Over a period of 8 years, the dominant single process was gravity transfer, including, in order of relative importance, earth slides, mudflows, rockfalls, and creep and solifluction. Earth slides and mudflows, as a group, moved on the average 70 metric tons of material per square kilometer per year as compared, for example, with 6 metric tons per square kilometer carried in solution by streams.

Quantitative measurement and differentiation of solifluction and of various creep processes is difficult, time consuming, and involves special equipment since the movements are only minute fractions of a centimeter. Such studies carried out over a period of many years in some test sites in Greenland and near Madison, Wisconsin, show that, generally, both creep and solifluction operate simultaneously, but one process dominates at any one time. Near Madison, wetting and drying produced swelling and shrinking just as effectively as freezing and thawing; hence, creep is especially effective in that area. The process lowers the steeper nonwooded slopes about 2 millimeters per year.

Measurement of many rockfalls, slumps, and earth flows have been made after failure, but few slopes have been, or are being, watched closely enough to predict when future failures will occur. Recently a few communities have enacted zoning laws requiring expert advice or certification of property prior to construction in slide-prone areas. Unfortunately, residents are rebuilding homes in the slide area in Anchorage affected by the earthquake of 1964, though geologists agree the area is likely to move again.

REFERENCES

Kerr, P. F., 1963, *Quick clay,* Scientific American, vol. 209, no. 5, pp. 132-142.

Sharpe, C. F. S., 1938, *Landslides and Related Phenomena,* Columbia University Press, New York.

Thornbury, W. D., 1969, *Principles of Geomorphology, 2d ed.,* John Wiley & Sons Inc., New York.

CHAPTER 20
STREAMS

The work of running water is important to man. River deposits provide fertile land for agriculture and level land for cities. Rivers are pathways for commerce and a source of water for everyday use. They sustain crops, run machinery, and provide, via their valleys, access routes to mountain ranges and recreational areas.

Rivers sculpture the land by picking up sediment material and transporting it from one point on the land surface to another. Landscape features result from erosion, deposition, or a combination of these methods.

PROCESSES OF RAINFALL EROSION

Raindrop Impact

Erosion begins as raindrops strike the ground. Perhaps you have noticed a bare earth surface after an intense rain, where each small pebble has protected the underlying soil and now rests on a pedestal of earth. The unprotected soil around it is gone (Fig. 20.1). Falling raindrops possess a surprising amount of kinetic energy that, upon impact, is expended in loosening and picking up fine materials from the unprotected soil.

Assuming all the droplets of a 10-centimeter rain to be 0.25 centimeter in diameter, the terminal velocity is on the order of 7 meters per second (23 feet per second.) The total kinetic energy of the raindrops in one such rainstorm is sufficient to raise 10 centimeters of the soil to a height of more than 2 meters (Linsley et al., 1949). The soil is not actually lifted 2 meters, but once soil particles are loosened, their resistance to further erosion has been overcome.

Sheet Flow, Sheet Erosion, and Gullying

When precipitation exceeds the infiltration capacity of the soil on slopes, the excess water becomes surface runoff. On smooth slopes, runoff may take the form of sheet flow, which is a thin film of water moving downslope, causing sheet erosion. Careful studies of topsoil loss from cultivated fields point up the fact that sheet erosion combined with raindrop impact is responsible for the downslope movement of tens of tons of soil per acre per year. Sheet flow is normally capable of eroding only fine silt and clay, leaving behind sand and pebbles as a residual deposit.

Water moving across an irregular land surface as

309

FIGURE 20.1 *After an intense rain, each small pebble has protected the underlying soil and now rests on a pedestal of earth. (R. K. Fahnestock photo.)*

sheet flow may be concentrated by irregularities on the land surface into small streamlike features called rills. The channel of water into rills, with their larger flows or more concentrated volumes and higher velocities, makes a more effective erosional agent than sheet flow. They modify the originally smooth slopes by enlarging their small gullies into miniature steep-walled valleys (Fig. 20.2). Continued growth of rills is accomplished by headward erosion and downcutting. Eventually, branching systems of drainage lines are established; each collecting the runoff from its own small drainage basin. The drainage basin is the surface area that contributes water and sediment to a stream. The boundary of this area is called the drainage divide, since it separates the drainage of one system from that of another.

The most severe erosion takes place on slopes lacking a protective cover of vegetation. Vegetation protects the land surface from erosion by breaking and stopping raindrops before they hit the ground. Roots serve to hold the soil against erosion by the raindrops not previously intercepted.

Under natural conditions and depending upon evaporation rates, the maximum sediment loss occurs with a precipitation of 25 to 35 centimeters per year. This is enough precipitation to produce significant runoff, but not enough to sustain a protective vegetative cover. With lower or higher rainfalls, the sediment yield is markedly less. A lower yearly precipitation means less runoff to cause erosion, while a higher yearly precipitation usually supports enough vegetation to protect the land.

Thus, in summary, initial lowering of the land surface is accomplished by weathering, gravity transfer, rainsplash, and slope wash. After the development of stream valleys, gradational processes continue to deliver sedimentary materials to streams. The stream itself may continue to deepen and widen its valley and increase its drainage basin by headward erosion. Valley development generates new slopes, thus inten-

sifying gradational processes. These erosional processes acting on slopes of the drainage basin are extremely important in lowering the land surface because
only about 5 percent of the land is covered with stream
channels. The soil and fragments delivered to a stream
by these processes are transported away.

The lowering of a land surface is illustrated in the
basin of the Mississippi River and its tributaries, an
area of approximately 3.2×10^6 square kilometers.
The Mississippi basin is being lowered at an average
rate of about 0.3 meter in 9,000 years. It must be kept
in mind, however, that a large part of the Mississippi
drainage basin is covered by easily eroded, unconsolidated glacial deposits and wind-deposited silt.
The rate of lowering of bedrock basins may be considerably slower.

MECHANICS OF STREAM WORK

Turbulence and Velocity

In natural channels, water molecules move in very irregular or criss-crossing paths that tend to mix the
water thoroughly. This type of movement is called
turbulent flow. Average velocities of stream flow, 1 to
3 meters per second, occur above the stream bed.

Velocities nearer the stream bed are much lower
because of the frictional effects. Bottom turbulence
stirs up sediment and helps keep sediment suspended
within the flow.

Turbulent flow is often marked by boils or other
irregularities of the water surface. Turbulence is especially noticeable at rapids, where an ever-changing
pattern of eddies, smooth and swirling water surfaces,
and extremely turbulent white water can be seen.

Stream velocity is determined by the slope, or
gradient, of the stream bed, the shape and roughness of
the channel, and the discharge.

Slope   Slope is the means by which the potential
energy of the stream is transformed into kinetic
energy. The slope is defined as the drop of the stream
divided by the horizontal distance over which the drop
takes place. Figure 20.3 shows how the acceleration of
gravity produces higher velocities on steeper slopes.

Channel Characteristics   Irregularities of the
channel walls and bottom cause friction and turbulence and act to reduce velocity. The amount of friction is proportional to the number and size of the irregularities on the walls and bottom. Friction is least in
semicircular channels because they have less wetted

FIGURE 20.2 *Gullies on an upper Tennessee Valley farm. (Tennessee Valley
Authority.)*

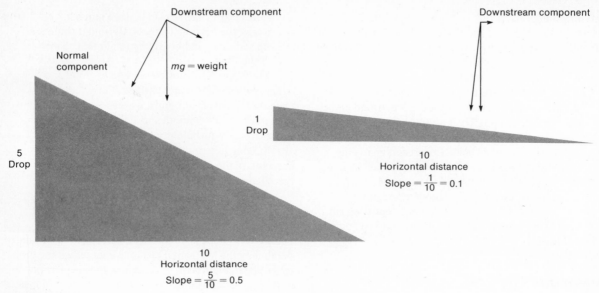

FIGURE 20.3 *Acceleration of gravity produces higher velocities on higher slopes.*

FIGURE 20.4 *Perimeters of channels having the same cross-sectional area. The semicircular channel has the smallest wetted perimeter.*

Shape	Cross-sectional area	Wetted perimeter
Rectangular	10 sq ft	12 ft
Square	10 sq ft	9.5 ft
Semicircular	10 sq ft	7.9 ft

surface area to interfere with the flow than do channels of other cross-sectional shapes (Fig. 20.4). Most natural channels are wider and shallower than this ideal semicircular channel. The type of sediment making up the channel also influences velocity. A gravel bottom offers more frictional resistance than sand.

Discharge   An increase in volume of water, or discharge, increases the depth of flow and in turn may increase the velocity. Any velocity changes are dependent upon a balance between discharge and the cross-sectional area of the stream.

Erosion

Erosion by flowing water occurs through a number of processes acting together. These are abrasion, the mechanical wear of the stream bed by the impact and friction of sediment in the moving water; hydraulic quarrying due to the drag and lift forces of moving water; and solution and chemical action of water on the materials forming the stream channel.

Abrasion   Clear water abrades very little. This is evident in many swift, clear, mountain streams where delicate plants such as mosses and algae can be seen growing on the stream bed. In contrast, mountain streams laden with sand or coarser sediment have no vegetation on their beds. When running water transports sediment grains, it becomes a powerful agent of erosion, capable of eroding deep canyons and gorges in solid rock. The downcutting of a stream is accomplished chiefly by the sediment that it carries along near its bed. These erosional tools chip, batter, and sandblast the rocks of the channel floor. The abrasive power of a river carrying a load of sand and pebbles varies approximately as the square of the velocity of the stream. For example, if the velocity of the current is doubled, many more and larger particles than before will be in motion, and each grain will possess more kinetic energy, thus eroding the rock exposed in the stream bed about four times faster.

Hydraulic Action   Erosion by stream water without the aid of sand and gravel is accomplished by the lift and drag forces of the stream current and by the impact of flowing water against a stream bank. Materials are dislodged and loosened by differences in the pressure of the water. They can range in size from sand to coarse gravel and boulders. Water pressure in a fracture can exert wedging forces large enough to move large blocks of bedrock.

Solution   Pure water does not exist under natural conditions. The water of the stream is constantly replenishing itself with impurities, which greatly increase its efficiency as a solvent. For example, stream water that flows through bogs and marshes will pick up carbonic acid and organic acids from the decaying vegetation. Silicate rocks dissolve at an almost imperceptible rate in stream water, but limestone may be readily dissolved if the stream water is acid. It is estimated that about 5 billion metric tons of solid earth material are dissolved from the continents annually. A large percentage of this material represents the dissolved substances in groundwater. In reality, therefore, most dissolved substances are delivered to the streams.

Transportation

Streams transport their sediment loads by sliding and rolling of the coarse materials along their beds, carrying fine sand, silt, and clay in suspension throughout the flow, and transferring the soluble compounds in solution. The transporting ability of a stream is enhanced by the fact that most of the mineral and rock fragments carried by a stream lose about 40 percent of their weight when submerged in water. Figure 20.5 shows the concentrations of several sizes of particles throughout the depth of flow.

Whether rock particles are transported as bed load or as suspended load depends on their size and mass and on the velocity of the stream.

The rate at which sediment particles settle depends on their size, the relative densities of the particle and the fluid, the viscosity of the fluid, and gravity. The relationship of these factors to obtain the settling velocity is known as Stokes' law.

The weight of a sediment particle is proportional to the volume (a function of the diameter cubed). The resistance to settling of the particles is a function of the surface area of the particle, or its diameter squared. Thus as size increases, the velocity of fall increases significantly.

Sediment particles larger than sand settle rapidly and are rarely carried in suspension. They move along the bottom in almost continuous contact with the bed and with other particles of similar size, moving about half as fast as the flow of the water.

Small particles settle slowly and thus tend to be carried in suspension. Almost any upward current is capable of lifting fine sand, silt, and clay and distributing them throughout the flow as suspended load. These materials are seldom on the stream bed and move as a suspension with about the same velocity as the surrounding water. Even smaller colloidal particles and ions in solution are kept dispersed throughout the flow by the turbulent movements.

The transport during floods of both large-sized particles and large volumes of sediment demonstrates that increases in discharge and velocity greatly increase the transporting power of streams. This additional transport capacity results from increases in turbulence as well as in volume. In general, for coarse loads, the dimensions of the largest particles transported as bed load vary directly with the 2.6th power of the velocity.

Figure 20.6 summarizes the results of several flume experiments and demonstrates a number of aspects of the erosion and transportation of sediment by running water. In the figure, the velocity character-

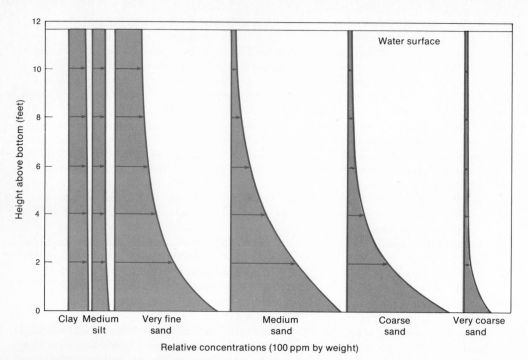

FIGURE 20.5 *Size of suspended sediment at various depths in the Missouri River at Kansas City, Missouri. Note that the coarser sediments are more concentrated near the bottom in contrast to the finer sediments, which are distributed uniformly throughout the flow. [After L. G. Straub in O. E. Meinzer (ed.), Hydrology, 1942.]*

istics of the stream are divided into three fields: the velocity of erosion, that of transportation, and that of deposition. Based on sizes of sediment, the graph indicates that it takes a greater velocity to put a given-sized material in motion than to transport that material once it is moving. For most sediments there is a velocity below which they will not be transported but will be deposited instead. For example, for cobbles, about 3.5 centimeters per second is required to set them in motion above the stream bed, while a velocity of about 2.4 centimeters per second is sufficient to keep them in motion along the stream bed. This is because lower velocities are needed at the stream bed as compared to the slightly higher velocities needed for a particle to bounce or roll above the stream bed. Moreover, bank materials are more easily set in motion since, in addition to the drag of the water, there is a component of gravity favoring motion of the bank material into the channel. As a result, the largest materials available for

transport from the bank may be considerably larger than those the stream can erode from its bed.

The easiest material to erode is not necessarily the smallest size. Silt and clay may be more resistant to erosion than fine sand because they have been compacted through burial or drying and possess strong cohesive qualities. Also, the velocity of the water in actual contact with the bottom may be extremely slow. Sand particles on the bottom, however, may be large enough to project upward through this boundary layer into regions of higher velocity, providing sufficient bottom roughness and thus turbulence to be picked up.

Figure 20.7 shows a variety of forms of the sedimentary material that occurs on stream beds. Each different form results from a particular flow regime. This fact was discovered by conducting laboratory flume experiments. These bedforms and the resulting deposits, if preserved, leave a record of the bottom conditions of the stream with different flow characteristics.

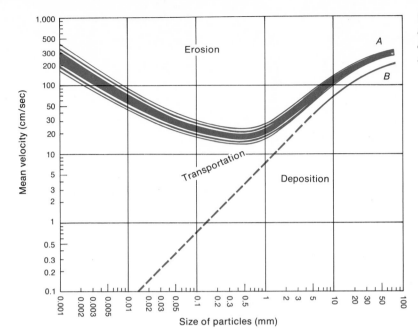

FIGURE 20.6 *Approximate curves for erosion and deposition of uniform material (logarithmic scale.) (From Filip Hjulstrom, 1939.)*

Lower regime

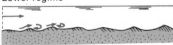

A. Typical ripple pattern

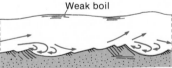

B. Dunes with ripples superposed

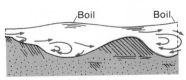

C. Dunes

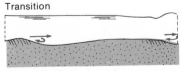

D. Washed-out dunes

Upper regime

E. Plane bed

F. Standing waves

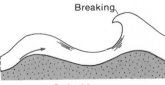

G. Antidunes

FIGURE 20.7 *Bedforms in sand channels. (From Simons and Richardson, 1961.)*

Ripples
Plane bed

Ripples

Dunes

FIGURE 20.8 *A record of successive bedforms developed in the Rio Grande. (From Harms and Fahnestock, SEPM Special Publication 12, 1965.)*

Different bedforms usually exist in different portions of the channel at the same time. A sedimentary record of dunes, followed by ripples, small antidunes, and then ripples develops as the velocity characteristics of the stream change (Fig. 20.8).

Deposition

During transportation of the bed load and suspended load, there is a constant exchange between materials in transport and materials on the bed. Particles are alternately picked up and dropped. The finer the sediment size, the more continuously it is transported in the flow from its initial pickup to its final deposition. Hjulstrom's curve (Fig. 20.6) shows that in portions of a channel where velocities are higher than average, these same-sized materials may be eroded.

In arid regions the dissolved materials in a stream may become concentrated enough by evaporation to be precipitated as a coating on pebbles during low-water stages. In humid regions, a stream fed by cold springs with a high content of dissolved calcium bicarbonate may deposit calcium carbonate along its course. The deposits form as a result of carbon dioxide being released by warming, agitation, or organic activity. Most stream waters are too dilute to form chemical deposits enroute to the sea.

Downstream Change in Size of Bed and Bank Materials

The material in bed, banks, and load of a stream decreases in size from the upstream portions toward the mouth. This is caused by selective sorting, wear on the particles, or a combination of the two.

Selective sorting takes place because finer grains are picked up more readily and are less likely to be deposited in a short distance than are coarser particles. Within a short distance along a channel in which deposition is taking place, selective sorting can produce striking changes in size of stream-bed materials. This mechanism can explain size change in the sediment downstream from a single source of sediment, such as a landslide or a deposit from a flooded tributary stream.

Wear on the sediment fragments is also important. Fragments are worn smaller as they are dragged, rolled, and pounded against one another and against bedrock exposed in the stream bed. The sharp edges and corners of angular fragments are rapidly smoothed, but after the initial rounding, additional wear takes place much more slowly. Abrasion experiments with fresh, angular rock fragments suggest that much longer distances of transport are necessary than are available in most stream systems to produce the decreases in size that are observed. Experiments have also shown that if the materials have time to weather between periods of transportation, later wear during transport is much more efficient. Such weathering occurs while the materials are temporarily deposited on river bars, flood plains, terraces, or slack-water reservoirs. There is, however, no single, simple answer for all streams.

STREAMS AS DYNAMIC SYSTEMS

Equilibrium

A stream whose hydraulic factors are constantly changing, shifting, and compensating to bring about a state of equilibrium is a graded stream. Hoover Mackin, a geomorphologist, first developed the idea of a graded stream. His concept follows:

A graded stream is one in which the slope is delicately adjusted over a period of years to provide, with available discharge and prevailing channel characteristics, just the velocity required for the transportation of the load supplied to it from the drainage basin. The graded stream is a stream in equilibrium. Its diagnostic characteristic is that any change in any of the controlling factors will cause a displacement of the equilibrium in a direction that will tend to absorb the effect of the change (Mackin, 1948).

Factors Controlling Equilibrium The primary factors of this equilibrium are stream discharge, sediment discharge and size of sediment particles, channel slope, and channel shape.

(a) Stream Discharge This is the quantity of water at any point and is the runoff from the drainage basin upstream of that point, usually measured in cubic meters per second. It is dependent mainly on climate. Its effect is expressed through the relationship:

$$Q \text{ (discharge)} = V_m \text{ (mean velocity)} \times A \text{ (cross-sectional area)}$$

(b) Sediment Discharge and Size of Sediment The quantity of the sediment discharge in tons per day and the size of sediment are determined by climate, weathering, and all processes delivering sediment to the stream. Other factors are the size of the grains and the composition and structure of the original rock material.

(c) Slope The slope adjusts automatically to provide the velocity necessary for transporting the amount and size of material being delivered to the stream. If the slope is too low for transportation of the sediment load, deposition results until the slope is sufficient for transportation (Fig. 20.9a). If the slope is so steep that it provides a velocity greater than necessary to transport the load, erosion reduces the slope. This new slope will provide the velocity needed to transport the sediment load (Fig. 20.9b).

(d) Channel Shape The ratio of width to depth is used to describe the channel shape. Channel shape is determined by the interaction of discharge, amount of sediment, slope, and local factors such as bank erodability and channel alignment.

Channel shape is important in determining stream velocity and, therefore, sediment transport. In general, the narrower the channel, the more efficient the sediment transport. When a narrowing of the channel occurs because of deposition along the bank, erosion of the bed is likely to result; or when deposition in the channel increases the erosive forces on the banks, widening will occur.

Base Level An important concept in stream equilibrium and stream and valley development is base level. This is the lowest level to which a stream can erode. Base level is reached when the slope of the stream approaches zero, checking or slowing the velocity. The stream is then incapable of eroding further. There are two types of base level, local or temporary, and ultimate or permanent. A local base level might be a resistant rock ledge or barrier, a lake, or the junction of a tributary. The ultimate base level of a stream at its mouth is slightly below sea level. Although it may not be permanent, it changes more slowly than most temporary base levels.

A stream will flow below base level to that depth appropriate to its size. For example, at New Orleans the level of the Mississippi River surface is only 3 to 6 meters above mean sea level, depending on the discharge, and the bottom of the Mississippi River channel at New Orleans is about 25 meters below mean sea level. When a stream enters a lake or the sea, sediment and water transport do not stop completely. It continues as a stream so long as there is a water-surface slope to provide the driving force.

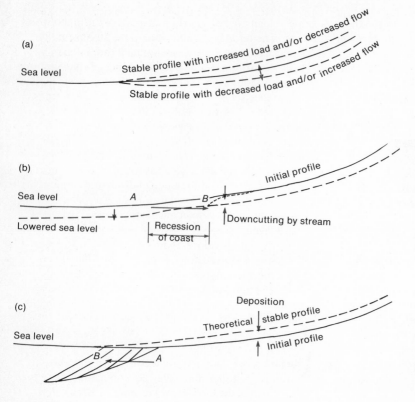

(a)

Sea level

Stable profile with increased load and/or decreased flow

Stable profile with decreased load and/or increased flow

(b)

Initial profile

Sea level   A   B

Lowered sea level

Recession of coast

Downcutting by stream

(c)

Deposition

Theoretical stable profile

Sea level

Initial profile

B   A

FIGURE 20.9 *Stream responses. Only named factors change; all others are assumed to remain constant. (a) Response to change in load and discharge. If slope is too low for transportation of the load, deposition results at the head of the reach until slope is sufficient. If slope is steep enough to provide a velocity greater than needed to transport the load carried into the reach, erosion results, until the slope is reduced to just that required to provide the velocity to transport the load. Note that increased load may offset, at least in part, the effects of increased flow. Decreased flow may similarly offset the effect of decreased load. Increased or decreased diameters of the load particles may substitute for increased or decreased quantities. (b) Response to either recession of coast or lowering of sea level. This is typical of changes resulting from continuing recession of portions of Lake Erie's north shore and lowering of sea level that occurred when much ocean water was held in the form of glacier ice on land. (c) Response to extension of delta. Typical of streams entering a quiet body of water, for example, in the vicinity of Los Angeles debris basins after flooding.*

Change in Equilibrium   Along the north shore of Lake Erie, wave erosion has caused the shoreline to retreat, shortening the small streams flowing into the lake (Fig. 20.9c). The response of the streams to the increase in slope in their downstream portions has been an increase in the sediment load by active downcutting. This has caused the undermining of bridges and culverts and has disrupted the road along the lake shore.

Several times during the Pleistocene epoch, sea level was lowered about 100 meters for thousands of years. This lowered sea level resulted when large quantities of water were withheld from the sea in the form of large continental glaciers. During these glacial periods the downstream portions of the streams usually adjusted to this lowering of their base level by downcutting. This response was like that of the streams on the north shore of Lake Erie. At the same

time the upper portions and perhaps the entire stream had to adjust to changes in discharge and sediment amounts brought on by the climatic changes that accompanied glaciation.

The changes in equilibrium cited here are relatively simple since only one or two factors changed. Even relatively minor shifts in climate can change discharge, sediment supply, and bank erodability, thus radically altering the form and character of the stream.

Greatly increased loads of sediment can be caused by glaciation, landslides, or man, all of which disturb the vegetation cover. Decreased sediment loads can result from the cessation of any of these activities or the establishment of a vegetation cover.

Changes in discharge can take place when a stream captures or appropriates the waters of another stream. Stream capture results in an increased drainage in one stream and decreased discharge to the

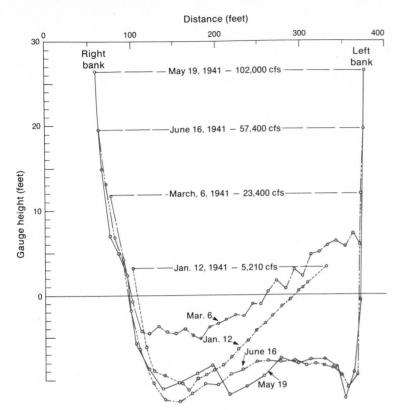

Right bank

Left bank

— May 19, 1941 — 102,000 cfs —

June 16, 1941 — 57,400 cfs

-March, 6, 1941 — 23,400 cfs—

— Jan. 12, 1941 — 5,210 cfs—

Gauge height (feet)

Mar. 6

Jan. 12

June 16

May 19

FIGURE 20.10   *Channel cross sections during spring high water, Colorado River at Grand Canyon, Arizona, 1941. (From Leopold and Maddock, U.S. Geological Survey Professional Paper 252, 1953.)*

other. A drainage rearrangement can also be caused by faulting, glaciation, vulcanism, or shifts in precipitation and evaporation.

Short-term changes in discharge or sediment load can cause adjustments that are similar to the longer term changes. During a flood stage, a river often erodes the bed as the flood waters rise and deposits material as it falls. Figure 20.10 indicates such an alternating pattern of erosion and deposition. Scour can occur only when there is more water available than required to carry the sediment load. Deposition can occur only when there is more sediment available than can be transported. One is often tempted to think of the entire length of a river undergoing scour during the passage of a large flood. If this were true, reservoirs would fill with sediment at a much faster rate than they do (Fig. 20.11). Along many portions of a stream during flood stages there is little or no deepening of the stream bed elevation; on the contrary deposition may occur.

FIGURE 20.11   *Possible responses to the changes produced by the construction of a dam and reservoir. The upstream response will depend on the amount of sediment being transported. If a large amount of sediment is transported, the response will be rapid deposition. The downstream changes may be simple erosion in response to the reduced sediment load, or they may be quite complicated, with reduced flood peaks completely offsetting the reduced loads. Flood deposited fans from tributary streams may block the decreased flow of the main stream and actually raise the water surface, drowning structures such as power houses and outlets.*

Tributary fans

Dam

Opposing tendencies may offset each other so that the result is only a small or slow shift in equilibrium. An example of this is found in the response of the Nile River to the closure of the Aswan High Dam. Before the Nile was dammed, it carried large quantities of sediment, especially during flood stages. Now the large sediment loads and the flood peaks of the Nile are gone. The water released from Lake Nasser, behind the dam, for irrigation purposes contains very little sediment, and one might assume that scour of the Nile sediments below the dam would be unavoidable. The irrigation releases are small in relation to the former annual floods and so far have not caused the erosion feared. The sediment contribution from tributary streams downstream from the dam equals or exceeds the transporting capacity of the irrigation releases.

Stream Profile

Most streams have long profiles that are steep in the headwaters and gradually become flatter downstream. There are, however, irregularities along this long profile.

If one considers equilibrium within a stream, it can be shown that slope adjusts to each change in the factors controlling equilibrium. Where a tributary joins the main stream, the additional discharge of water may allow transport of the sediment load on a flatter slope, or an increase in amount or size of sediments may require a steeper slope.

On resistant bedrock a different type of equilibrium may result in a similar profile. The rate of downcutting of a stream depends on the character of its bed. Where the rock is hard, slopes develop and continue to steepen. The capacity for downcutting increases until there is an equilibrium of action.

Where the slope of a stream is adjusted to the resistance of the bedrock, the work of the stream is not directed to the transportation of sediment but to the removing of the bedrock. In this case, the stream can usually transport far more sediment than is supplied, and the profile, although it may be similar in form, is not the same profile required for transport of the load.

Channel Characteristics

The variables involved in stream channel flow are width, depth, velocity, discharge, channel roughness, and slope.

The width, depth, and velocity at a cross section of a stream are functions of the discharge. Width, depth, and velocity all increase with increasing discharge (Fig. 20.10).

How does a stream change along its course from its headwaters to the sea? The discharge normally increases downstream as more tributaries join the stream. Compare a small-headwater mountain stream and the deep quiet waters of a major river many kilometers downstream as it approaches the sea.

The way scientists study the effects of an increased volume of water in a stream channel is to compare characteristics of channels where discharges occur with equal frequency. For example, if a discharge of 14 cubic meters per second occurred in the headwaters the same number of times per year that a discharge of 140 cubic meters per second occurred on the main stream, then the channel characteristics at these differing discharges can be compared.

If width, depth, and velocities at a position in the headwater region and a position on the major stream are compared at mean annual discharge, velocity can be noted to increase downstream, the slope and grain size decrease, and the width and depth increase. Recall the relationship between discharge mean velocity and cross-sectional area cited earlier.

Two causes of the downstream decrease in grain size of materials were also discussed earlier. An examination of Fig. 20.6 shows that bed materials of the size found in headwater tributaries would require velocities of at least 2 meters or 200 centimeters per second for erosion and transportation. Bed materials in the down-stream reaches would require velocities of 1 to 2 meters per second for transport and erosion of gravel and 0.3 to 0.8 meters per second for sand. This means that the mean annual discharge at the downstream point would be capable of eroding the channel, but the mean annual discharge at the upstream point would have little or no effect on its channel.

VALLEY DEVELOPMENT

Every river appears to consist of a main trunk, fed from a variety of branch streams, each running in a valley proportional to its size. All of them form an interconnecting system.

Irregularities on a surface newly exposed to

stream erosion, such as an uplifted continental shelf, concentrate runoff into a series of lakes and wet areas connected by streams. Where these streams are steep, there is sufficient energy available for erosion, and they cut downward and headward, forming isolated valley segments. When the lakes are filled with debris or their basins become breached by headward erosion of the outlet stream, a continuous valley is formed.

As a valley is deepened, the floor of its channel approaches a level below the water table. After a valley penetrates this level, this new source of water supports a stream even during dry seasons. Such streams are permanent streams. Streams that have not reached the water table and flow only after a period of precipitation are called intermittent streams.

Widening of a valley goes on in conjunction with its deepening. Most valleys are much wider than the streams that carved them. Widening is caused by gravity transfer, rainwash, and lateral erosion of the bank or the bedrock by the stream. Lateral cutting occurs at the outside of a bend in the channel. When the bend is against the valley wall, the erosion tends to widen the valley and steepen the wall. This steepening makes the processes of weathering, gravity transfer, and rainwash more effective. The character of the rock over which the stream flows may produce local variations in the width so that narrow portions alternate with wider ones. Where a stream crosses a tilted bed of hard resistant rock lying between softer materials, the valley widens faster in the soft material upstream and downstream, leaving a narrow channel in the resistant rock.

Canyons and Gorges

Valley walls in arid areas are comparatively steep because chemical weathering and gravity-transfer processes that would make them more gentle operate very slowly. Also, in some regions, conditions for downcutting are more favorable than for lateral erosion. Examples can be seen along the Gunnison River in Colorado and along the upper portions of Zion Canyon, Utah. In Zion Canyon, near-vertical walls over 600 meters high are only a few hundred meters apart at the brink of the canyon.

The Colorado River has a total length of about 3,200 kilometers and drains an area of about 580,000 square kilometers, most of which is a high plateau 1,800 to 2,400 meters above sea level. In the high plateau of northern Arizona, the Grand Canyon is about

350 kilometers long, 16 to 25 kilometers wide, and a little more than 1.6 kilometers deep. At this depth, the total width from rim to rim is 13 to 20 kilometers, but its width at the bottom is only slightly greater than that of the stream. If the slopes of the canyon were uniform, they would have an angle of less than 15°, but the variations in resistance to weathering and erosion of the sedimentary strata have produced large steplike slopes, or rock terraces. The steep faces of some of these terraces drop vertically for more than 300 meters. Most of the rock terraces are formed by slope processes and small tributary streams rather than by the main stream. The upper series of rocks in which the canyon is cut is composed of flat-lying beds of limestones, and sandstones, and shales (Fig. 1.1). Below this series, the stream has cut into Precambrian crystalline schists and granites.

Classification of Streams

The course followed by a stream in its journey to the sea results from the original slope and irregularities of the land surface and the relative ease with which the different rocks encountered can be eroded, as well as from a complex history of erosion and deposition brought about by climate change, uplift, subsidence, faulting, and folding.

A consequent stream is one whose course has been determined by the original slope and the irregularities of the land surface. Such streams are characteristic of coastal-plain areas, such as the Atlantic and Gulf Coasts. Here, the surface is comparatively uniform and regular, with a gentle slope to the sea. Other consequent streams flow off volcanic cones, across lava fields, or over irregularly rolling plains of glacial deposits.

Some stream courses are directed by differences in the structure and character of the bedrock formations. Erosion proceeds more readily in the less resistant beds so that eventually a stream may undergo marked changes in position and direction from its original consequent course. Rivers formed in this way are called subsequent streams (Fig. 20.12).

As a result of the downcutting by a consequent stream, tributaries develop down the new slope to the main stream, independent of the original topography. If there is no directional control, as in areas of flat-lying sediments or unfractured crystalline rocks, the headward growth and multiplication of tributaries

FIGURE 20.12 *Split Mountain, viewed upstream, showing the nearly symmetrical form of the anticline, breached by the Green River. The river is thought to have been superposed from a sedimentary cover. (Hal Rumel photo.)*

produce a dendritic drainage pattern (Fig. 20.13*a*). This pattern resembles the branches of a tree or the veins of a leaf.

Often the direction of a valley is controlled by the strike of the fractures in the bedrock of the area which it drains. Guided by such zones of weakness, the major streams and small tributary streams develop angular drainage patterns (Fig. 20.13*b*). Such patterns are particularly evident in the Colorado Plateau, Connecticut, and Ontario, where large areas of strongly fractured rocks are exposed at the surface.

In regions where vertical and horizontal movements along faults have taken place on a large scale, many valleys follow the fault zones for great distances. The fractured rock of the zone is relatively erodable, and the relief produced by the faulting movement may alter the course of the stream. At some places, long narrow blocks of the earth's crust have been depressed into valleylike basins, or grabens, that may be occupied by streams (Fig. 20.14). The Dead Sea Basin and the Jordan Valley are typical examples. Owens Valley in eastern California has a similar history, as do many large valleys in the southwestern United States.

Although mountains would appear to be an impenetrable barrier to a river, rivers now cut through most of the major ranges of the world. There are three ways that streams can cut across the resistant rock exposed by erosion of folded or tilted beds (Figs. 20.15 and 20.12). First, an antecedent stream may predate the formation or uplift of the structure by eroding downward as rapidly as the structure is raised in its path (Fig. 20.15). Such an origin has been proposed but not proven for the Colorado River in the Grand Canyon.

FIGURE 20.13 *Stream drainage patterns. (a) Dendritic drainage pattern in central Africa. This type of drainage is found where directional control of erosion is lacking. (United States Air Force.) (b) An adjusted stream pattern characteristic of the mature stage of an erosion cycle in a region of folded or tilted strata of different degrees of resistance to erosion. The arrangement resembles a trellis.*

(a)

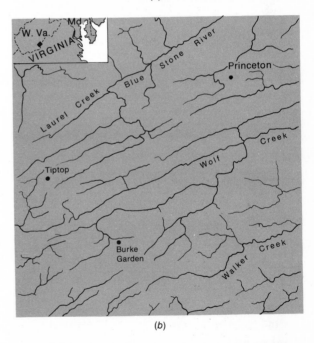

(b)

Another way is where a superposed stream develops on an erosion surface or sedimentary fill that at one time extended over the structure. The stream cuts downward across the tilted rocks. Portions of the Hudson, Delaware, Susquehanna (Fig. 20.16), and Potomac Rivers in the Appalachian region, are considered to be superposed from a former cover of sediments that were subsequently removed. In the Rocky Mountains region, many rivers, including the Bighorn, Madison, Platte, Arkansas, Green, and Snake, are superposed very impressively in spectacular canyons across mountain ranges.

A third and much less common method by which a stream may develop a course across resistant beds is stream piracy. As streams extend or modify their drainage basins, stream piracy can occur when one stream cuts back by the headward extension of its tributaries until it captures some of the headwaters of another stream beyond a divide. The stream whose drainage basin has been invaded is said to have been "beheaded" by the pirate stream (Fig. 20.17). Conditions that may cause a stream to erode faster than an adjoining one are a greater discharge, less resistant rocks in which to excavate its channel, and a higher gradient due to a shorter distance to the sea.

If a stream flows across tilted strata, a narrows called a water gap (Fig. 20.18) is developed where the valley crosses the harder beds. If such a stream is diverted by piracy, the abandoned water gap becomes a wind gap common in the Appalachian region. These wind gaps served as passes through the mountains for the early settlers traveling by wagon to Kentucky and Tennessee and became strategic points during the campaigns of the Civil War.

Channel Pattern

Meanders Streams are rarely straight; instead, they develop a series of winding curves, or meanders.

FIGURE 20.14 *Valley following fault zones for great distances. Here the valley is occupied by an arm of the sea. (NASA photo.)*

FIGURE 20.15 *The Carson River and Prison Hill in Nevada, looking south. The Carson River is an antecedent stream. The river leaves the Carson Valley and cuts a gorge through the south end of Prison Hill. (Spence Air Photos.)*

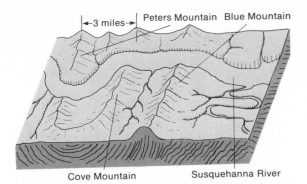

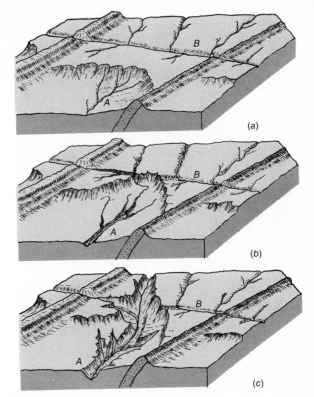

FIGURE 20.16   *A superposed stream flowing over truncated folds; the Susquehanna River, near Harrisburg, Pennsylvania. (After D. W. Johnson.)*

FIGURE 20.17   *Diagram illustrating stream piracy. (a) The tributaries at A are advancing by headward erosion toward the valley of the stream B. (b) The stream B has been beheaded or captured, and its headwaters are diverted to the pirate stream A. (c) The valley of A is extended and deepened. (Based on a drawing by Davis.)*

FIGURE 20.18   *The Delaware Water Gap, where the river cuts across the level-topped ridge known as Kittatinny Mountain, on the Pennsylvania-New Jersey border. This type of narrows is found where a stream flows across a resistant ridge of tilted rock. (Aero Service Corporation; Division of Litton Industries.)*

FIGURE 20.19   *Meanders and oxbow lakes along Mudjatic River in northern Saskatchewan, Canada. The scrolls of former channels on the flood plain are conspicuous. (National Air Photo Library, Canada, Photo A1814-27.)*

The curves tend to enlarge by erosion on the outside of bends and by deposition on the inside. Even in straight portions of the streams, bars are usually attached alternately to opposite sides of the channel, causing the main current to meander (Fig. 20.19). The vertical cross section of the channel at the bend is asymmetrical, with the deepest portion toward the outside of the bend. The size of meanders is proportional to the size of the stream. They are confined within the meander belt, which ranges from 8 to 18 times the width of the stream (Fig. 20.20). Meander forms occur in sediment materials of all sizes, although streams with large amounts of silt and clay in their banks have the tightest meanders.

A deposit called a point bar generally forms at the inside of a bend (Figs. 20.21*A* and 20.21*B*). The main current makes a rapid sweep along the outer bank, and in the slower water on the inside of the bend, deposition of the coarser portion of the load takes place.

Once started, meanders enlarge and become more sinuous, until a series of loops separated by narrow necks of land is developed. Eventually, the stream may cut through the narrowed neck of land between two loops and abandon the former channel. Where this happens, the river shortens its course, and the enlargement of the bend begins again. Deposition occurs in the abandoned channel openings adjacent to the main stream, blocks the ends, and creates an oxbow lake (Figs. 20.19, 20.20, and 20.22).

Meander cutoffs occur every few years along active channels. Such changes on the present Mississippi (Fig. 20.22) are prevented in places by artificial bank protection costing upward of $1 million per kilometer. In other places, artificial cutoffs are created by dredging.

Mark Twain in 1883 remarked on cutoffs altering the length of the Mississippi River: "the Mississippi between Cairo and New Orleans was twelve hundred and fifty miles long one hundred and seventy-six years ago. It was eleven hundred and eighty after the cutoff in 1722. It was one thousand and forty after the American Bend cutoff. It has lost sixty-seven miles since. Consequently its length is only nine hundred and seventy-three miles at present.... In the space of one hundred and seventy-six years the lower Mississippi has shortened itself two hundred and forty-two miles. That is an average of a trifle over one mile and a third per year. Therefore, any calm person, who is not blind or idiotic, can see that in the Oolitic Silurian Period, just a million years ago next November, the Lower Mississippi River was upwards of one million three hundred thousand miles long, and stuck out over the Gulf of Mexico like a fishing rod. And by the same token any person can see that seven hundred and forty-two years from now the Lower Mississippi will be only a mile and three quarters long, and Cairo and New Orleans will have joined their streets together, and be plodding comfortably along under a single mayor and a mutual board of aldermen. There is something fascinating about science. One gets such wholesale returns of conjecture out of such a trifling investment of fact."

A visual imagination like Mark Twain's often gets dampened by scientific facts. Nineteen natural cutoffs since 1765 shortened the river a total of 400 kilometers. Yet the length of the river between Cairo and Baton Rouge was approximately 1,400 kilometers in 1765, 1,500 kilometers in 1825, 1,350 kilometers in 1882, and 1,410 kilometers in 1930. Clearly, shortening is not the only active process; the river is also being lengthened by the enlargement of all active bends.

Braided Streams A stream with velocities capable of transporting a large bed load of sand and coarser materials also places a large erosive stress on the banks. If the banks collapse or are buried completely, broad flat channels develop in which numerous sand bars deflect the current. Then, at low-river stages, flow is not in single channel but in many dividing and rejoining interconnected streamlets (Fig. 20.23). These channels, together with the bars that separate them, are continually shifting, creating a multichannel, or braided, pattern. Braided patterns occur in streams that are downcutting, are in relative equilibrium, and perhaps most commonly, are depositing. The common section of a braided reach is broad and shallow.

Braided streams are common in areas where the river water evaporates rapidly, sinks into the ground, or is diverted for such uses as irrigation. These conditions diminish all but flood flows. One example is the Platte River in Nebraska, which flows in a broad alluvial valley nearly 1.6 kilometer wide. During most of the year a small volume of water flows through a series of interlacing channels that shift positions at every flood.

Glacial streams frequently are braided. At times of peak melt, the flow patterns change rapidly. A shift may take place in a matter of minutes as fluctuations in flow and minor changes in bed elevation due to erosion and deposition cause some old channels to be reoccupied and some of the existing ones abandoned (Fig.

FIGURE 20.20 *Stages in the formation of cutoffs and in the development of new meanders. (After A.K. Lobeck, 1939.)*

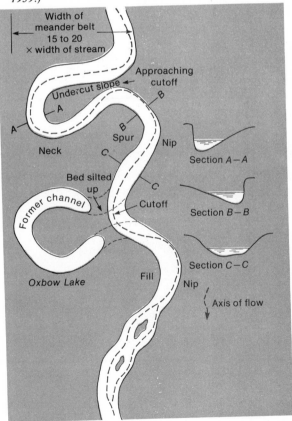

FIGURE 20.21A *Point bar and gravel veneer, Greybull River, Wyoming. (R. K. Fahnestock photo.)*

FIGURE 20.21B *A series of point-bar developments along the Ohio River in Indiana. The river flows from right to left. U. S. Geological Survey photo.)*

20.24). An average of about 15 centimeters of debris was removed from the valley floor during the month of July 1958, when the photos of Fig. 20.24 were taken.

OTHER EROSIONAL FEATURES ASSOCIATED WITH STREAMS

Mesas and Buttes

In regions where resistant horizontal sedimentary strata, gravel terraces, or lava flows cover soft clays or partially indurated sediments, flat-topped tableland areas called mesas become isolated by the headward cutting of tributary streams. The level top of a mesa consists of a resistant bed that tends to protect the less resistant strata below it. Buttes are small isolated hills, some of them remnants of former mesas.

Badlands

This is a topographic term given to an extensive gully system most often developed on relatively soft, poorly consolidated, flat-lying beds of silt and clayey sediments (Fig. 20.25). Badlands are well developed in the northwestern Great Plains region of the United States and Canada, especially in the Dakotas. A semiarid climate, in which rainfall is concentrated in a few heavy showers, is favorable to the development of badlands. The dry climate is responsible for a lack of protective vegetation on the slopes, encouraging erosion.

Cuestas and Hogbacks

On gently tilted rock strata, asymmetrical ridges called cuestas are formed, with steep erosional escarpments cutting across the bedding on one side and gentle slopes parallel to the bedding on the other side. Where the beds tilt about 30° or more, the resulting hogback has slopes equally steep on both sides (Fig. 20.26).

Waterfalls and Rapids

There is no sharp distinction between a waterfall and a rapids; steep rapids are commonly called falls. Whenever erosion proceeds at different rates in different reaches of a river, falls and rapids develop on the more resistant rocks. The slant of the resistant rocks affects the form and life of a waterfall.

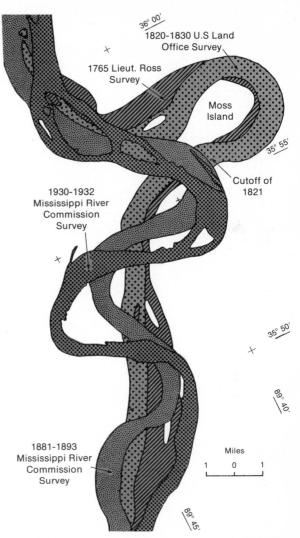

FIGURE 20.22 *Progressive changes in the channel of the Mississippi River above Memphis, during a period of more than 150 years. The sketch shows meanders and the meander belt of a major stream on its flood plain. (United States Army Corps of Engineers.)*

FIGURE 20.23 *Braided pattern of the Slims River, Yukon Territory, Canada. (Walter A. Wood photo.)*

Niagara Falls offers an excellent example of a cap-rock falls (Fig. 20.27). This cataract plunges about 55 meters over a dolomite cap 25 meters thick that overlies a soft shale. It has carved out a gorge 60 meters beneath the level of the water below the falls so that the total depth of the gorge from the rim to the river bottom is about 110 meters (Fig. 20.28). As the turbulent flow at the base of the falls loosens the soft shale, it undermines the capping dolomite. Large masses of rock then plunge into the pool at the bottom. Meter by meter the falls have receded up the river a total of about 11 kilometers from the Niagara Escarpment, which is 13 kilometers south of Lake Ontario. Measurements made from time to time indicate that the average rate of recession since 1875 is approximately 1 meter a year (Fig. 20.28).

At Yellowstone Falls, a vertical dike crosses the weaker rocks in which the deep canyon of the Yellowstone River was cut (Fig. 20.29). Falls of this sort cannot recede upstream because the softer formations up-

stream can be eroded only as fast as the resistant rock mass is cut down ahead of it.

Where main stream valleys have been eroded deeper than their tributary valleys by glaciers, the waters of the tributary streams fall into the main stream over steep precipices. Yosemite Falls in California plunges over a granite cliff into the glacially deepened Merced River valley with an initial drop of 440 meters (Fig. 20.30). It then cascades for about 240 meters over a jagged surface with a steep slope and finally falls 100 meters more over a vertical cliff to the valley floor. The numerous falls of the tributary streams in the Finger Lakes area in New York were also formed in this manner.

Along the east margin of the Appalachians, the resistant crystalline rocks of the Piedmont are bordered by a coastal plain composed of relatively unconsolidated sands and clays. Streams flowing from the Appalachians to the Atlantic developed rapids and falls along the line of contact between the two rock

FIGURE 20.24 *Braided channel changes, White River, Mt. Rainier, Washington, 1:00 to 2:00 P.M., July 18,1958. (a) 1:00 P.M.; (b) 2:00 P.M. Discharge is about 9.7 cubic feet per second. Arrows indicate new channels or bars that have appeared since the previous photograph. (U.S. Geological Survey, Professional Paper 422-A.)*

FIGURE 20.25 *Badlands carved by rainwash and gullying near Scotts Bluff, Nebraska. In semiarid climates, badlands are formed when a maze of gullies is cut into poorly consolidated sediments of varying resistance. (Darton, U.S. Geological Survey.)*

FIGURE 20.26 *Hogbacks formed by differential erosion of upturned beds of unequal resistance along the Rocky Mountains front near Morrison, Colorado. Unlike cuestas, hogbacks are underlain by rather steeply dipping strata. Therefore, the two sides of the ridge have slopes that are nearly equal. (Lovering, U.S. Geological Survey.)*

FIGURE 20.27 *Niagara Falls from the air. The river begins its drop in the rapids at the upstream end of Goat Island (upper center) and finally plunges over the dolomite brink, most of the water going over the Canadian Falls (right). Resistant dolomite caps softer shale. The Falls are receding because of sapping underneath the dolomite capping. Note the fallen blocks. (The Photographic Survey Corporation, Ltd.)*

FIGURE 20.28 *Canadian Falls of Niagara, showing its known retreat in about 280 years by sapping of relatively weak rocks beneath a resistant cap rock of dolomite. (After Gilbert.)*

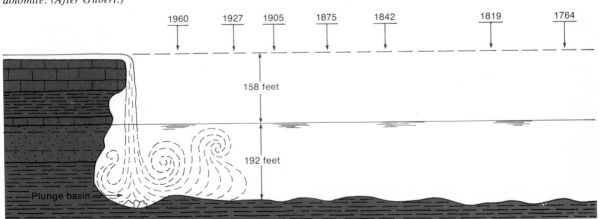

FIGURE 20.29 *Lower Falls and Grand Canyon of the Yellowstone River, Yellowstone National Park, Wyoming. The greater than 100-meter waterfall is held up by a vertical mass of resistant rock. The river has been able to entrench itself downstream in comparatively weak rock that has been altered and softened in the past by hot water and steam. (Burlington Northern Railway.)*

FIGURE 20.30 *Yosemite Falls, Yosemite National Park, California. This tributary to the Merced River discharges from a hanging valley. Hanging valleys are formed where the principal stream or a glacier has deepened the main valley far below the level of the tributaries. (Southern Pacific Railroad.)*

types. Because of the great number of falls along a relatively narrow zone, the region is referred to as the "fall zone " or "fall line."

Under exceptional conditions, rapids may originate where damming has been caused by the deposition material in the stream bed by landslides, glacial action, tributary streams, or lava flows. The rapids of the Colorado River in the Grand Canyon occur at the junctions with tributaries where huge boulders have been deposited by floodwaters from these steep intermittent streams.

Potholes

If water-carrying sediment is given rotary motion by eddies, it tends to grind out round or kettle-shaped excavations in the bedrock of the valley floor or at the base of waterfalls. These are potholes, or plunge pools (Figs. 20.31 and 20.32). The rate of deepening is accel-

FIGURE 20.31   *A portion of Dry Falls of the Columbia River, nearly 150 meters high and almost 5 kilometers wide, higher and much wider than Niagara Falls. Plunge pool, Fall Lake, is in the foreground. (Spence Air Photos.)*

FIGURE 20.32   *Potholes on the bed of the Susquehanna River at Conewago Falls, Pennsylvania, as exposed during unusually low-water stage in 1947. These remarkable pits have been abraded by swirling waters of the rapids at high-water stages.* (Lancaster Intelligencer Journal.)

erated during periods of high water, when stream velocities are increased. Some potholes are spiral shaped, with a larger diameter near the bottom than at the top. They may be formed in massive igneous rocks such as basalt or granite or in softer sedimentary strata such as limestone or shale. The swift currents of waterfalls and rapids favor their development. Potholes range from a few centimeters to 4 or 6 meters and also vary in depth. Some of the larger potholes in Interstate Park at Taylor's Falls, Minnesota, are 6 meters in diameter and have been eroded 15 meters or more deep in solid basalt. Pothole formation aids in the enlargement of steep-walled valleys, such as the valley of the inner portion along the Colorado River above Hoover Dam, the canyon of the ancient outlet of Owens Valley, near Little Lake, California, or the plunge pool at Dry Falls, Washington (Fig. 20.31).

Natural Bridges

A natural bridge or arch may be formed in several ways. If the rock of a stream bed is fractured above a waterfall, some of the water may descend through a fracture and then follow a bedding plane until it issues into the main channel below the falls or behind the curtain of water forming the falls. This fracture may be enlarged until the interior channel can accommodate all the water, leaving the former lip of the falls as a natural bridge.

Where stream meanders become deeply entrenched, the lateral erosion of a river against the rock walls may undercut the neck of a meander at the level of the water. The process continues from both sides until a hole is cut through and the stream flows through the perforation, leaving the arch of rock as a bridge over the stream. The famous Rainbow Natural Bridge in San Juan County, Utah, was formed in this way (Fig. 20.33).

Other natural bridges have been formed by the incomplete collapse of the roofs of lava tunnels, by incomplete collapse of caves, wave erosion, sandblasting by wind, and by differential weathering.

Flood Plains

A stream in flood stage overflows its normal channel and may spread from valley wall to valley wall, depositing fine silt, mud, and sand on the new valley floor. The height of this flood plain may be increased until it is at such an elevation above the normal stream level that only during very large floods is it ever overspread by the river. Wide flood plains may be formed

FIGURE 20.33 *Rainbow Natural Bridge, southern Utah. The bridge, 92 meters high, was formed by the stream undercutting the narrow neck of one of its entrenched meanders and leaving a resistant layer of sandstone spanning the 82-meter gap. (National Park Service photo.)*

by the long continued lateral erosion of a stream in equilibrium or by the filling of a broad valley by an aggrading stream. Flood plains are also formed as a result of meandering and point-bar deposition on the inside of the meander bend.

The flood plain of the Mississippi River between the junction with the Ohio River at Cairo, Illinois, and the Gulf of Mexico varies from 50 to 100 kilometers in width and is approximately 1,000 kilometers long. Most flood plains are bounded on either side by relatively steep valley walls. Where the rocks bordering a valley are weak and easily eroded, the slopes are so gentle that it is difficult to detect where the flood plain ends and the valley sides begin.

During flood stage, the current has the highest velocity in the deep water of the main channel. Along the margin of the channel where the rapid current mixes with the slowly moving shallow water on the flood plain, the velocity is sufficiently checked to cause the coarser part of the suspended load to be deposited. In this way the flood-plain deposits are built up highest on the immediate border of the channel and slope gradually toward the valley sides. These embankments of aggraded material are called natural levees because of their resemblance to man-made levees constructed to confine a stream to a narrow channel. Sediments in levees are coarser than those deposited on the outer

parts of a flood plain. Levees are low ridges, seldom more than 2 meters higher than the plain toward which they descend. Some levees, however, are high enough to stand out as long low islands. The main channel of the river flows on one side and the floodwaters of the flood plain and uplands on the other (Fig. 20.34) during all but the highest floods. During moderately high water, natural levees may serve as protection for the river flats since they tend to retain the waters in a definite channel. However, during higher floods, a river may break through the levees or back up the tributaries and flood the lowlands.

The worst flooding in a generation occurred on the Mississippi River during March-April 1973. Estimates of acreage covered by the flooding waters ranged from 3 million in Mississippi, 2.3 million in Louisiana, to 1.3 million in Kentucky, Illinois, and Missouri. Levees and dikes were breached in many areas along the river, causing the evacuation of entire communities, the flooding of farmlands with newly planted crops, and over $72 million in damage.

Along many rivers, the natural levees are so high that tributaries are forced to flow parallel to the main stream for considerable distances before they join (Fig. 20.35). The Yazoo River, which parallels the main channel of the Mississippi River for about 320 kilometers, is a typical example.

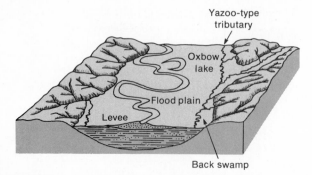

FIGURE 20.35 *Natural levees. Note that many small tributaries may run for some distance parallel to the main stream behind the natural levees before they are able to join the main stream. Such tributaries are called Yazoo-type rivers after the Yazoo River, which flows for 320 kilometers along the east side of the Mississippi River before joining. (After A. K. Lobeck, 1939.)*

In contrast to these flood plains and features formed largely by overflow deposition, a vigorous stream in equilibrium carrying coarse gravels through a region of relatively erodable rock may cut laterally and deposit on its bars a thickness of gravel covered by a few meters of finer overbank deposits. The total thickness of these deposits is approximately the same as the depth of the river in flood (Fig. 20.10). This type of flood plain is essentially a veneer of gravel over a bedrock surface and may be developed in a single lateral swing of the river, although the river usually swings from side to side in its valley more than once.

Terraces

Stream terraces are the remnants of former flood plains, abandoned when the streams actively downcut and develop new flood plains at lower levels (Fig. 20.36).

The concept of flood frequency is useful in defin-

FIGURE 20.36 *Alluvial terraces along the Esk Valley, Canterbury, New Zealand. Former flood plains have been left behind as discontinuous benches by subsequent downcutting by the streams. (V. C. Browne, Christchurch.)*

ing the difference between a flood plain and a former flood plain, or terrace. A flood plain is inundated on the average once every 2 to 5 years, while a terrace is flooded rarely or not at all. Many terraces are too high above the river to be flooded at any time, since they were left behind early in the history of the valley as the stream cut down. For example, the Greybull River of Wyoming has several terraces from about 400 meters to less than 16 meters above the present river. The usual cause of stream-terrace formation is the disturbance of equilibrium by the lowering of base level. Many terraces also develop as a result of a climatic change.

Terraces are normal features of most stream valleys. They may be cut into alluvial valley fill. When cut into bedrock they may consist only of a veneer of flood-plain deposits over bedrock. They may be cut out of alluvial fill. Terraces protected by spurs of hard rock at their bases or on the upstream side are said to be rock-defended terraces. When terraces exist along both sides of the valley and extend along the stream for many kilometers, they are called paired, or cyclic, terraces (Fig. 20.37). They usually represent a long period when equilibrium was not disturbed.

Small isolated remnants of flood plains may be left at various levels along the valley walls when rivers cut down rapidly from one equilibrium position to another (Fig. 20.37). These remnants are usually unrelated to other remnants along or across the valley. These flood-plain remnants are true terraces, but they do not represent long periods of stability as do cyclic, or paired, terraces.

FIGURE 20.37   *(a) Paired, or cyclic, terraces are indicative of successive stable or equilibrium stream profiles during downcutting. Terraces occur on both sides or along long reaches of the stream. Their large extent is evidence of long pauses with lateral cutting during times of a stable stream slope. (b) Nonpaired, or noncyclic, terraces. Downcutting is continuous. Only small remnants of former flood-plain positions remain. Both paired and nonpaired terraces may occur along the same stream, the nonpaired terrace remnants forming as the stream downcuts from one stable position to another.*

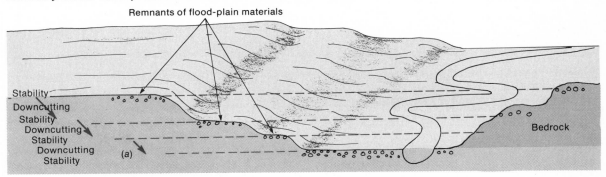

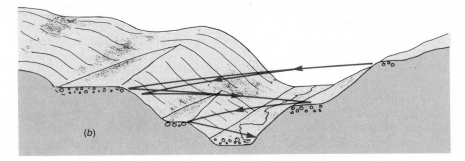

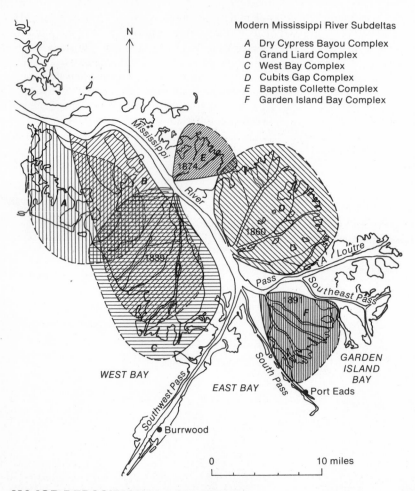

Modern Mississippi River Subdeltas

A   Dry Cypress Bayou Complex
B   Grand Liard Complex
C   West Bay Complex
D   Cubits Gap Complex
E   Baptiste Collette Complex
F   Garden Island Bay Complex

FIGURE 20.38   *The subdeltas of the modern birdsfoot delta of the Mississippi River. (From Coleman and Gagliano, 1964.)*

MAJOR DEPOSITIONAL LANDFORMS

Deltas

Where rivers laden with sediments flow into a body of quiet water, such as a lake or ocean, the velocity is checked and deposition results. Where these deposits are not removed by shore or tidal currents, they form a delta, so called because of similarities in shape with the Greek letter Δ. As the stream current reaches the quiet water, the bulk of its coarser load is deposited, and the finer silt and clay are carried farther out. The river water tends to float on the heavier salt water of the ocean until the fresh and saline waters mix. With mixing, the salts of the seawater cause the fine clay particles to coagulate into clumps large enough to settle out. The development of a succession of deltas and

distributaries of the Mississippi River is shown in Fig. 20.38.

With the growth of new distributary systems and the abandonment of others, many partially filled areas remain as lakes on the delta. Lake Pontchartrain is one of the best known (Fig. 20.38).

Delta deposits have all the variations of normal fluvial deposits, in addition to the variations caused by the character of the lake or ocean into which they are deposited. Oceans or lakes may vary in level because of tides or seasonal variations in runoff. A large tidal range can profoundly alter the form of a delta by distributing sediments tens of kilometers from the mouth of the river. The deltas of the Colorado River and the rivers flowing into Cook Inlet, Alaska, each area with large tidal ranges, differ radically in form and deposits

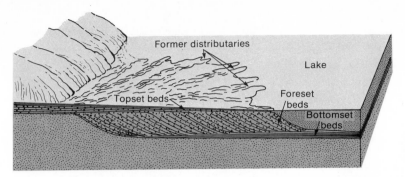

FIGURE 20.39 *Diagram illustrating the form, growth, and internal structure of a delta. Materials are added on the frontal slope as foreset beds, beyond as bottomset beds, and across the top as topset beds. Each line represents a surface during the growth of the delta. In subsiding areas, the topset beds may attain considerable thickness. (From Strahler, 1969, Physical Geography, 3d ed., copyright © 1951, 1960, 1969, John Wiley & Sons, Inc.)*

from the Nile and Mississippi deltas (Fig. 20.38).

The size, shape, location, and alignment of the ocean or lake can have major effects on the size and character of the waves and currents that rework the deltaic deposits of a river. Vigorous wave or tidal currents may redistribute along the shore the materials delivered by the river and prevent the formation of an extensive delta protruding prominently from the shoreline. An example is the combined deltas of the Colorado and Brazos Rivers of Texas. More often the waves and currents will rework only a part of the delta materials, redistributing them and filling areas between distributaries. In the process of reworking, silts and clays are carried out to deeper water, and the coarser materials are left on the delta margin. From the distribution of the various sizes of sediment in ancient delta deposits, the nature of both the river and the body of quiet water can be inferred.

Deltas show clearly three types of bedding (Figs. 20.39 and 20.40). The flood-plain sediments on the

FIGURE 20.40 *Topset and foreset beds of a large delta deposited in glacial Lake Hitchcock in the Connecticut Valley, Note man for scale. (R. K. Fahnestock photo.)*

FIGURE 20.41 *Coalescing fans in Death Valley. View is from the southeast looking toward Black Mountain and Furnace Creek. (Spence Air Photos.)*

delta surface are called topset beds; they overlie steeply dipping foreset beds that rest on gently dipping bottomset beds. The bed load of the stream accumulates just below the brink of the delta front until the slope becomes too steep and this deposit slumps to a stable angle of repose (20 to 35° from the horizontal). It is this slumping that results in the steep foreset beds. The top of the foreset beds marks the approximate elevation of the bottom of the inflowing stream. The bottomset beds are fine sands, silts, and clays that were carried in suspension by the stream and settled on the bottom beyond the delta front. As the delta front continues to be built forward, foreset beds are deposited over previously deposited bottomsets, and in time, the topset beds of the flood plain are deposited on the foreset beds.

Subsidence of large deltas has been noted at the mouths of many rivers. In general, deposition has kept pace with the subsidence, but in some delta deposits, marine limestones and shales are found interbedded with fresh-water sediments and soils. From the structural relations of the sediments, it is evident that, at times, subsidence gained on upbuilding and that the delta surface again was covered by the sea. Deep borings in various deltas reveal similar successions. At New Orleans, driftwood deposits were found by boring at a depth of 350 meters. Subsidence is due primarily to the downwarping of the ocean floor on which the sediments are accumulating. Compaction of the soft underlying sediments is also a factor.

Deltas can be recognized in ancient rocks by the interbedding of point bar, natural levee, swamp deposits, lake deposits, wave-worked beach sands, and foreset and bottomset beds.

One of the proposed explanations for some of the extensive deposits of coals and associated marine and fresh-water deposits of the world is the existence of ancient deltas. Coal is thought to represent the plant remains deposited in the back-swamp areas. The deposition of plant materials ended when subsidence shifted the swampy areas to below sea level and began again when fluvial deposition once more brought the areas above sea level.

Alluvial Fans

Where a heavily sediment-laden stream flows from a narrow valley in the mountains onto a plain or wider valley floor, its velocity suddenly decreases, and a large part of its load is deposited in the form of a fan (Fig. 20.41). The velocity may decrease because slope decreases, discharge decreases as the water sinks into the fan deposits, or the resistance to flow increases as the stream divides into many channels on the fan. In these channels, the stream encounters banks with lower resistance to erosion than that of the bedrock channel of the mountain valley. Any or all of these factors may be important in the formation of a single fan. The fan deposit is cone-shaped and is usually thickest at the mouth of the valley.

Streams that discharge near each other onto the same plain may deposit fans which coalesce and form a continuous sheet of sediments along the base of the mountain range (Fig. 20.41).

Alluvial fans are usually composed of stream deposits, but where mud and silt are available in great abundance, mudflows may be important contributors to the volume of the fan. For example, in 1963, as a result of a prolonged and intense storm, mudflows provided the bulk of the sediment deposited on alluvial fans of half of the tributaries to the Shoshone River of Wyoming.

Alluvial fans do not occur in valleys that have resulted from the uninterrupted processes of normal erosion. Uplift, faulting, and glaciation can initiate the building of fans by disrupting the existing equilibrium of sediment and water, or both, or by changing the base level or gradient of a stream.

Streams and Lake Development

The deposition at the junction of streams can form lake basins. A main stream and a tributary may not be matched in load or transporting ability for a period of time. If the tributary carries too much gravel or sand for the main stream to carry away, an alluvial fan may be deposited across the main valley, damming the main stream to form a lake. Examples are Lake Pepin on the Mississippi River and lakes in the Illinois River Valley. Conversely, a main stream, because of a greatly increased load, may build its bed above the level of its tributaries, the stream deposits then block the mouths of the tributaries and form lake basins in the tributary valleys. Examples are lakes in Washington and Idaho north and east of Spokane and ancient lakes in the tributaries to the Allegheny and Wabash valleys (Fig. 20.42).

On river flood plains, meander cutoffs create oxbow lakes, and deposition of natural levees raises the river margins above the level of the adjacent land and form lake basins. Deltas will often enclose lake basins.

Filling of Lake Basins   An inflowing stream commonly deposits a delta at the point where it enters a lake. Many lake basins are partially filled by deltaic deposits. The town of Watkins Glen, New York, is built on a delta at the head of Seneca Lake. This delta is nearly 5 kilometers long, more than 1.6 kilometers

FIGURE 20.42   *Lakes formed when tributary valleys were blocked by glacial outwash deposits in the main valley. Similar conditions produced the ancient glacial lakes of the Allegheny valley in western New York. (From A. K. Lobeck,* A Popular Guide to the Geology and Physiography of Allegany State Park, *1972, New York State Museum.)*

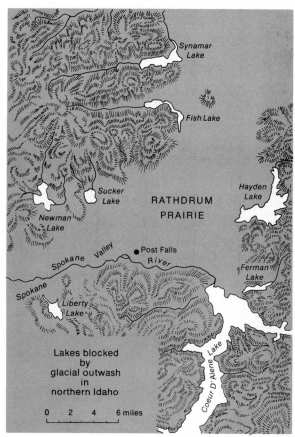

wide, and at least 150 to 180 meters thick. Ithaca, New York, occupies a similar setting on a delta in Lake Cayuga, another of the Finger Lakes (Fig. 20.43). Hanging deltas above the present level of the Finger Lakes were deposited in higher lakes formerly ponded in the valleys between glaciers on the north and moraines on the south. These deltas occur at succeedingly higher elevations to the north, and so they also provide an indication of the isostatic rebound of the earth after the weight of the continental glaciers was removed by melting.

When a stream enters a lake, the sediment-laden water may form turbidity currents that flow because

FIGURE 20.43 *The shape and orientation of the Finger Lakes of western New York. These lakes lie in preglacial valleys that were greatly deepened by glaciation. The bottoms of the bedrock troughs of Lakes Seneca and Cayuga lie below sea level.*

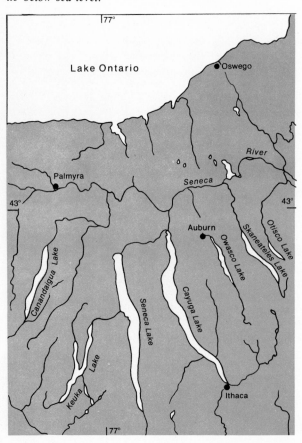

they are denser than the lake waters. Oceanic turbidity currents are discussed in Chap. 23. The turbid stream waters sink and then flow on any submerged surface of greater density, whether it is a lower, denser mass of water or the lake bottom itself. These turbidity currents transport silts and clays to distant parts of a lake or reservoir. Usually, this fine material will eventually settle out in quiet water, and so the outflowing stream from the lake may be largely free of sediment. In Lake Mead, the lake behind Hoover Dam, some turbidity currents laden with silt and clay particles brought into the lakes by the Colorado River, travel along the bottom or on a density-surface thermocline for the 200-kilometer length of the reservoir. They rise to the surface when they strike the dam, muddying the otherwise clear surface water.

SEQUENTIAL DEVELOPMENT OF A REGIONAL EROSIONAL TOPOGRAPHY

As a drainage system is developed in a newly uplifted or rising land area such as an emergent coastal plain, the progress of erosion produces a sequence of changes in the topography. These changes are determined by the altitude of the uplifted area and the resistance to erosion and structure of its subsurface rocks. Streams increase the surface relief of the area by excavating gullies and valleys as they cut downward toward base level, leaving remnants of the initial surface as divides. As streams approach base level, lateral cutting becomes more dominant, the widening of the stream valleys accompanied by the degradation of the interstream areas eventually results in a relatively level surface. Hence, stream erosion over geologic time would tend to eventually produce a plain. A cycle of erosion can be established and defined as the time required for a region to pass from low relief to high relief and to return to low relief again.

For purposes of the comparison and study of land surfaces, the erosion cycle of a region can be divided into three stages—youthful, mature, and old.

Regional Youth

During the youthful phase of a region, most streams flow in V-shaped canyons or valleys with steep sides. The side slopes are steep because sufficient time has not elapsed for gravity transfer to alter the slopes and widen the valleys. If the area has been recently uplifted, streams are not numerous, and they have few tributaries. Since they have not had time to erode extensively, they may still have rapids and waterfalls along their courses. The extensive, noneroded, poorly drained divide areas are the dominant landform of a youthful region. Upland lakes and swamps may exist.

Regional Maturity

As erosion continues, the features of the landscape characteristic of youth change. Valleys with flaring sides and gently rounded upper slopes take the place of the sharp straight lines of the youthful landscape. Because the tributaries have cut headward, the divides are narrowed, and the region becomes thoroughly dissected by a complex network of valleys. This type of topography is very rugged (Fig. 20.44). Typical mature topography now exists in the region of the Allegheny and Cumberland Plateaus, west of the Appalachian Mountains (Fig. 20.45).

Regional Old Age

By continued erosion, the rugged landscape of the mature stage of stream erosion is gradually reduced, forming broad valleys with gentle slopes and low divides. The sediments supplied to the streams are finer, and this, combined with seaward extension of deltas, lowers the gradient until the streams deposit along much of their courses rather than erode. They meander over the deposits of their own flood plains. A land area thus worn down nearly to a plain, with the gentle slopes of extreme old-age topography, is called a peneplain. The most prevalent landform at this stage is simply lowland (Fig. 20.46). Frequently, hills of resistant rock rise above the general level of the peneplain.

The amount of time required for each stage is measured in hundreds of thousands to millions of years but varies with the particular area; one area may illustrate the characteristics of maturity in less time than an adjoining one, especially if the former is cut in

FIGURE 20.44   *Mature topography, Kern County, California. Streams are numerous, the land is all in slope, and the divides are narrow. (Spence Air Photos.)*

FIGURE 20.45 *The Allegheny Plateau at Jacobsburg, Ohio.*

FIGURE 20.46 *The skyline of the Laurentians eastward at Long Lake, upper portion of the Manigotagan River, southeastern Manitoba near the Ontario border. (Geological Survey of Canada, Ottawa, No. 85301.)*

relatively unresistant rocks and the latter in more resistant rocks.

As the drainage basin of a river system passes through successive stages of a normal cycle of erosion, it may be interrupted at any stage by glaciation, vulcanism, renewed uplift, or a change in sea level. A lava flow may fill a stream channel or it may bury the entire region and profoundly modify the drainage system. Where the erosion cycle is interrupted, a new cycle is initiated on the surface of the lava field.

Movements of the earth's crust or changes in sea level may modify the cycle of erosion locally and temporarily, or the streams may begin the cycle anew. If an area is elevated, the gradients of the streams are increased and they begin cutting gorges and canyons in the bottoms of their old valleys, with the result that the region takes on some of the characteristics of the youthful stage. Such a region is said to be rejuvenated, and the streams become rejuvenated streams. After such a rejuvenation, the river carves its valley within the new upland; and if the old streams had meandering courses before rejuvenation, the winding channel is deepened and the old meanders become entrenched (Fig. 20.47).

Because there are so many interruptions of the erosion cycle, many geologists question whether a cycle is ever completed and a true peneplain developed. Many also question the strict sequential development of landforms.

STREAM SCULPTURE IN ARID AREAS

Although less effective, stream sculpture is not lacking in deserts (Fig. 20.48). In semiarid regions it is easily the dominant gradational process, even though long periods go by with little or no precipitation. Some deserts may have no rainfall for several years and then are subjected to torrential rainfalls that result in erosive sheet floods and valley torrents.

Many of the deserts and semideserts of the United States are internal, or closed, drainage basins, separated by mountain ranges from an outlet to the sea. The scarcity of water is the reason that through-flowing drainage has not developed. Some of these desert basins extend below sea level, and sediment can escape from them only by wind activity. Such basins, however, may eventually fill to the point that through-drainage can evolve (Fig. 20.41).

FIGURE 20.47 *Entrenched meanders ("goose necks") in San Juan Canyon, Utah. The meandering course of the stream was developed during a previous erosion cycle, when the river was flowing over a graded plain. The meanders became entrenched when the stream was rejuvenated by an uplift of the area. (National Park Service Photo.)*

FIGURE 20.48 *Receding cliffs and detached remnants rising above the pediment in an arid region. Monument Valley, Arizona-Utah. (The Atchison, Topeka and Santa Fe Railway Company.)*

REFERENCES

Bloom, A. L., 1969, *The Surface of the Earth,* Prentice-Hall, Inc., Englewood Cliffs, N. J.

Eagleson, P. S., 1970, *Dynamic Hydrology,* McGraw-Hill Book Company, New York.

Ellison, W. D., 1948, *Erosion by raindrop,* Scientific American, vol. 179, no. 5, pp. 40-45.

Gilbert, G. K., 1914, *Transportation of Debris By Running Water,* United States Geological Survey Professional Paper 86.

Hansen, W. R., 1969, *The Geologic Story of the Uinta Mountains,* United States Geological Survey Bulletin 1291.

Harms, J. C., and Fahnestock, R. K., *Stratification, Bed Forms, and Flow Phenomena (with an example from the Rio Grande),* American Association of Petroleum Geologists, Special Publication No. 12.

Hjulstrom, F., 1935, *Studies of the morphological activity of rivers as illustrated by the River Fryis,* University of Upsala Geological Institution Bulletin, vol. 25, pp. 221-527.

Lane, E. W., 1955, *The importance of fluvial morphology in hydraulic engineering,* American Society of Civil Engineers Proceedings, vol. 81, no. 745.

Leopold, L. B., Wolman, M. G., and Miller, J. P., 1964, *Fluvial Processes in Geomorphology,* W. H. Freeman and Company, San Francisco.

Linsley, R. K., Kohler, M. A., and Paulhus, J. L. H., 1949, *Applied Hydrology,* McGraw-Hill Book Company, New York.

Mackin, J. H., 1948, *Concept of the graded river,* Geological Society of America Bulletin, vol. 59, pp. 463-512.

Matthes, F. H., 1951, *Paradoxes of the Mississippi,* Scientific American, vol. 184, no. 14, pp. 18-23.

Morisawa, M., 1968, *Streams, Their Dynamics and Morphology,* McGraw-Hill Book Company, New York.

Playfair, J., 1802, *Illustrations of the Huttonian Theory of the Earth* (reprinted 1956), University of Illinois Press, Urbana, Illinois.

Sundborg, A., 1956, *The River Klaralven,* Geography Analer, vol. 38, nos. 2-3, pp. 127-316.

Thornbury, W. D., 1969, *Principles of Geomorphology,* 2d ed., John Wiley & Sons, Inc., New York.

Twain, Mark, 1883, *Life on the Mississippi,* James R. Osgood and Company, New York.

CHAPTER 21
SUBSURFACE WATER

Perennial streams, springs, geysers, wells, and limestone caves have a common source—subsurface water. Beneath the ground surface and protected from evaporation, large quantities of water fill the small spaces and fractures of the soil and rock. Slowly, this water flows from the entry areas, where it is supplied by rainfall, to the discharge areas. About 37 times as much water occurs underground as is contained in all the lakes, rivers, and swamps of the world. This chapter will consider the geologic conditions that control this important resource and relate its effects on the rocks in which it is stored and through which it moves.

WATER IN THE GROUND

Three genetic types of water occur beneath the ground surface: meteoric, connate, and magmatic. By far the most abundant is meteoric, water derived from the atmosphere. Meteoric water makes up nearly 100 percent of the water in the hydrologic cycle (Chap. 9) and is the source of almost all subsurface water.

Water trapped in the spaces between sediment particles as they were deposited as rocks on the ocean floors is connate water. It contains dissolved salts but usually differs in composition from present seawater. During the long periods of geologic time that it has been in the rocks, it has dissolved additional available salts, as well as depositing some mineral matter in the surrounding rocks.

Magmatic water is that water derived from deep-seated magmatic bodies which may originally contain as much as 10 percent dissolved water. Most of this water is forced into the surrounding rocks when the magmas crystallize to form igneous rocks. During this process, the waters may deposit mineral ores, such as copper, lead, zinc, and silver. Although all water on the earth may have originated from igneous activity, the amount now added in this manner is quite small compared to the amounts of meteoric water available from the atmosphere. The water of hot springs in areas of recent volcanic activity may be partly of magmatic origin.

The occurrence and movement of subsurface water is largely controlled by the porosity and permeability of various rocks.

Porosity

Openings in rocks include fractures, joints, and the

349

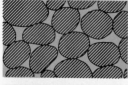
Well-sorted sedimentary deposit has a high porosity

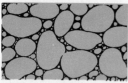
Poorly sorted sedimentary deposit has less porosity because the openings between the larger particles are filled with finer materials

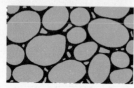
Deposition of a mineral cement may reduce porosity

Fractures and bedding may result in porosity

Solution of soluble rock may enlarge fractures increasing porosity

FIGURE 21.1 *Factors affecting porosity or rocks.*

openings between individual mineral or sediment grains called pores. The porosity of rocks or soils is the ratio of volume of the pore space to the total volume of the material including its pores. Thus, if 1 liter of sand will hold 0.3 liter of water when saturated, its porosity is said to be 30 percent, since three-tenths of its total volume is made up of space between the grains. The porosity of different types of rock varies from less than 1 percent in unfractured granite to more than 40 percent in a poorly cemented sandstone. Weathered, highly fractured, and layered rocks are usually more porous than massive igneous rocks.

The porosity of a sedimentary deposit depends chiefly on the shape and arrangement of the constitu-

ent particles, the degree of sorting of its particles, the cementation and compacting to which it has been subjected since its deposition, the removal of soluble mineral matter through solution, and the fracturing of the rock, resulting in joints and other openings. Figure 21.1 summarizes the effects of all these factors.

Rock and Soil Permeability

The permeability of rock or soil is a measure of its capacity for transmitting a fluid. Permeability varies with the size and shape of the pores and the size, shape, and extent of their interconnection. Permeable rocks are always porous, but a rock with high porosity is not necessarily highly permeable. For example, frothy volcanic pumice is a rock material with very high porosity but very low permeability. A pumice block may float in water for days because its pores are not interconnected and water cannot penetrate it. A bed of pumice fragments, however, is highly permeable because of the many interconnected openings between fragments.

In the discussion on streams it was demonstrated that the rate at which water flows is related to the friction it encounters—the greater the friction, the slower it flows. Friction is generated by the interaction of the water and the surfaces it encounters; consequently, the more surface area, the more friction.

Coarse-grained sand and gravel, which do not have fine-grained particles between the coarser grains, are permeable earth materials. On the other hand, fine-grained materials like clays and shales, although very porous, have large surface areas and are relatively impermeable. A volume of 28,000 cubic centimeters of spherical particles 1 millimeter in diameter has an internal surface area of about 3,300 square meters. The same volume of 0.02-millimeter particles (silt size) has a surface area of 165,000 square meters (more than 1 acre), and 0.001-millimeter particles (clay) make up a surface area of more than 3×10^6 square meters, or more than 70 acres. The fineness of clays and shales provides a large surface area to which a thin film of water clings by molecular force so tightly that the water will not flow. These rocks have a low permeability, and so even though their porosity may be quite high, they are poor sources of water. Rock units that are both porous and permeable and contain abundant water are called aquifers.

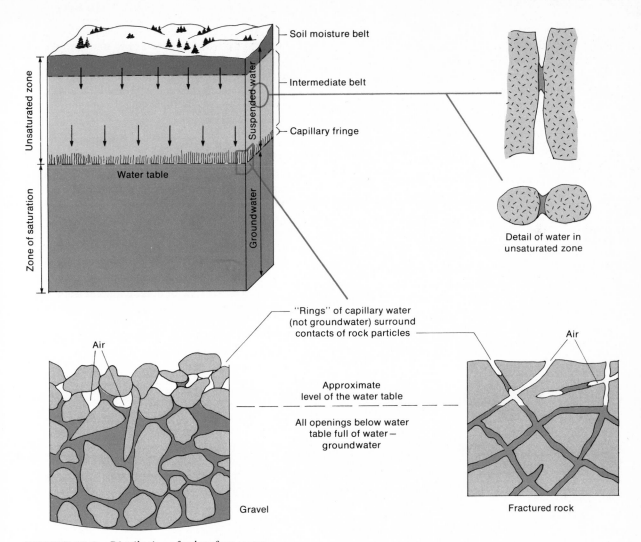

Detail of water in
unsaturated zone

"Rings" of capillary water
(not groundwater) surround
contacts of rock particles

Approximate
level of the water table

All openings below water
table full of water—
groundwater

Air

Gravel

Fractured rock

FIGURE 21.2 *Distribution of subsurface water.*

ZONES OF GROUNDWATER

Two zones of subsurface water are usually encountered when a well is dug (Fig. 21.2). First, there is an unsaturated zone, in which the pores in the rock or soil may contain both air and water. Any water in this zone is suspended water. This zone varies in thickness, since its lower limit is determined by the position of the water table. Below it occurs a second zone, the zone of saturation, in which the pores are filled with groundwater.

The unsaturated zone is divided into three belts: a belt of soil moisture, an intermediate belt, and a capillary fringe. The boundaries of these belts are indistinct. The belt of soil moisture is located just below the ground surface. It is of particular importance since this is the zone where most plants get their water. Some soil moisture is used by plants and returned through transpiration to the atmosphere. Some soil moisture evaporates directly from the soil.

Plants greatly increase the surface area available for evaporation. A large fraction of the snow and rain falling on a forest canopy evaporates from leaf and limb surfaces and never reaches the ground.

Vegetation removes larger quantities of water from the upper part of the soil. This water is evaporated from the plant cells into the atmosphere by the process known as transpiration. Transpiration may equal or exceed the evaporation from a free water surface; in well-vegetated areas it will exceed the evaporation from the belt of soil moisture. The roots of some plants may reach the capillary fringe or the water table and may actually withdraw groundwater from storage and cause a lowering of the water table. Common plants of this type are willows, cottonwoods, and salt cedar. The greatly diminished flow or drying up of some streams in the Southwest has been ascribed to the spreading of salt cedar, a shrub introduced to the area for ornamental purposes more than 100 years ago. Eradication of such plants holds some promise of increased yields of groundwater.

Water in the intermediate belt is below the roots of most plants and the effects of surface evaporation. The intermediate belt transmits water to greater depths. At its base, additional moisture occurs in a belt called the capillary fringe. In this belt, a few centimeters to a few meters thick, water from the saturated zone below is drawn up into the pores of the rock by the molecular attraction between the rock material and water.

Below the capillary fringe, water fills most of the pores of the soil or rock, and water will stand in a well that reaches this depth. This is the zone of saturation, and the water in it is called groundwater. This zone will yield well water if the rock is permeable enough. The surface of the zone of saturation is called the water table, or groundwater level.

The Water Table

The water table tends to follow in a subdued way the land surface, thus it has higher elevations under hills than under adjacent valleys.

In hilly regions with average rainfall, the unsaturated zone may be a few to a hundred meters thick (Fig. 21.3). In arid regions, where rainfall is small and evaporation high, the unsaturated zone may be much thicker, since the water table is at greater depths. Near permanent streams, lakes, or other bodies of water, the water table is close to the surface.

A body of water in permeable material may be perched, or suspended, within the unsaturated zone above the main water table. Perched water tables may be important sources of water in arid and semiarid regions.

Perched Water Tables   These are generally found above an irregularly shaped mass of impermeable rock or in lenses or wedge-shaped masses of sand and gravel, which catch and hold the downward moving water (Fig. 21.3). Deepening a well in such a situation can have a disastrous effect if the impermeable

FIGURE 21.3   *Diagram showing the relation of the water table to hills and valleys, and lakes and swamps.*

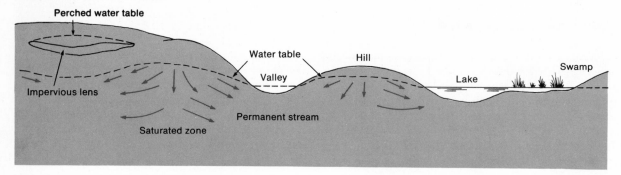

layer is breached. The result can be the drainage of the perched water table.

In parts of the glaciated region of northeastern North America, where large areas are covered by a thick mantle of clayey glacial drift, lake basins may lie above the regional water table. The perched water in such lakes is prevented from moving down to the main water table by impermeable clay layers between the floor of the lake and the underlying regional water table. Accumulation of organic debris on the bottom of a depression may also decrease the permeability of the sediments to the point that a lake may be formed and maintained.

Infiltration   Infiltration, the movement of water downward through the unsaturated zone to the water table, is controlled by the amount and intensity of rainfall, the topography of the region, the permeability of the rock or soil, and the vegetation.

At the beginning, when a slow rain begins to fall on permeable material, such as a sandy soil, there is no surface runoff. Runoff does not occur until the uppermost layer of material becomes fully saturated. However, rainfall on impermeable material runs off promptly. This difference depends on the permeability of the soil or other surface material.

The steeper the slope of the ground, the greater the amount of runoff. On nearly level ground, surface-flow velocities are negligible and water has a longer time to soak in. Plants, plant remains, and minor irregularities of the surface interrupt the runoff of surface water and increase the time available for infiltration.

Inclined strata may allow more water to penetrate into the subsurface than do flat-lying beds. Water passing through inclined beds will follow the most permeable layers. If rock beds or soil zones are horizontal, the infiltration rate is limited by the least permeable layers through which the water must pass.

GROUNDWATER MOVEMENT

The slow movement of water through interconnecting openings in the zone of saturation is called percolation. Groundwater percolation depends on the slope of the water table, or the hydraulic gradient (the fall in elevation of the water divided by the distance over which

the fall takes place), and the permeabilities of differing rock layers, as well as the inclination of the sediment or rock layers.

The pressure at a point in a fluid produced by the weight of overlying fluid is hydrostatic pressure. The hydrostatic pressure is greater under the higher parts of the water table than it is at the same elevation under lower parts of the water table. This pressure difference produces a movement of the water (Fig. 21.3). Groundwater follows the paths of least resistance from high-pressure areas under the hills to low-pressure areas under valleys. The change in elevation of the water table between a wet year and a dry year is normally greater under the hills than under the valleys. As a result, the water supply from shallow wells is less dependable under hills than under valleys.

The depth to which surface waters penetrate varies with the porosity and permeability of the rocks. In some permeable rocks, surface waters reach depths of a thousand meters, whereas in other impermeable rocks, very little water collects at depths of more than a hundred meters. If permeable paths are directly downward because of the inclination of rock layers or fracture zones of high permeability, the water may move downward to considerable depths before it rises through some permeable formation or fracture to the surface.

In massive igneous and metamorphic rocks, permeability is mainly a function of the number and size of the fractures. If fractures are absent, significant supplies of water are infrequently found at depths of more than ten meters. In the copper-bearing rocks of Keweenaw Peninsula, Michigan, shafts have been sunk more than 1.6 kilometers below the surface. In many of the mines, the rocks are dry near the lower end of the shafts. In other mines and tunnels, however, appreciable quantities of water have been found at great depths. Traces of calcium chloride dumped into a lake 430 meters above the Moffat Tunnel in Colorado appeared in the water entering the tunnel via fractures within 2 hours.

Calculations of Groundwater Flow

A major step in the understanding of groundwater movement was made in 1856 by Henry Darcy. Darcy found that the discharge of groundwater is directly proportional to the hydraulic slope and the cross-sectional area of the aquifer transmitting the flow. These relationships can be expressed in the form of an equation:

$$Q(\text{discharge}) = V(\text{velocity}) \times A(\text{cross-sectional area})$$

$$V = P \ (\text{coefficient of permeability})$$
$$\times I \left(\frac{\text{change in elevation}}{\text{distance traveled}} \right)$$

Therefore

$$Q = P \times I \times A$$

The coefficient of permeability P depends on all the aquifer properties affecting permeability and the fluid properties affecting flow. The flow through small pores of most aquifers is laminar rather than turbulent; therefore, viscous forces are quite important. The most important factor affecting viscosity of groundwater is temperature. Under most conditions, temperature and viscosity in a given setting are nearly constant. When groundwater temperatures differ significantly from the standard conditions (17°C), however, changes in viscosity must be taken into account in determining the value of P and therefore the velocity.

Figure 21.4 shows the range of values for P in a variety of materials. An idea of natural flow rates can be obtained by using coefficients of permeability P. For example, a water-table slope of 1.88 meters per kilometer (10 feet per mile) for values of P of 10^{-1} (fine sand) and 10^5 for clean gravel gives velocities of 0.0007 meter (0.0025 foot) per day and 0.7 meter (2.5 feet) per day, respectively. Similarly, a slope of 18.8 meters per kilometer (100 feet per mile) will yield 10 times these velocities:

$$V = PI = 10^{-1} \times \tfrac{1.88}{1,000} \times \tfrac{1}{264}$$
$$= 0.007 \text{ meter per day}$$

where $\tfrac{1}{264}$ is the conversion factor for gallons to cubic meters.

These velocities are based on the cross-sectional area of the aquifer and not on the cross section of the pores actually transmitting fluid. The actual velocity in the pores must be several times as great as the velocity given by the Darcy equation. This is still extremely slow when compared with the velocities of surface flow in streams or the flow in large underground solution conduits. The movement of water in large openings, such as solution-enlarged fractures, is turbulent flow and is analogous to the flow in streams or pipes.

Groundwater and Streams

Most rivers receive groundwater from springs and general seepage. Such rivers, known as effluent (gaining) streams, occur in low places on the water table, and hence, groundwater flows into them. In areas where the water table occurs well below stream elevations, streams lose water downward into the earth. Such rivers are known as influent (losing) streams. They cause a ridge in the underlying water table (Fig. 21.3). Losing streams are often intermittent because of the absence of surface runoff during dry periods. A stream may be gaining along part of its course and losing in other parts. Many rivers are losing during floods, when they raise the water table in adjacent flood plains and other aquifers, and gaining during dry seasons. Streams that continue to flow during long dry seasons may be maintained by the groundwater they receive. Other sources include artificial means, melting snow, or glaciers.

Infiltration from a stream is greatly influenced by the permeability of the stream bed. If a stream has bed and banks of silt and clay, infiltration losses may be small. Large amounts of silt and clay in the stream load

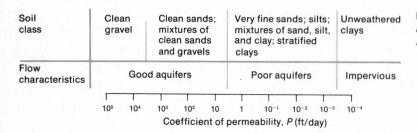

FIGURE 21.4 *Coefficient of permeability for different classes of unconsolidated materials. (After D. K. Todd, 1959.)*

Soil class	Clean gravel	Clean sands; mixtures of clean sands and gravels	Very fine sands; silts; mixtures of sand, silt, and clay; stratified clays	Unweathered clays
Flow characteristics	Good aquifers		Poor aquifers	Impervious

Coefficient of permeability, P (ft/day): 10^5 10^4 10^3 10^2 10 1 10^{-1} 10^{-2} 10^{-3} 10^{-4}

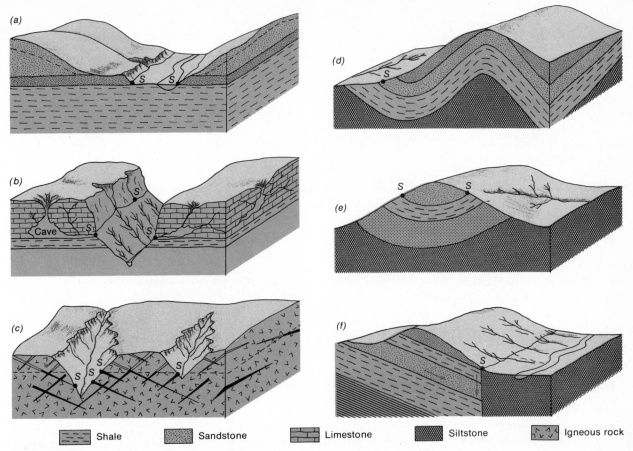

FIGURE 21.5 *Subsurface rock structures favorable for the formation of springs. Locations of springs are marked by (S).*

serve to seal stream channels against large water losses. In a similar manner, clay is used to seal irrigation ditches and channels from losses by infiltration.

Springs

Springs occur wherever underground water flows to the surface through natural openings in the ground. Their rate and manner of flow are regulated by the available water and by the permeability and structure of the rocks (Fig. 21.5). Where an aquifer rests upon impervious beds and the water table intersects the surface, the water may move out in the form of hundreds of small seepage springs. Over a period of time, well-

defined underground water courses may develop in some rocks where fractures, solution-enlarged fractures, or permeable zones concentrate the flow into a few larger springs. Springs may issue along fault planes that cut the impervious strata. Some springs discharge fresh water on the floor of the sea, where it may rise through the denser salt water before mixing. In water-deficient areas, the location and interception of spring flows going to the sea is a valuable source of fresh water.

It has been calculated that there are in the United States 65 springs of first magnitude, that is, with an average daily discharge of not less than 2.8 cubic meters per second. Thirty-eight occur along openings

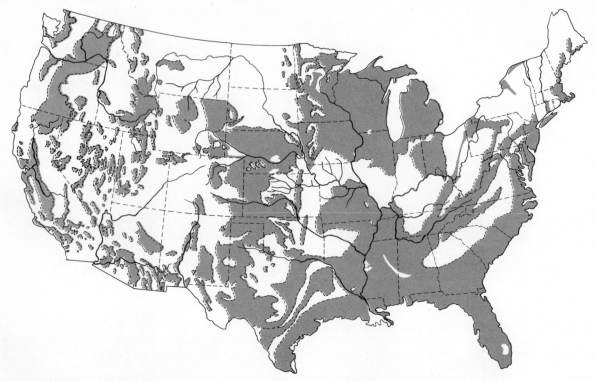

FIGURE 21.6 *Groundwater areas in the United States. (After H. E. Thomas,* The Conservation of Ground Water, *McGraw-Hill Book Company, New York, 1951.) Patterns show areas underlain by aquifers generally capable of yielding to individual wells about 189 liters per minute or more of water containing not more than 2,000 parts per million of dissolved solids (includes some areas where more highly mineralized water is actually used).*

of fractures and irregular flow surfaces of lava flows and in associated gravel deposits in Idaho, Oregon, and California. Twenty-four occur in solution-enlarged fractures in limestone in the Ozark area of Missouri and Arkansas, along the Balcones fault zone of Texas, and in Florida. The other three springs of first magnitude issue from a sandstone in Montana, and their great discharge is believed to be due to faults or other special features.

GROUNDWATER RESOURCES

Groundwater is a natural resource on which a large part of our population now depends. It supplies many municipalities and is widely used in agriculture and in-

dustry. Wells and springs are estimated to furnish an average of 113,700,000 to 132,650,000 cubic meters (120 billion liters) per day, roughly one-fifth of the total water used for irrigation, industry, and domestic purposes.

Most of the groundwater used comes from unconsolidated sediments, from permeable solid sedimentary rocks, and some from lava flows. In unconsolidated sediments, the main sources are alluvial sands and gravels, glacial outwash, stream terraces and flood plains, and sandy coastal-plain sediments. Gravels give the highest yields, sands somewhat lower yields, and silts, clays, mudflows, and glacial tills rather low yields. In consolidated rocks the greatest yield may be expected from permeable sandstone, porous limestone, closely jointed rocks, such as quartzite and granite, and jointed, vesicular, or cavernous basalt flows.

Explosives and acid are used to increase permeability in the vicinity of some wells. Shattering of solid rocks by atomic explosives may serve to increase infiltration, percolation, and recovery at wells if the problems of radioactivity can be solved. Shale commonly yields very small quantities of water, unless it is highly fractured. Water yields from limestones vary from large where solution has enlarged fractures to much less where fractures in the limestone are infrequent and solution enlargement has not been effective.

In most igneous rocks, permeability depends on the number and size of the fractures within the rocks. Some lava flows and deposits of pumice fragments are so permeable that surface streams are infrequent. On the other hand, some granite terrains are so impermeable that surface runoff dominates. Figure 21.6 summarizes the distribution of aquifers that yield significant quantities of good water in the United States.

Wells

Digging or boring for water dates back to very early historic times. Some of the earliest wells drilled were in China and India. Irrigation works using well water were constructed in Babylonia as early as 2,000 B.C. In present-day India and the United States, more land is irrigated from wells than from streams.

Many wells are merely shallow holes dug or bored into the earth to depths below the water table. They serve as chambers into which groundwater flows and from which it can be pumped. Most new well systems producing large water supplies are located with great care after careful hydrologic analysis of such factors as the rainfall on intake areas, lithology of the rocks, and their structure and permeability. Many individual wells, however, are located by other means. Water witchers (dowsers) claim to be able to locate water and many other valuable minerals by a variety of techniques, the most common of which involves the use of a forked stick. Geologists and water-supply engineers are almost unanimous in their skepticism. Whatever success is achieved by water witching results from the fact that, almost everywhere, water exists at some depth below the surface. A young boy, a future geologist, whose grandfather showed him the location of a water line by using a forked stick or welding rod may not be so skeptical.

Even in highly permeable rocks, the removal of water through a well depresses the water table centrally toward the well (Fig. 21.7). In materials of low permeability, such a cone of depression is steep and may have a radius of no more than a hundred meters. In uniformly permeable sands and gravels, however, the cone of depression from a heavily pumped well may extend a thousand meters from the pumping well and interfere with neighboring wells. When a field of wells is developed, the wells must be carefully spaced so that the maximum yield can be obtained from the smallest number of wells.

Artificial recharge of wells has been practiced in some parts of the United States since the early 1800s. Water can be put into the ground, thus increasing the hydrostatic pressure, infiltration rate, and length of time available for recharge to take place. It serves two purposes—water conservation and improvement of quality. It allows increased recharge rates at times of surplus surface water, dilution of mineralized groundwater, and modification of groundwater temperature.

Artesian Wells   Artesian wells received their name from Artois, a province in France, where the water in many wells rises above the surface of the earth, similar

FIGURE 21.7 *The cone of depression of the water level around a well that is being pumped.*

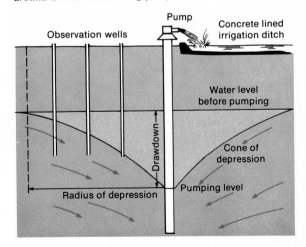

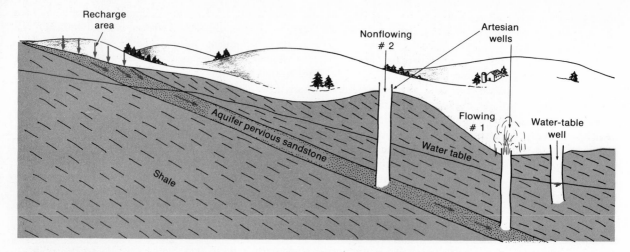

FIGURE 21.8 *Diagram showing subsurface conditions favorable for artesian wells. The water in the artesian wells rises above the top of the sandstone aquifer where it was first encountered while drilling. Well (2) is an artesian well, even though it does not flow at the surface as well (1) does. The water in the sandstone aquifer is under pressure at all points lower in elevation than the recharge area. The water in the water-table well comes from fractures.*

to a fountain. Today the term artesian well is applied to any well in which groundwater rises above the point at which it was initially encountered; it need not flow out at the surface.

Artesian flow takes place because of differences in hydrostatic pressure that exist in different parts of a confined aquifer. In an artesian system, the water-bearing bed may be compared to a water-filled tube of permeable rock in which the intake is higher than the outlet. The following conditions are essential for a flowing artesian well (Fig. 21.8) in sedimentary strata:

1. There must be an aquifer that can hold and furnish water.
2. There must be impermeable beds above and below the aquifer to confine the water and to prevent it from escaping.
3. The place at which water enters the aquifer must be higher than the place where the well penetrates the water-yielding bed.

These requirements apply only to sedimentary strata. Artesian flows may be obtained from other kinds of rock, even from unconsolidated sediments,

but where encountered by the well, the water must be under hydrostatic pressure.

At many places the sources of artesian water in the intake areas are scores or even hundreds of kilometers distant from the wells. Most wells in the famous Dakota Sandstone basin of North and South Dakota are 300 meters deep and they may be 250 kilometers from the intake area. This aquifer supplies a major part of the water used in the Dakotas.

Although many artesian wells in glacial drift are less than 30 meters deep, artesian water in solid rock has been encountered at depths of more than 1 kilometer.

Groundwater Conservation

In the past, most residents of the United States have never worried much about their water supply. A fairly high average rainfall over the well-populated areas and the presence of large inland rivers and lakes are responsible for this attitude. Most people thought of water as one resource over which they needed to have little concern because they believed it to be everlastingly abundant. During the past few decades, with increasing population and industrial uses, we have been forced to recognize that difficulties with our water supply are increasing, both as to quantity and quality, and that we must eliminate wasteful or harmful practices and adopt sound conservation methods.

To conserve groundwater, it is possible to (1) restrict the rate of withdrawal by pumping at the natural rate of recharge, (2) avoid waste, especially from flowing or leaking wells, (3) prevent seawater from moving landward in aquifers that outcrop on the ocean floor, (4) control the use of aquifers for waste disposal, (5) reduce evaporation losses by using only the right amounts of water for the correct times for each irrigated crop and by changing industrial cooling practices, (6) assist replenishment by artificial recharge and by preventing excess runoff of surface water, and (7) return used water to the ground from industrial cooling, air conditioning, and treated waste water. One of the primary uses of water is the treatment of sewage and other wastes by dilution, which renders the water unfit for reuse. Applying methods of waste treatment other than dilution would make reuse possible and increase the amount of water available.

GRADATION BY GROUNDWATER

Water sinks readily into the ground in regions where there are numerous cracks or joints in the rocks. If the rocks are easily soluble, such as limestone, dolomite, gypsum, and rock salt, the joints are gradually enlarged by the descending water. At the intersection of joints and easily dissolved layers, horizontal chambers of considerable size may be developed (Fig. 21.9). Areas in the United States where solution has been important in sculpturing the land are shown in Fig. 21.10. The chemistry of the solution of limestone by naturally carbonated waters is treated in Chap. 10.

One of the most noticeable results of solution is a sinkhole, or sink. A sinkhole is a surface depression formed by solution of susceptible rock without any physical disturbance of the surrounding rock or by the collapse of the rock over a small chamber or part of a cave. The latter mechanism, collapse, is most important and may result in the loss of property (Fig. 21.11). In a 2.6-square-kilometer area in Orange County, Indiana, more than 1,000 sinkholes were recorded. Occasionally, sinkholes become sufficiently plugged with

FIGURE 21.9 *Diagram of Mammoth Cave, Kentucky, showing different "levels" of the cave that follow the horizontal bedding planes of the rock and "domes" that follow the vertical joints. (After A. K. Lobeck, 1939.)*

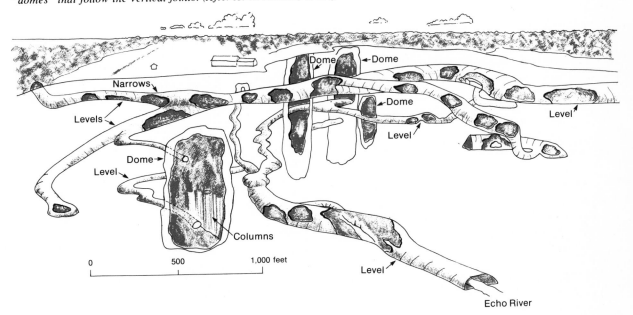

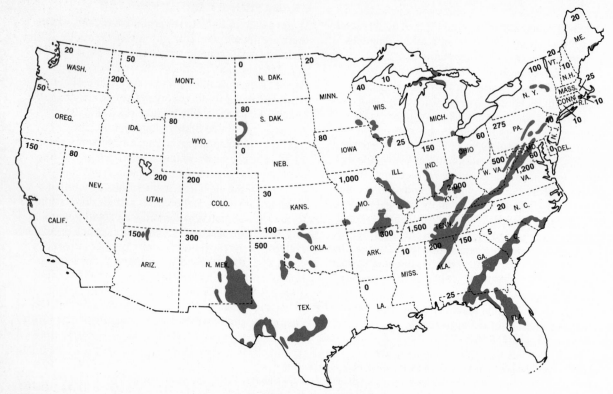

FIGURE 21.10 *Areas in the United States where solution has been important in sculpturing the land. Numbers indicate estimated total of caves in a state. Less than five caves not indicated. [After maps by Wm. E. Davies, U.S. Geological Survey, and* Map of Cavern Areas in the U.S., *in C. E. Mohr and H. N. Sloane (eds.),* Celebrated American Caves, *Rutgers University Press, 1955.]*

organic debris, rock fragments, and soil materials to form ponds or lakes. Alachua Lake, Florida, is an example. Prior to 1871, the surface drainage of the Alachua Prairie emptied into a large sinkhole. That year the outlet of the sink was clogged, and a lake nearly 13 kilometers long and 6 kilometers wide formed. About 20 years later the outlet opened again, and the lake drained underground.

The formation of one of the largest sinkholes in Alabama, and perhaps in the United States, apparently occurred in late December 1972. On December 2, residents of Shelby County, Alabama, were startled by a home-shaking rumble, followed by the snapping and breaking of trees. An investigation of the area of the source of the sound resulted in the discovery of a sinkhole, 140 meters long, 115 meters wide, and 50

meters deep. The sinkhole is within a 16-square-kilometer area, containing over 1,000 sinkholes and subsidence areas. The bedrock of the area is composed of limestones and dolomites that are covered with a deep weathered zone. The sinkhole was formed in the deep weathered zone. The cause may be in part due to an abnormally high rainfall in November and to groundwater withdrawals in quarries to the east.

Caves

Along fracture zones and bedding planes where limestone is most readily attacked, nearly all the rock is dissolved, and only small amounts of sand and clay remain. Thus, in the course of time, an elaborate system of narrow to spacious tunnels and open chambers may be formed in solid rock (Figs. 21.9 and 21.12).

FIGURE 21.11 *Home lost in land collapse in Bartow, Florida, May 22, 1967. Sinkhole that formed was 158 by 38 meters and 18 meters deep. (U.S. Geological Survey.)*

FIGURE 21.12 *Idealized diagram of a portion of the karst region of southern Indiana. (After Wm. J. Wayne,* The Compass, *vol. 27, p. 218, 1950.)*

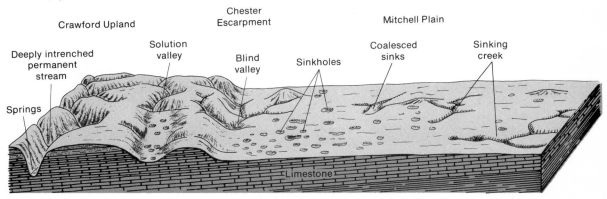

Most solution takes place at or just below the water table. Many caves show features such as ceiling pockets and intricate network patterns, which strongly suggests that the passages were filled with water during their formation. Probably a network of similar-sized passages such as those in Fig. 21.9 results from pressure flow beneath the water table. The layout resembles in many respects the water-distribution system of a city.

Most large caves are nearly horizontal even in inclined soluble rocks, suggesting control by the position of the water table. Solution is most effective near the top of the zone of saturation because, as is shown in Fig. 21.13, mixing of downward percolating waters with those from below always produces conditions under which more calcium carbonate can be dissolved.

A lowering of the water table occurs after the solution of the main cave. This lowering of the water table normally results from the downcutting of the local surface-stream network over a long period of time. Caves produced by this sequence of events would then provide evidence both of the history of

FIGURE 21.13 *Mixing of groundwater and unsaturated-zone water always produces water undersaturated in calcite. The zone of mixing just below the water table is therefore the zone of greatest solution.* [*After G. W. Moore,* Limestone Caves, *in Rhodes W. Fairbridge (ed.),* The Encyclopedia of Geomorphology, *1968.*]

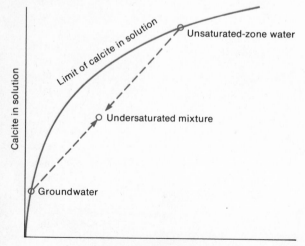

land surface development and the subsurface development of a region. Carlsbad Caverns and many of the caves of Missouri are thought to be examples of this type of cave formation.

Big Spring in the Missouri Ozarks is cited as an example of a sub-water-table cave system in the process of formation. This spring discharges 947,500 cubic meters (1 billion liters) of water per day, containing 192 metric tons of dissolved materials, into the Current River. This escape of spring water under hydrostatic pressure from a water-filled cave system probably began when, in deepening its valley, the Current River broke through the roof of a conduit. Openings of former groundwater courses can be seen in the face of the cliff above the spring. The sources of Big Spring's water are rainfall and minor stream-water leakages. The source of the dissolved material is the limestone through which the stream flows. Not until the Current River deepens its valley below the present level of the main conduit will this underground system of openings be drained.

One of the largest known cave systems in the world is that of the Carlsbad Caverns National Park, New Mexico. It is located in limestone and gypsum rocks of the Guadalupe Mountains. Because of its huge size and excellent display of subsurface water deposits, it was made a national park in 1930. One gallery, the Big Room, is about 1,200 meters long, with a maximum width of 90 meters and a maximum ceiling height of about 90 meters.

Another well-known American cavern is Mammoth Cave in Kentucky (Fig. 21.9). Some portions of it have been studied and mapped in detail, but others have never been explored. This vast labyrinth consists of several hundred kilometers of connected galleries, with underground lakes, rivers, and waterfalls included in the system. These galleries range in height from less than 1 meter to more than 30 meters. In some parts of the cave, one gallery is located above another. Mammoth Dome, an enlarged portion of the cavern, is about 120 meters long, 50 meters wide, and 25 to 75 meters high.

In some areas, little evidence of any network of cave passages exists in the surface. A network of this type is found near Hannibal, Missouri (perhaps the locale of Tom Sawyer's adventures), which has nearly 6 kilometers of passages in an area of not more than 30 acres. There are only two unobstructed entrances.

About 24 cave passages are blocked by rubble where they meet the hill slopes. A possible mechanism for the breakdown and concealment of the cave entrances is shown in Fig. 21.14.

Incomplete collapse of the roof of an underground cavern may result in a natural bridge. Natural Bridge, Virginia, is a good example. Natural Tunnel, Scott County, Virginia, is a cave remnant some 270 meters long that is now occupied by Stock Creek and the Southern Railroad.

Karst

Irregular topography that is abundant where sinkholes occur is called karst, or karst topography. This unusual type of topography gets its name from the Karst Mountains northeast of the head of the Adriatic Sea, where this type of topography is well developed. In the United States, somewhat similar topography occurs in the limestone areas of Florida (Fig. 21.15), central Tennessee, Kentucky, and parts of southern Indiana (Fig. 21.12).

With the development of karst, the slopes become steep and clifflike. The surface may also contain a network of numerous short gullies and ravines, which terminate abruptly where they discharge their waters into subterranean channels; consequently, the land surface is very rough. The progressive solution of limestone in a karst area is shown by sinks and solution basins that enlarge to form solution valleys (Fig. 21.12). Long-continued destruction of a limestone region may leave isolated limestone hills or groups of hills called haystack hills. In Puerto Rico, hills formed in this way are called Cockpit Country and rise up to 100 meters above the surrounding plain.

AGGRADATION BY GROUNDWATER

Most spring water contains some ions in solution. The ions most commonly present are bicarbonates and sulfates of calcium, magnesium, and sodium, with small amounts of chlorides, silicates, and phosphates of these and other elements. Carbon dioxide from the atmosphere is picked up by rainwater to form a weak acid, carbonic acid. In combination with humic acid from organic matter in the soil, carbonic acid dissolves the minerals. The soluble products are carried off as ions in solution in ground and stream water.

FIGURE 21.14 *Stages in cave-passage truncation in central Kentucky karst. (After R. W. Brucker, National Speleological Society Bulletin, vol. 28, pp. 171-178, 1966.)*

Spring Deposits

The amount of ions carried in solution by some spring waters is enormous. When these waters emerge from the ground, they may be subject to evaporation, temperature changes, loss of carbon dioxide through agitation, or release of pressure, as well as the work of bacteria or algae. All these changes tend to cause deposition of some of the dissolved substances.

Calcium carbonate is the material deposited by both hot and cold springs. The calcium carbonate deposited by springs around their vents is called travertine. Because of the long-continued overflow of springs, most travertine occurs in the form of terraces (Fig. 21.16). Falls Creek, Oklahoma, receives water from springs that carry so much calcium carbonate that natural dams of travertine have been deposited across the stream valley. Water from the Silver Springs in central Florida releases about 660 metric tons of dissolved solids daily. A banded-crystalline variety of travertine from Baja California, Mexico, is "Mexican onyx." An odd type of breccia is formed when rock fragments are cemented together by travertine. Both materials are highly prized as ornamental stone.

Thousands of tons of gypsum are deposited annually at the springs of Leuk in Switzerland. Other springs deposit such substances as iron hydroxide,

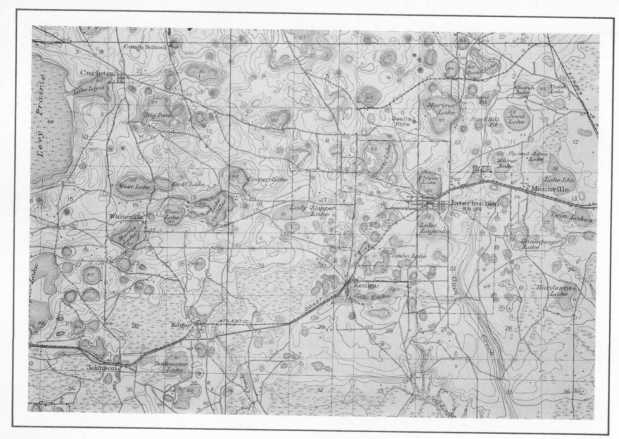

FIGURE 21.15 *Sinkhole lakes in Florida. (U.S. Geological Survey Topographic Map.)*

manganese oxide, sodium chloride, sodium carbonate, sulfur, and metallic sulfides.

Deposition in Caves

Although some cave deposits may have been formed when the cave was filled with water, most deposits have been formed above the water table after the cave was drained. A common name for cave deposits, which are usually composed of $CaCO_3$, is dripstone. Dripstone assumes such forms as stalactites, which are attached to the roof of the cavern or to some projecting ledge, and stalagmites, which form on the limestone floor of the cavern and build upward to make mounds and cones (Fig. 21.17).

Stalactites originate on the damp roof of a cave,

where drops of water gather and begin to evaporate and lose carbon dioxide. The drops can then hold less calcium carbonate and deposit any excess as rings at their margins. Drop after drop lengthens the ring into a long pendant with a hole in the middle like a pipestem. An iciclelike stalactite broken across shows a radial structure, with fibrous crystals around the hole passing concentric zones of growth. A growing stalactite, kept moist by calcium-bearing water trickling over its surface, is lengthened at the lower end and thickens outward.

Stalactites that reach the floor of a cave become solid stalks, which may thicken into massive columns or pillars. Many pillars are formed by the union of stalagmites that grow upward from the floor merging

FIGURE 21.16 *Travertine deposits named Opal Terraces at Mammoth Hot Springs, Yellowstone National Park, Wyoming. Hot mineral-spring water carrying calcium carbonate in solution rises to the surface and slowly overflows these shallow pools. Cooling, loss of carbon dioxide, evaporation, and algae cause precipitation of travertine on the bottoms and rims of the pools. The light-colored embankment in the middle distance is a large deposit of similar origin. (Burlington Northern Railway.)*

FIGURE 21.17 *Dripstone below a joint crack in the roof of Mayfield Cave, Texas. Its forms include (1) slender, hollow, pendant tubes that resemble soda straws; (2) stouter, iciclelike stalactites; (3) stalagmite mounds on the floor; and (4) continuous columns or pillars where stalactites and stalagmites join. This array, here less than 1 meter high, shows in miniature the different stages of development that are equally representative of similar features many tens of meters long. (James F. Quinlan, Jr.)*

with their counterparts from the roof (Fig. 21.18).

Stalactites assume many shapes as determined by the manner in which the water trickles over them and by the amount of water present. Beautiful forms fringed with crystals of calcite, draperies hanging from the roof, grotesque shapes that extend from floor to ceiling, and ornamental pillars may be observed in the same cavern.

FIGURE 21.18 *Stalactites and stalagmites, some of them joined to form pillars in Natural Bridge Caverns near San Antonio, Texas. (Natural Bridge Caverns, Inc.)*

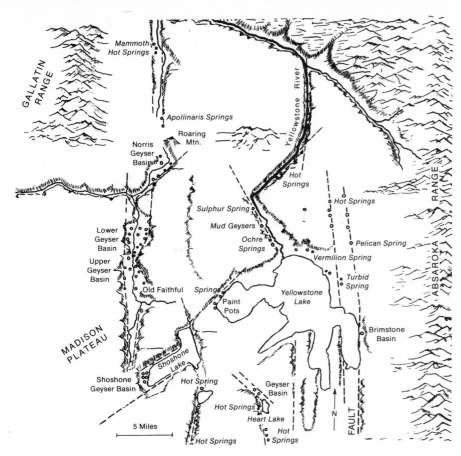

FIGURE 21.19   *The Yellowstone Basin, showing the localization of geysers and hot springs along fracture zones. (After Lobeck,* Atlas of American Geology.)

THERMAL GROUNDWATER PHENOMENA

Hot Springs

Hot springs have an average temperature noticeably above the mean annual air temperature at the same locality. They are of great interest because they reveal the movement of groundwater. Most hot springs are found in areas of relatively recent volcanic activity. This association suggests a connection between certain hot springs and igneous rocks that are cooling underground. Circulation of the groundwater is evidently made possible in places by deep-reaching, large-scale faults, along which hot springs are aligned (Figs. 21.19 and 21.20).

Some hot springs and a considerable number of warm ones, however, are found in areas remote from igneous centers, as at Hot Springs, Arkansas. It is probable that the water of these springs is rain water (meteoric water) that has been heated by circulating to great depths. Groundwater at a depth of 1 kilometer is about 40°C hotter than the average surface temperature. If such water were to rise without much dilution or cooling, it would be noticeably warmer than surface water.

FIGURE 21.20 *Lower Geyser Basin, Yellowstone Park. (Spence Air Photos.)*

It has been suggested that only those United States springs having temperatures at least 9°C above the mean annual air temperature be classed as thermal. In the contiguous United States there are more than 1,250 areas in which thermal springs are found. In many of these areas there is a multitude of springs. Hot Springs, Virginia, and Hot Springs, Arkansas, are well-known health resorts. Thermopolis, Wyoming, is the site of the largest hot spring in the United States. It flows an average of 350 cubic meters per second at a temperature of 57°C.

Gases of probably magmatic origin are also present in many hot springs, and certain vents that emit gases in dry seasons become hot springs in wet seasons. Boiling springs are those in which the gases erupt vigorously and agitate the water. Paint pots and moon pools are springs that contain abundant fine-grained rock particles, particularly oxides of iron, which color the thickened water yellow, blue, or red. Algae, which are simple forms of plant life, thrive in

the warm waters of certain springs and add color to them. Light-colored algae (i. e., pink or light gray) live in fairly hot water; darker-colored types live in tepid water.

Geysers

Geysers are hot springs from which the water is expelled vigorously at intervals. The name geyser comes from an erupting hot spring (geysir) in Iceland near Langervaten. Geysers are less common than ordinary hot springs, but abundant geysers occur in Yellowstone National Park in the United States (Figs. 21.19 and 21.20) and in Iceland, Chile, and New Zealand. All are in areas of recent volcanic activity.

The hot water emitted from a geyser is mainly surface water that has soaked into the ground and has been heated by magmatic material or hot igneous rocks. The output of some geysers has been directly related to wet and dry years. Certain geysers eject hot water 60 to 80 meters into the air, while others eject only a few meters above the ground surface. Most geysers erupt at irregular intervals. The eruptions of some are finished within a few minutes, while others last an hour or more. One of the best-known geysers in the world is Old Faithful at Yellow- stone Park (Fig. 21.21). Since 1870, Old Faithful has erupted regularly at an average interval of about 65 minutes.

However, during this period the interval has actually varied from 34 minutes to 91 minutes.

A geyser eruption is usually preceded by rumblings and violent boiling. Water flows over the top of the vent followed by strong jets of water, which are thrown scores of meters into the air. The theory of geyser eruption generally accepted is that of Bunsen, who studied the geysers of Iceland. The theory is based on the fact that the temperature at which water boils increases with pressure. This temperature is more than 100°C at 50 meters of water depth and 180°C at 100 meters. A research hole drilled at the Norris Geyser Basin at Yellowstone showed a temperature of 240°C at a depth of 330 meters, and the temperature was still rising. If the water flows into a fissure or conduit (Fig. 21.22) and becomes warm at depth by absorbing hot gases or by contact with hot rocks, it will remain as water until it exceeds the boiling point for the existing pressure. When the temperature rises at some point in the system so that steam forms, the steam may expand faster than it can rise, lifting up the column of water. As some of the

FIGURE 21.21 *Old Faithful Geyser in eruption, Yellowstone National Park, Wyoming. (Burlington Northern Railway.)*

FIGURE 21.22 *Diagrammatic section illustrating geyser eruption, according to the theory of Bunsen. Because of the restricted circulation, groundwater in fractures and vents below the surface is heated to the boiling point at depth near the main source of heat before it reaches the boiling point in the upper part of the vent. For simplicity, only a few fractures are shown.*

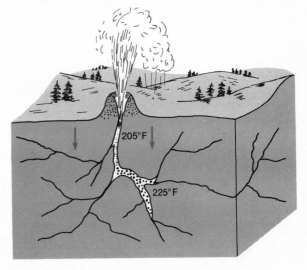

water spills over at the top, the pressure in the column is reduced. The water, well above its new boiling point, converts to steam and drives both water and steam from the vent in an eruption.

The collection of steam and other gases in the high parts of crooked tubes of some geysers may also cause eruptions. Any constriction of the tubes tends to prevent convection of the heated water. This hastens local boiling.

Geyserite, or siliceous sinter, is a term applied to siliceous hot-spring deposits, particularly those made by geysers. Where its deposits are abundant, a geyser may build a mound around the orifice. Silica occurs abundantly in hot waters and is the chief material deposited by the geysers and hot springs of Yellowstone National Park, with the exception of the calcareous Mammoth Hot Springs.

Geothermal Energy

While geysers and hot springs are rather spectacular geologic occurrences, they are also an important potential energy source. Geothermal energy is the energy produced by releasing steam and hot water from natural hot areas through drill holes. This is not a new source of energy. Electric power has been produced from steam and hot water vents at Larderello, Italy, since 1905; hot water and steam have been used for heating in parts of Iceland since 1925; and electric power has been produced from the Geysers area of northern California since 1960. The renewed interest in using geothermal energy has developed because re-

in using geothermal energy has developed because reserves of fossil fuels are dwindling and 10 million kilowatts of electricity generated from geothermal energy is equivalent to the energy in 1 billion tons of coal; it is a nonpolluting source of power, although the disposal of hot, brine-laden waters is a problem; and present cost analyses indicate it may be an inexpensive power source if compared to present sources. Potential reserves from depths of up to 10 kilometers total 6×10^{24} calories, equivalent to 9×10^{14} short tons of coal.

The use or projected use of geothermal energy is not without its problems. Geothermal wells have their own peculiar noxious qualities due to gases contained in the steam and hot water. Environmentalists claim the use of geothermal power would set a precedent for private power-plant development on public lands. Steam cannot be transported long distances; therefore, a power plant must be constructed at the well. The construction of power plants also means other supporting construction. Others claim that the venting pipes for the wells create a noise that resembles a 747 jet. The possibility exists of surface sagging or cave-ins, due to liquid and gas withdrawal.

Despite objections, and the prospect of solving the problems, in 1974 the federal government opened 58 million acres of public land for exploratory geothermal drilling. It has been predicted that by 1985 the production of geothermal power could replace the equivalent of $8.9 billion worth of oil. It was also predicted that by 1976 geothermal power could meet all electrical needs of the city of San Francisco.

REFERENCES

Horberg, L., 1949, *Geomorphic history of the Carlsbad Caverns area: New Mexico,* Journal of Geology, vol. 57, pp. 464-476.

Thornbury, W. D., 1969, *Principles of Geomorphology,* 2d ed., John Wiley & Sons, New York.

Todd, D. K., 1967, *Ground Water Hydrology,* John Wiley & Sons, New York.

CHAPTER 22

GROUND ICE, GLACIERS, AND GLACIATION

A world of snow and ice occupies Antarctica and a great deal of the northernmost Northern Hemisphere. The far north is characterized by cold temperatures, a vast expanse of frozen ground, heavy snowfall in the coastal mountains, and a magnificent display of glaciers. The area of northern Alaska is receiving increased attention because of its petroleum reserves.

At times it is an exhilarating region, but it can be harsh and beset with unique difficulties. One major problem is the abundant ice in the ground, which leads to complications in construction and land use. Perennial and seasonal ground ice greatly affects gradational processes, groundwater, and vegetation. Glaciers are of special interest because they represent a dynamic erosional agent.

Growing ice crystals in a confined space develop pressures up to 2,100 kilograms per square centimeter, more than 8 times the tensile strength of any rock. Hence, the disruptive effects of ice can be enormous, and the perennial snow and ice can superimpose on the landscape their own distinctive and strikingly different features (Fig. 22.1).

At present, about 10 percent of the land area of the globe is under glacial ice (Fig. 22.2). At times within the last 1 to 3 million years, different areas totaling about 30 percent of the land area of the earth were inundated by ice up to thousands of meters thick (Fig. 22.2).

In the context of this discussion geologists have recently found evidence of Ice-Age glaciers in North Carolina. It has long been believed that in eastern continental United States, glaciers had been no farther south than what is now the New Jersey-Pennsylvania area. The evidence is a glacially carved U-shaped valley 1,430 meters up the side of Grandfather Mountain. At lower elevations, rock outcrops show a polished effect, as well as carved parallel grooves. If this evidence holds up, it will force a revision of geology textbooks. Many geologists, however, doubt the validity of the evidence.

This chapter will consider the role of snow and ice in geology, with emphasis on frost action, glaciers, and glaciation.

OCCURRENCES AND PROPERTIES OF ICE

Ice forms wherever water freezes—in streams, lakes, the sea, the atmosphere, and in the ground. Approxi-

FIGURE 22.1 *Stripping of tundra over permafrost has caused melting of ground ice and collapse of the surface near Barrow, Alaska. Caterpillar tractor trails provide scale. (R. F. Black photo.)*

mately 1.3 percent of the total water on the earth occurs as ice. Its most familiar form—snow—is commonly composed of skeletal hexagonal crystals of great delicacy and beauty. Ice also occurs as frost crystals and in needlelike, dendritic, and feathery forms, similar to the aggregates seen on window panes.

In the ground, ice forms single crystals and aggregates in voids, films on soil particles, veins, and dikes, and irregular masses of all sizes and shapes. Large wedge-shaped masses in polygonal patterns are unique to the cold polar regions.

Ice is among the commonest of substances on the earth, and few realize that it is a mineral, or rock. Its simple formula of H_2O belies its complex behavior. Mineralogists speak of its physical properties, such as its hexagonal crystal structure (similar to quartz), its conchoidal fracture, its brittleness at low temperature, and its specific gravity of 0.917. Unlike almost all other naturally occurring minerals and rocks, ice expands on freezing and shrinks on melting. The expansion, a 9 percent increase in volume, permits ice to float on water instead of sinking. This is due to the fact that the maximum density of water is 1.000 at 4°C and decreases thereafter to the freezing point at 0°C, when its density becomes 0.917.

GROUND ICE

All soils are porous, and rocks near the earth's surface that appear to be solid have innumerable pores and cracks which contain water in warm humid climates or ice in cold climates.

Permafrost

Perennially frozen ground, or permafrost, occurs in about one-fourth of the land area of the globe and was formerly much more widespread. In the Northern Hemisphere it is about equally divided between Eurasia and North America (Fig. 22.3).

Permafrost provides an impervious substratum over wide areas in polar regions, which indirectly produces especially striking effects in an active layer. This layer is the surface zone that freezes and thaws seasonally. The top of the permafrost is never flat for any appreciable distance, and water will collect in the hollows and saturate the active layer. The gravity-transfer processes of soil creep and soil flow occur widely in this area each summer and are the most important associated gradational processes. They quickly destroy small stream channels or rills on the

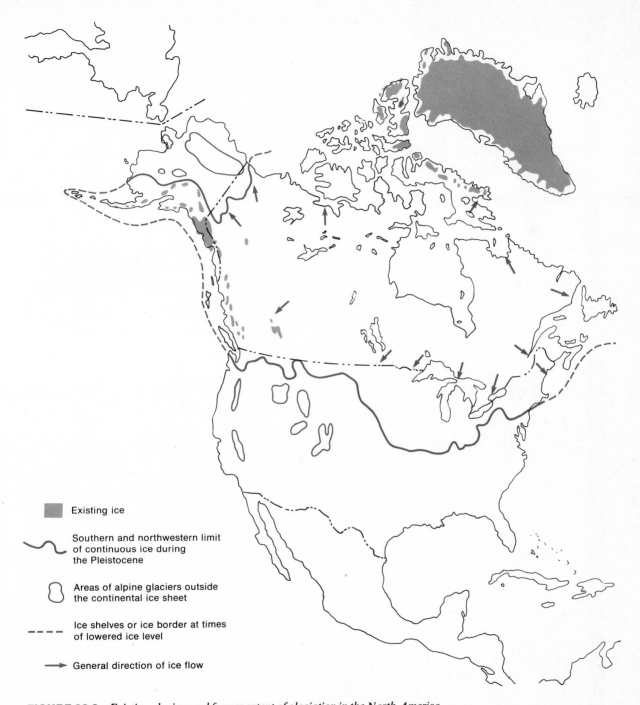

Existing ice

Southern and northwestern limit
of continuous ice during
the Pleistocene

Areas of alpine glaciers outside
the continental ice sheet

Ice shelves or ice border at times
of lowered ice level

General direction of ice flow

FIGURE 22.2 *Existing glaciers and former extent of glaciation in the North America.*

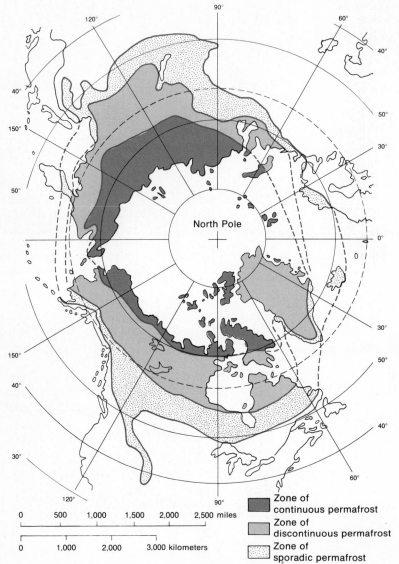

FIGURE 22.3 *Extent of permafrost in the Nothern Hemisphere. (After Black, 1954, Permafrost—A Review,* Geological Society of America Bulletin, *vol. 65, pp. 839-856.)*

North Pole

0   500   1,000   1,500   2,000   2,500 miles

0   1,000   2,000   3,000 kilometers

Zone of continuous permafrost

Zone of discontinuous permafrost

Zone of sporadic permafrost

tundra, and so, long, smooth, gentle slopes are commonplace (Fig. 22.4). These may be interrupted by small terraces or benches from combined downslope movements and frost processes.

Permafrost is continuous and is many hundreds of meters thick in the high polar regions. It becomes dis-**continuous, sporadic, and thinner and has a warmer temperature toward** lower latitudes until it eventually **disappears.**

Permafrost effectively seals off groundwater infiltration and often preserves organic remains indefinitely. At a depth where no annual temperature change occurs (about 20 to 30 meters), the temperature of permafrost approximates the mean annual air temperature of the locality. It is the temperature of the permafrost at the depth of no annual change, together with its distribution, that is used in classifying it for geologic, biologic, and engineering purposes.

Permafrost results when the net total heat balance over a period of several years produces a temperature continuously below 0°C. In polar regions, low temperatures and permafrost have existed for thousands of years. As climates have warmed during the past century, some isolated bodies of permafrost at the southern limit have been or are being destroyed. Where glaciers have recently melted or where new sand bars, deltas, and lake bottoms have been exposed to cold climates, new permafrost is forming today.

Permafrost Frost Action   The growth of ice lenses as the result of water movement to freezing centers and consequent expansion cause the disruptive effects of frost action and frost shattering in unconsolidated rocks and minerals. Capillary action of water in fine-grained materials can lift that same water many meters. As growing ice crystals attract water to their surfaces, they build ice lenses that buckle the surface. Ice-crystal growth is limited by the sizes of the pores, the confining pressure, the thermal gradient, and the availability of water. Pores too large or too small cut off the upward flow of water. Large pores provide very little capillary attraction, and small pores hold the water too tightly for it to pass. Ice lenses form best in mixtures of clay and silt. Ice lenses increase the water content of the frozen material and lift the surface as they enlarge. Thus, roads are commonly buckled during freezing, then collapse and break up when t[ ] melts.

The coefficients of expansion of ice are 4 to 6 times that of most other rocks. Soils with a great deal of contained ice are prone to vertical cracking because of the contraction of the ground surface when temperatures drop suddenly. The patterns are polygonal like those in mud cracks or in cooling lava.

This stirring and sorting action on earth materials by frost in the active layer of the permafrost produces a variety of patterned ground features. These features are broadly classified into varieties of circles, nets, polygons (Fig. 22.5), steps, and stripes. Most do not require permafrost for their formation, but commonly they are better developed, larger, and more widespread in the discontinuous zone of permafrost. Open cracks in permafrost partially fill with snow in winter, with frost in spring under reversed temperature gradients, and with water from spring thaw. The result is a narrow vertical dike of ice a fraction of a centimeter wide that penetrates 3 to 10 meters into the permafrost. The crack, or fracture, is a zone of tensional weakness, and so cracking the following year commonly follows the original ice dike.

In time, a considerable width of ice is introduced into the ground to form an ice wedge (Fig. 22.6). In dry areas, as in Antarctica, similar cracks take place, but drifting sand partly fills them to produce sand wedges. Sand and ice wedges grow to widths of 6 or 10 meters

FIGURE 22.4   *Smooth tundra slopes on Seward Peninsula, Alaska. Bedrock residuals rise above the surrounding slopes of accumulation of waste derived by frost action. (R. F. Black photo.)*

FIGURE 22.5 *Sorted polygons surrounded by troughs with vegetation in the Colorado Front Range. Frost stirring prevents vegetation from colonizing the bare mud and small stones. (R. F. Black photo.)*

over thousands of years and produce distinctive surface relief (Fig. 22.7).

Abundant ground ice upon thawing promotes solifluction, collapse of the ground, and faster erosion (Fig. 22.8). On exposed Arctic coasts, wave erosion has been measured for decades at average rates of 10 to 30 meters per year. In the last century Alaska has lost several kilometers of Arctic coast near the North Slope, a rate that will probably continue for many decades. When houses and other buildings rest on permafrost without adequate insulation, they upset the thermal regime. Thawing of the ice-saturated

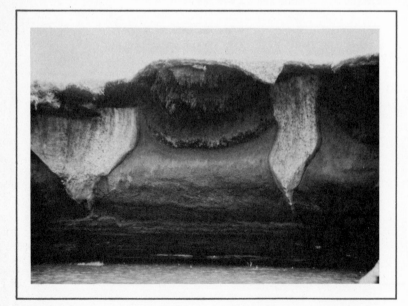

FIGURE 22.6 *Ice wedges exposed in a 5-meter bank along the Arctic coast of northern Alaska. The oldest wedge is the largest. Sedge peat laid down in small ponds between ice wedges festoons the upper part of the bank. (R. F. Black photo.)*

FIGURE 22.7   *Ice-wedge polygons in northern Alaska. The wedges underlie the raised turf surrounding the ponds. The ponds average about 7 meters across. (R. F. Black photo.)*

ground causes collapse and destruction of the structures (Fig. 22.9).

Where permafrost has been thawed to considerable depth, refreezing produces interesting pressure effects, following the freezing-milk-bottle principle. Ice mounds, called "pingos" (Fig. 22.10), grow for months or years. They attain heights up to 50 meters and diameters of hundreds of meters. Some pingos are known to be thousands of years old.

Tundra environments on the Arctic coastal plain of Alaska, Canada, and Siberia are characterized by tens of thousands of oriented thaw lakes (Fig. 22.11).

FIGURE 22.8   *Thaw of permafrost at water level along the Arctic Ocean allows the ice-rich bank to collapse rapidly. Rates of coastal recession average many tens of meters per year. (R. F. Black photo.)*

FIGURE 22.9 *Settling and cracking of post office building at Nome, Alaska, from melting of permafrost. (R. F. Black photo.)*

Most of these lakes seem to owe their origin to the thawing of permafrost and their orientation to a persistence of prevailing winds normal to their long axes. Many thaw lakes are found in the interior of the continents, in the discontinuous zone of permafrost, but they have little if any orientation.

The rate at which the various frost processes operate differs markedly. Ice wedges require from hundreds to thousands of years to show marked surface expression, since the growth of wedges is only about 0.05 to 0.10 centimeter per year. Some patterned ground in the active layer on level ground may

FIGURE 22.10 *A pingo, an ice-cored mound, over 45 meters high in the McKenzie Delta of Canada. (R. F. Black photo.)*

require only a few years, whereas other forms generally need centuries. Heaving and settling movements range from 1 to 2 centimeters annually.

Seasonal Frost   In areas where the ground freezes only seasonally, bedrock or impervious clay can have an effect similar to permafrost in trapping moisture. A saturated soil, even in an open system when freezing from the top down, can develop pressures from 0.7 to 5.6 kilograms per square centimeter.

The freezing of cultivated land with boulders in a fine-grained soil or substrate brings new boulders to the surface each year. Stones, fence posts, pilings, and Carex (sedge) tufts are buckled when they are frozen within a soil matrix that in turn is heaved by ice segregation. Ice crystals can grow below a boulder that conducts heat more rapidly than the adjacent soil, pushing it upward by the force of crystallization of the growing ice. Once buckled upward, a boulder rarely settles back into its original position when the ground thaws. This is because fine soil or smaller pebbles and sand usually fall into the cavity below it.

FIGURE 22.11   *Oriented lakes of northern Alaska. Wind blowing normal to the long axes of the lakes is believed to have oriented them as they grew during thaw of permafrost. View south at Barrow, Alaska. (U.S. Navy photo.)*

GLACIERS

Geologists study glaciers not only because they are striking features of the landscape (Fig. 22.12) but

FIGURE 22.12   *Alpine glaciers with sharp, jagged peaks rising above them in the Alps. (C. Tairra photo.)*

because of their tremendous erosional capabilities. They are a vital source of moisture for streams that man uses, and their deposits have provided a ready supply of coarse aggregate for construction (Fig. 22.13). Some glaciers directly threaten or have already overrun settlements in alpine areas, whereas others have provided lakes (Fig. 22.14). More subtly, glaciers exert some control on the weather. They are measures of present and past climates. Indirect effects are extended far beyond glacier borders by streams and by cold ocean-bottom currents of the oceans.

Glaciers hold vast amounts of water. If all present-day ice were to melt, sea level would rise about 65 meters and inundate all major coastal cities.

Waxing and waning ice sheets controlled the migrations of man and other life during the Pleistocene epoch, a time characterized by massive glaciers and the development of man in the evolutionary sequence.

Not all glaciers have been continuous from the Pleistocene to the present. For example, the glaciers in Glacier National Park probably came into existence only about 4,000 years ago, after a warm, dry period that preceded their formation by 2,000 years. They have fluctuated in size since this time, attaining a maximum size around the middle of the last century.

All in all, the formation and beauty of glaciers, the glaciated landscape (Fig. 22.15), and the influence of glaciers upon events and processes in the earth's history provide an exciting geological study.

Origin of Glaciers

Glaciers originate by conversion of perennial snow to ice. They are classified on the basis of geometry, flow, and temperature. Although ice illustrates an elastic response to sudden stresses, ice is viscoplastic (both viscous and plastic, interchangeably or together). In response to sustained stresses, it consequently flows readily at temperatures near the melting point.

The ice below perennial snow fields, tens of meters thick, can spread outward in radial flow, if on a level surface, or in directed linear flow, if in a valley.

The Development of Glacial Ice

Snow Fields A snow field is an area of perennial snow. In high latitudes or at high altitudes or in areas with sufficient snowfall where the mean annual temperature is below the freezing point of water, a great deal of the snow remains unmelted from year to year and accumulates to great depths. Certain regions in Antarctica and northern Greenland lack snow fields

FIGURE 22.13 *Bedded sand and gravel overlain with unsorted drift or till.*

FIGURE 22.14 *Cirque Lakes, Lake Ellen Wilson and Lake Lincoln, with Mount Jackson in the background, at Glacier National Park, Montana. (Burlington Northern Railway photo.)*

because the infrequent very light snowfalls are sublimated or melted on warm rocks even though air temperatures remain below freezing. Snow fields are common in all high mountains of the world, even on peaks like Kilimanjaro and Ruwenzori, both of which are located near the equator. The two largest areas of perennial snow are Antarctica and Greenland. Together they contain more than 96 percent of the total ice and snow on the earth.

The lower limit of a snow field or that limit where a net accumulation is balanced by a net melting is the snow line. The snow line is usually a fairly regular regional or climatic line, but locally it shifts up or down because of topography, exposure, and wind. The temperature regime and amount of snow are the most important factors, with greater snowfalls partly offsetting warmer temperatures. Snow lines are higher in the interiors of continents away from sources of moisture.

Cloudy mountain belts with maritime air and a very high precipitation have the lowest snow lines.

FIGURE 22.15 *Aurlandsfjord near Naroyfjord, Norway. Precipitous walls drop hundreds of meters below sea level. (R. F. Black photo.)*

Annual snow lines fluctuate somewhat from year to year. Snow lines during the Pleistocene epoch were hundreds of meters lower than at present.

In snow fields, the most recent snow grades downward into older snow, which is progressively more dense (Fig. 22.16). Seen in finer detail, the snow layers of slightly different density and texture reflect each snow storm and warm spell, winter versus summer accumulation. Dust, volcanic ash, pollen, micrometeorites, and small organisms may accumulate in the various layers to comprise a record of a part of the earth's history.

Névé New snow has a low specific gravity, perhaps as low as 0.05. Further density changes convert old snow to névé, or firn, by melting and recrystallization. Further increases in density involve sublimation, melting and refreezing, and recrystallization into ice. At specific gravities between 0.82 and 0.84, the remaining pores are isolated and the mass becomes impermeable. Further reductions in pore space increase the density to 0.90, a common specific gravity for glacial ice (Fig. 22.16).

The time required for conversion of névé into ice ranges from less than 1 year in temperate regions to many centuries in polar regions. Cold inhibits the

process, while a heavy accumulation of snow increases loading and speeds up conversion. Accumulation is high in a maritime climate, whereas it is lower in a continental climate. For example, the Vatna Glacier in Iceland, at an elevation of approximately 1,300 meters, receives a mean annual precipitation of about 400 centimeters, mostly as snow, whereas the Antarctic ice sheet at the South Pole receives only 6 to 9 centimeters of precipitation annually. Near Juneau, Alaska, over 35 meters of snow fall each year on the upper Taku Glacier.

Glacial Movement

Under very low stress rates a single ice crystal glides readily in any direction along its basal plane. The crystal lattices of an aggregate of ice crystals deform less easily but similarly as viscoplastic flow. The main effect of a great thickness of ice on its flow characteristics is to lower the pressure melting point and make the ice respond as though it were warmer; then it flows more readily even at cold temperatures.

Recrystallization occurs readily with flow, but cold temperatures inhibit both flow and recrystallization. The actual mechanisms are not well understood, nor can the behavior of glacial ice always be predicted.

The crystal fabrics of ice may reveal something of its history of deformation. The hardness of glacial ice increases as its temperature drops, being soft enough at the melting point to be scratched with one's fingernail but as hard as iron at temperatures of −60 to −75°C which are common on the Antarctic ice cap.

Types of Glaciers

A radially deploying glacier is considered an ice cap (Fig. 22.17) if it is on top of highlands and an ice sheet if it approaches continental dimensions.

Mountain, or alpine, glaciers (Fig. 22.18) and cirque glaciers (in the headward bowl-shaped part of a valley that the glacier itself eroded, Fig. 22.19) are typical forms of glaciers. Upon leaving these high source areas the glacier becomes a valley glacier. Later, the valley glacier may move out onto a plain and spread out radially from the valley mouth. Here it may develop several forms called foot glaciers (Fig. 22.20), or tongue glaciers (Fig. 22.21). Valley glaciers may extend into lakes or the ocean, where they coalesce into ice shelves or break off in huge icebergs (Fig. 22.21). Where two or more valley glaciers join and spread out at the base of a mountain range, they form a piedmont glacier (Fig. 22.22).

Distribution and Size   Glaciers occur in the same latitudes and altitudes as the perennial snow fields from which they form. The termini of very active glaciers generally extend much lower than perennial snow lines.

The huge ice sheets of Antarctica and Greenland are continental-sized accumulations of ice covering an entire land area, except for a few mountains. The glaciers in mountain ranges vary in size from short, wide patches of ice and snow on the narrow benches of a cliff, such as the snowbank glaciers of the Big Horn Mountains of Wyoming, to the long tongues of ice extending tens of kilometers down valleys, as in the mountains of Alaska or the Himalayas in Asia. Glacier National Park has about 60 glaciers, all small. Only two approach 1.3 square kilometers in area. Mt. Rainier has 63 square kilometers of ice contained in 20 glaciers. The Seward Glacier in Alaska is 80 kilometers long and 5 to 8 kilometers wide. The Rennick Glacier in Antarctica is at least 260 kilometers long and up to 80 kilometers wide. The glaciers on the

slopes of Mt. Everest are more than 50 kilometers long. Those in the Alps are numerous but are usually not more than a few kilometers in length. The Juneau ice field of southern Alaska and western Canada is an intermontane system of valley glaciers, snow fields, and small ice caps all connected in a complex encompassing an area of 3,800 square kilometers.

(a) Antarctica   The vast ice sheet blanketing Antarc-

FIGURE 22.16   *Bulk density of snow and ice on Upper Seward Glacier, Alaska.*

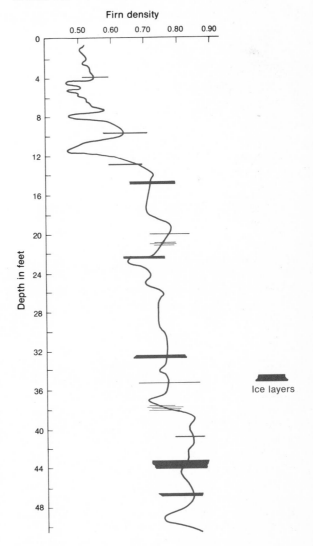

FIGURE 22.17   *Penny Ice Cap on Baffin Island, Canada. Two outlet glaciers join to form Coronation Glacier, in the foreground. (National Air Photo Library, Canada. Photo TE14-L82.)*

FIGURE 22.18   *An Alaskan glacier with numerous tributaries forming a series of cirques, or blunt, steep-headed valleys. (U.S. Air Force photo.)*

FIGURE 22.19   *Cirques enclosing small crescentic glaciers in the high Sierra Nevada near Mount Whitney, California. (Spence Air Photos.)*

FIGURE 22.20   *The Canada Glacier, Taylor Valley, Antarctica, a foot glacier. The distal part has expanded outward in a bulb on the valley floor. (R. F. Black photo.)*

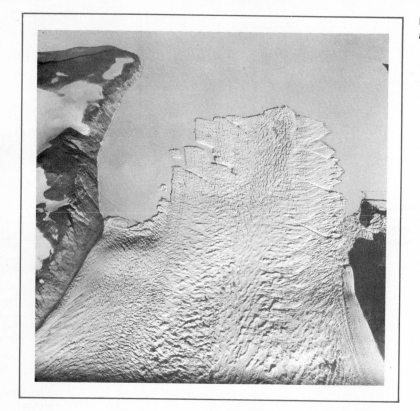

FIGURE 22.21 *MacKay Glacier tongue, Antarctica. (U. S. Navy)*

FIGURE 22.22 *Malaspina Glacier, southeastern Alaska, showing accordionlike pleating of moranial bands parallel with the margin. (Austin Post, U.S. Geological Survey.)*

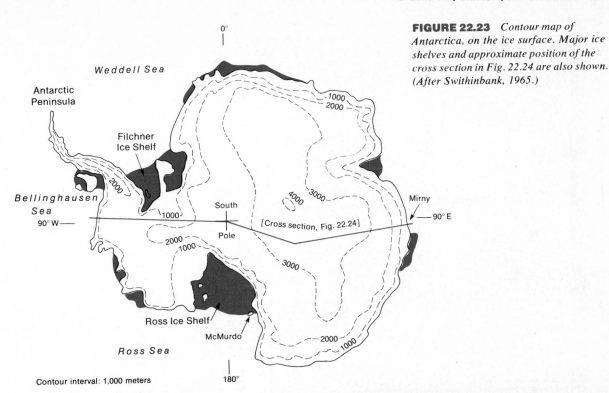

FIGURE 22.23   *Contour map of Antarctica, on the ice surface. Major ice shelves and approximate position of the cross section in Fig. 22.24 are also shown. (After Swithinbank, 1965.)*

Contour interval: 1,000 meters

tica contains about 90 percent of the world's ice. It covers an estimated 13 million square kilometers, exceeding 2,000 meters in thickness over 75 percent of the area (Figs. 22.23 and 22.24). For comparison, the area of coterminous United States is only 7.7 million square kilometers; Canada is 9.7 million. The deepest hole drilled in the Antarctic ice penetrated 2,164 meters to bedrock near Byrd Station in 1968. Seismic evidence indicates that the maximum thickness exceeds 4,000 meters.

The Antarctic continent contains enough ice, if melted, to raise sea level about 60 meters. However, the ice sheet is colder than is necessary to perpetuate itself and would require thousands of years to change its mean temperature, even following a global temperature rise of 1-2°. Only 2 percent of Antarctica is ice free (Fig. 22.25). The Antarctic ice sheet is comparable in size to the last ice sheet that covered most of Canada and the northern United States.

Antarctic precipitation is in the form of snow, with frost in more humid areas. The mean annual accumulation for Antarctica is only 15 to 18 centimeters—the lowest of any continent.

The depth of transformation of snow to glacial ice

FIGURE 22.24   *Cross section of part of Antarctica, showing ice thickness and underlying bedrock topography. (After Bentley et al., 1964.)*

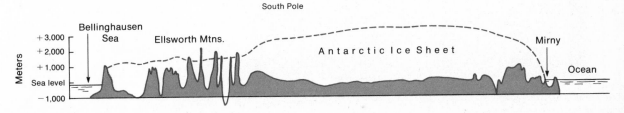

FIGURE 22.25   *Taylor Valley, one of several dry valleys in the Ross Sea area of Antarctica. Small alpine glaciers flow into the valley from the adjacent highlands; Taylor Glacier, an outlet glacier from the Antarctic ice sheet, terminates in Lake Bonney, a proglacial lake. (R. F. Black photo.)*

FIGURE 22.26   *Crevassed folds in the Excelsior Glacier, Alaska. (Spence Air Photos.)*

on Antarctica ranges from 30 to 60 meters near sea level at the edge of the ice sheet to more than 150 meters in the high inland plateau. The time required to convert the snow to ice is 150 years at Little America on the coast and 1,000 years at the South Pole.

As the inland ice flows toward the edge of the continent, it thins and is influenced by the bedrock topography. In East Antarctica, discharge is directly into the sea along a broad front from which relatively fast-moving ice tongues protrude. In West Antarctica, much of the inland ice is fed into huge floating ice shelves that fringe more than one-third of the coastline of the continent. They are floating sheets that flow under their own weight.

Floating ice shelves cover more than 1.3 million square kilometers, the Ross Ice Shelf (500,000 square kilometers, four-fifths the size of Texas) and the Filchner Ice Shelf (380,000 square kilometers) being the two largest. Their thickness varies from about 120 meters at the seaward ice front to some 1,200 meters at the junction with land ice. Where the ice shelves abut land, folds or large fractures are produced (Fig. 22.26).

Rock-walled valley glaciers are well developed along the ranges of Victoria Land and southward. Eight major outlet glaciers from the ice sheet supply vast quantities of ice to the Ross Ice Shelf. Those outlet glaciers are the largest valley glaciers in the world. The Beardmore (Fig. 22.27) and Rennick Glaciers are, respectively, 200 to 260 kilometers long and between 30 and 80 kilometers wide. The former is on the route used by Scott to reach the South Pole.

Wastage of ice from Antarctica seems largely to result from calving of icebergs into the sea. Sublimation and local basal melting are only locally important. Estimates of accumulation versus wastage suggest but do not prove a net positive balance.

The weight of this vast quantity of snow and ice on the Antarctic continent depresses the land surface 600 meters on the average. Removal of the snow and ice would result in that much rebound and bring the rock surface of East Antarctica in line with that of other continents. West Antarctica would still remain a partially submerged series of mountainous islands. The bulk of relief of the continent is due to the accumulation of snow and ice.

Ice movement is measured in meters per year in the interior. However, movement of ice in outlet glaciers like the Amundsen, Scott, and Byrd is on the order of 1 to 2 meters per day.

The ice-free areas of Antarctica are particularly interesting because they record ice fluctuations, give

FIGURE 22.27 *Beardmore Glacier, one of the larger outlet glaciers of the Antarctic ice sheet. Mountains in the distance are more than 160 kilometers away. Gibraltar Rock on the left rises more than 1 kilometer above the ice. (R. F Black photo.)*

insight into the bedrock geology beneath the ice, and provide examples of geologic processes in extreme polar deserts.

(b) Greenland   Greenland is a large island with an area of somewhat more than 2 million square kilometers (Fig. 22.28). Its central domelike glacial plateau reaches an elevation of about 3,000 meters. The coastal area is very irregular and is fringed with rugged mountains through which outlet glaciers from the cen-

FIGURE 22.28   *The continental ice sheet on Greenland covers all but a narrow border of the landmass.*

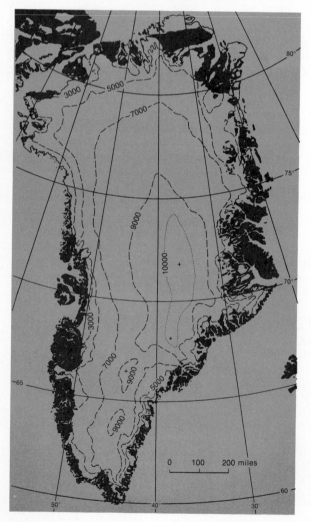

tral ice cap reach fjords and the sea. About 80 percent of the island is covered by a vast continental glacier; only the northern cold desert area and the coastal mountains are comparatively bare. The ice sheet occupies a large depression, which in places is below sea level. Maximum thickness of ice is about 3,000 meters, the same as its maximum elevation. The ice thins markedly northward, where colder temperature and low rates of precipitation produce a polar desert. Most snow originates from the North Atlantic and is fed to the ice sheet along its southeastern side.

The Glacial Budget

The mass balance of glaciers refers to the budget of nourishment and wastage, the economy, for an entire glacier for one year—that which comes in versus that which goes out of the system (Fig. 22.29). Total annual accumulation of snow and ice must be balanced against total annual wastage to determine the status of a glacier. An unbalanced budget can lead to an advance or a retreat of a glacier.

In recent years, mass-balance studies of some of the thousands of small glaciers in western North America reveal interesting patterns. Former positive budgets for small high-altitude glaciers in the northern Coast Ranges have tended to develop markedly negative ones. Of some 200 of Alaska's glaciers measured during the past 25 years, 63 percent have been retreating, 7 percent advancing, and 30 percent unchanged. The advancing glaciers contain more ice than all the others combined.

The activity of a glacier, and hence the work it can do, is determined in large part by the size of its budget. Small budgets for large glaciers in cold polar areas mean little movement of ice and little work accomplished directly by ice or indirectly by runoff. In contrast, warm glaciers with big budgets move rapidly over their beds, doing much more work both by the ice and by its meltwaters.

Nourishment   Glaciers are nourished by direct fall of snow, by wind-driven snow, by avalanching from mountain slopes above them (Fig. 22.30), by hail or rain that is frozen to the glacier, and even in part by the condensation of moisture directly from the air. In cold polar environments, the accumulation of snow and ice need only be in small amounts to balance losses,

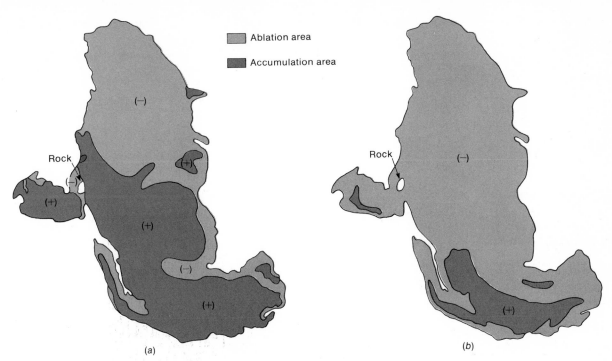

Ablation area

Accumulation area

(a)

(b)

FIGURE 22.29   *Total mass net balance (a) for 1965 and (b) for 1966 of the South Cascade Glacier, North Cascade Range, Washington. (After Meier et al., 1971.)*

FIGURE 22.30   *Avalanche or rock slide onto the Schwann Glacier, Alaska, resulting from the Good Friday earthquake of 1964. (M. M. Miller.)*

but in warm areas much more accumulation is needed to avoid having a glacier destroyed during the long summers.

Wastage Wastage of glaciers is the total loss of snow and ice by direct calving of icebergs into water bodies, by melting at the surface and at the base of the ice, by sublimation, and by deflation, or blowing away, of snow and ice. Wastage in all its forms commonly is referred to as ablation. Wastage of dead ice leads to knob-and-swale surfaces of great complexity.

Movement of Glaciers

Glacial motion was recognized as early as 1705. The first estimates of the amount of movement were made by noting the changes that took place in the surface debris. Conspicuous boulders on the ice surface were seen to move downglacier slowly from year to year. Later, careful measurements of the rate of change were made with surveying instruments and stakes placed on or imbedded in the sides of glaciers. They easily demonstrated that the rate of movement varies considerably from time to time, not only in different glaciers but in different parts of the same glacier (Fig. 22.31).

Many large alpine glaciers move from 0.3 to 1 meter per day. Calculations show that an ice particle would require approximately 500 years to move from the summit of the Jungfrau to the end of Aletsch Glacier, Switzerland, an ice stream about 16 kilometers long. Many of the large Alaskan glaciers flow more rapidly. The Muir Glacier commonly moves as much as 2 meters per day, and the Childs Glacier flows nearly 10 meters per day during the summer months. Extraordinary velocities have been recorded for the ice tongues that descend to the fjords along the coast of Greenland. Rates of nearly 30 meters per day have been observed, but in the same region, the inland ice at some distance back from the fjords moves only a few centimeters per day. The smallest glaciers in Glacier National Park move only 2 to 3 meters per year, and the largest, probably 8 to 10 meters per year.

The distribution of velocity in many glaciers resembles that of a river. The center generally moves more rapidly than the sides because of resistance against the walls of the valley. The surface moves more rapidly than the deeper parts of the ice because the deeper portion bears a greater load of debris and encounters more irregularities on the valley floor. At curves, the outside bend of a glacier moves more rapidly than the inside of the bend.

The velocity of glacial flow is influenced by (1) the gradient of the surface under the ice, (2) the surface slope of the ice, (3) the roughness of the surface over which the ice flows, (4) the ice temperature, (5) the amount of water available at the base of the ice, and (6) the quantity of debris in the ice.

Glacier movement involves sliding, shearing, and plastic flow. Warm glaciers move mainly by sliding and shearing over their beds. Minute amounts of free water at the base of the ice speed up sliding. In the deeper portions, or zone of flowage, glaciers move as viscoplastic solids, partly by slippages along planes parallel with the bases of the hexagonal ice crystals and partly by molecular exchange of ice across grain boundaries, permitting recrystallization. Large shears and tension fractures (crevasses) also occur within glaciers. The upper brittle zone rides along the top of

FIGURE 22.31 *Cross section of the Saskatchewan Glacier, Alberta, Canada, showing measured velocity vectors and calculated velocity-depth profiles at representative places. (After Meier, 1966.)*

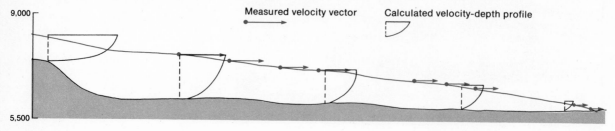

Measured velocity vector   Calculated velocity-depth profile

the zone of flowage, and the top moves faster than the base. The movement within a valley glacier is obliquely downward in the accumulation zone and obliquely toward the surface in the ablation zone.

Because the ice in glaciers is always moving in a fluctuating manner, the terminus of a glacier is rarely stationary. When ablation exceeds the forward movement or the resupply of ice, the terminus of the glacier retreats; when more ice moves down than is lost, the margin advances.

Most oscillations follow climatic cycles, but some advances follow earthquakes, while others come without any recognized warning or obvious relationship to climatic factors. The Black Rapids and Muldrow Glaciers of the central Alaska Range and the Walsh and Steele Glaciers of Yukon and Alaska, all valley glaciers a few kilometers wide, began to advance or surge years ago at rates of 8 to 30 meters per day—many times their normal rates. Their termini advanced several kilometers and then stagnated. Over 200 surging glaciers are known in western North America that move erratically at very rapid rates, and then lie almost dormant for many years before surging again.

Reconstruction of the rates of movement of former continental ice sheets is exceedingly difficult because the ice removed almost all unconsolidated material it passed over, except near the fluctuating margin. What evidence remains, however, suggests that advances were relatively steady over periods of many thousands of years to an outer fluctuating limit where vast stagnation of the outer periphery took place. The ice ablated in a wide outer zone, while it continued forward in an inner zone that also stagnated later. Destruction of ice sheets involved some thousands of years, but generally, this is only a fraction of the time it took the ice sheet to grow to its maximum dimensions.

Structures of Glaciers

Internal adjustments by flowage and recrystallization accommodate movement in much of the ice, but velocity differences in the upper brittle zone lead to various tension and compression fractures.

At the upper end of a mountain glacier where the ice breaks away from the snow fields, a great concentric crack or a series of open fissures is generally seen

and is known as a bergschrund. The displaced portion of the glacier has moved downward and outward like the head of a slump block. The bergschrund commonly extends downward to the rock wall at the head of the valley. Meltwater and snow later enter the bergschrund to form solid ice. As the ice pulls away from the headwall, an enormous quarrying force is exerted on the headwall, approaching the tensile strength of the ice itself. Plucking enlarges the headwall into the steep-walled, blunt-ended, amphitheaterlike valley head called a cirque (Figs. 22.18 and 22.19).

Under compressing flow, where velocity decreases downstream and thickening of a glacier takes place, plates or units of ice commonly shear obliquely upward over the slower ice in front. Shearing also characterizes the terminal area, where debris dumped along the margin acts as a barrier. Advancing ice rides up and over, by repeated shear, leaving a deposit of dirty ice stacked like an inclined deck of cards. In parts of some valley glaciers where ice descends over steep drops in the underlying bedrock surface, a large crevasse or a series of crevasses similar to a bergschrund may form. It extends straight across the valley or curve, depending on the particular configuration of the valley and the flow pattern of the glacier.

In response to widening, a glacier often develops longitudinal crevasses along its center that fan out down ice along flow lines. Many blocky and irregular patterns also form. Since the surface ice moves more rapidly than ice lower down, and other differential movements on the surface take place, any crevasse may be distorted and reoriented with time. All crevasses narrow downward and die out in the zone of flowage, which is deeper in cold ice than in warm. Crevasses generally range in depth from 10 to 50 meters. In clean ice, the crevasses fill partly with snow and hoarfrost. Debris and water fill others. Where the glacier is thin, surface water plunges through crevasses and forms subglacial streams.

GLACIAL DRIFT

All debris in transit in glaciers or deposited from them is called drift. Long before the glacial theory was proposed, surface rock waste in parts of Britain was thought to have "drifted" in by floods or by floating ice. Although those particular deposits are now at-

FIGURE 22.32   *The "Two Creeks" buried soil in drift near Two Creeks, Wisconsin. Roots of trees are seen in the black soil at the trowel, which have been radiocarbon dated at 11,850 years B.P. (before present). Till sheets in the lower right and upper left corners of the photograph are separated by lacustrine sediments and in turn are separated by the soil. The soil was contorted by the overriding glacier. (R. F. Black photo.)*

FIGURE 22.33   *Scratched and striated opposite sides of a cobblestone from glacial drift.*

tributed to mass movements and frost action, the term is still used for glacial deposits. Drift may be either stratified or nonstratified, sorted or unsorted (Fig. 22.32).

Acquisition

Glaciers acquire debris by direct erosion by ice and by gravity transfer of earth material. The debris in mountain glaciers generally concentrates along valley walls, but high-velocity avalanches may carry debris for kilometers across glaciers (Fig. 22.30). All glaciers erode, although their efficiency varies with the temperature of the ice, its thickness, its rate of flow, the hardness of the bedrock, and the characteristics of the fragments used as abrading tools. If a glacier is frozen to its bed and is very cold, the flow of ice is slow and little work is accomplished. In contrast, ice at the pressure melting point is very effective in quarrying and abrading. Rapidly sliding debris-laden, warm ice is effective in polishing and striating the bedrock.

Ice sheets that cover all topography can acquire material only from below and transport it as basal drift. Hence, drift on top of ice sheets can only be found at or near the terminus, where upward-directed flow and rapid ablation can bring it to the surface. Some drift is found down-ice from a nunatak, which is an isolated mountain top rising above an ice sheet (Fig. 22.17). On the other hand, merging valley glaciers bring their supraglacial and marginal drift zones in contact with each other and literally spread drift in zones throughout the glacier from head to toe (Figs. 22.17 and 22.18).

Debris is acquired (1) by freezing onto pre-existing soil and loose rock in the path of an advancing glacier, (2) by quarrying action involving freezing onto jointed rocks and subsequent plucking, and (3) by abrasion of bedrock by rock waste held in the glacier against its bed. Quarrying produces large angular rock fragments; abrasion produces silt- and clay-size mineral fragments, or rock flour. Abrasion of rock by drift in the ice ranges from fine polishing to coarse grinding and grooving. Only a small part of the drift of a glacier comes in contact with the bed or with other fragments in the ice. Generally, less than 10 percent of the rocks in glacial deposits are striated and grooved by such contact (Fig. 22.33). Under favorable circumstances, big boulders may deeply groove the bedrock below (Fig. 22.34). The polished, striated, or grooved surface of bedrock shows clearly the direction of ice flow and is called a glacial pavement.

If loose material is too abundant below a thin warm glacier to be picked up and transported, it may

FIGURE 22.34 *Glacial grooves in bedrock, Kelleys Island, Ohio. The former glacier scoured off all the loose rock material, leaving a clear surface.*

FIGURE 22.35   *Drumlin, Fond du Lac County, Wisconsin. Ice moved from left to right. (R. F. Black photo.)*

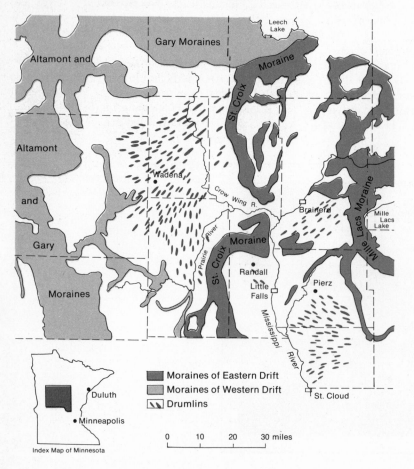

FIGURE 22.36   *Map of drumlin fields in central Minnesota. (After Allen Schneider.)*

be molded into fluted forms or drumlins (Fig. 22.35), which are streamline forms like inverted spoons. Drumlins and fluted forms have various sizes and shapes—from a few meters to 70 meters high and up to several kilometers in length. Both erosion and deposition take place in a complex sequence of events. New York, Wisconsin, and Minnesota contain excellent drumlin fields (Fig. 22.36).

Transportation

Glaciers are able to transport huge boulders and large loads across distances of hundreds of kilometers and even move material thousands of meters up in elevation. Material acquired at the base of glaciers stays along the base until it becomes so abundant that it interferes with glacier flow and causes oblique upward shearing of ice over it. Debris acquired along the outer margin from adjacent hillsides tends to remain on the margin. However, where two valley glaciers or distinct lobes of ice come together, their corresponding marginal zones of debris are joined in a medial position. Such debris may lie between two juxtaposed masses of ice and extend from the surface to the bed of the glacier, or it may be superimposed or inset to only shallow depths. The drift in transit as distinct bodies is commonly called moraine, described as medial, lateral, or end moraine according to its position.

Valley glaciers and ice sheets handle their loads somewhat differently. Debris acquired at the head of valley glaciers descends to the bottom and is carried farthest—to the very terminus. That acquired below the head is carried proportionally less distance below the firn line before returning obliquely toward the surface. Since large continental ice sheets are fed snow most copiously from 10 to over 100 kilometers in from their margins, their basal debris is carried most actively in that peripheral zone. No indicator minerals or rocks are known to have been carried all the way from the center of an ice sheet to the outer margin.

Surficial drift has special effects. Thin drift or isolated stones carried on the surface of glaciers may melt their way a short distance down into the ice; drift 15 centimeters or more thick and large stones protect the surface from ablation (Figs. 22.37 and 22.38). Such drift increases relief and subsequent sliding around on the surface during ablation. In stagnating ice, the surface drift thickens and supports vegetation.

Deposition

All material transported by glaciers ultimately is dropped when the ice melts. Part may be laid down directly from moving ice or from sublimating ice in cold regions, part may be carried away from the glaciers by wind and running water, and part may be dropped

FIGURE 22.37 *Debris cones produced by differential ablation, or surface melting, in 12 days, midsummer 1950, on Emmons Glacier, northeast side of Mount Rainier, Washington. The amount of ablation is shown by the difference in level between the point the man indicates on the rod and the ice surface on which he stands. The vertical rods had been set in holes drilled nearly 2 meters into the ice and were further supported by sand piled around them, which now protects the surface. (George P. Rigsby.)*

FIGURE 22.38 *A glacier table, or rock-capped ice pillar, on the surface of Gornergrat Glacier, Switzerland. The column of ice is protected from melting by the shadow of the slab of rock. (E. S. Moore.)*

from floating glaciers and icebergs into rivers, lakes, and the ocean. The drift may be deposited particle by particle or in huge masses.

Drift deposited directly from ice characteristically is not stratified and contains abundant coarse angular rock fragments imbedded in a finer matrix. Such drift is called till. Descriptive terms, such as clayey, sandy, or stony are used to help classify its texture. Generally, till is poorly sorted and heterogeneous—representing all rocks over which the glacier rode but mostly those of a nearby source (Fig. 22.36). Bedded water-laid and nonbedded ice-laid deposits merge and mix so intimately in places as to be practically inseparable.

However, any deposit of unconsolidated clay, silt, sand, or gravel over which the ice moves can be incorporated in a glacier, transported, and deposited directly from ice, as is the poorly sorted till just cited. Any previous stratification, though, will be destroyed in the process. Till is easily mistaken for mudflow debris and other mass-moved material, and vice versa.

Basal drift in glaciers can be plastered on the surface below by pressure melting of the lowermost ice, even though the glacier is advancing. Such drift can be somewhat bedded or fissile, and if identifiable, it is called lodgment till. In contrast, till on top of the ice or let down from the surface by ablation is much looser,

more modified by water, and is called ablation till (Fig. 22.37). In New York, for uncertain reasons, basal drift is more common in the lee of obstacles than in front.

Glacial drift deposited in discrete zones marginal to a glacier builds up forms called moraines, which are classified by descriptive adjectives according to that position the debris was transported in the ice. The term moraine has been subsequently expanded to include any drift having constructional topographic form closely associated with direct glacial action. For example, an end moraine is that ridge of drift marginal to and at the end of a glacier; a lateral moraine is on the side. Ground moraine is defined loosely as haphazardly dumped thin drift behind end moraines. End moraines in complex systems have been traced from the Dakotas to the Atlantic Ocean and for hundreds of kilometers in Canada. An extensive complex of end moraines has been mapped in the Great Lakes area (Fig. 22.39).

The amount of drift laid down by continental ice sheets is staggering. Average thickness of drift in central New York is 17 meters, in New Hampshire 10 meters, and in central Ohio 27 meters.

When large portions of ice sheets stagnate and waste away, a chaotic variety of dead ice or ice-disintegration features is produced. Meltwater is plentiful in lakes and streams on and between ice blocks of a stagnating glacier. The drift generally is moved, sorted, and deposited by water, but it is also shifted around by mass movement on the ice as relief is modified by ablation. In warm or wet glaciers during deposition, little till remains—instead, most drift is deposited by water, whether outside the margin (Fig. 22.40) or within the limits of ice (Fig. 22.41). Kames (water-laid ice-contact deposits, Fig. 22.42), eskers (subglacial stream deposits, Fig. 22.43), and crevasse fills (till or drift in surface fractures, Fig. 22.44) are common.

During stagnation of a glacier, drift accumulates over and buries ice blocks. A hollow called a kettle forms as the ice slowly melts. Kettles range from a few meters to many thousands of meters in circumference and may be circular to very irregular in outline (Fig. 22.45). Some contain water, but others are too shallow to intersect the water table. Tens of thousands of depressions, ponds, and lakes in Wisconsin, Minnesota, and Michigan owe their origin to this process (Fig. 22.46).

In lakes and ponds associated with active glaciers, diurnal, annual, and aperiodic variations in discharge and sediment load lead to rhythmically stratified, fine clastic sediments. Annually deposited couplets of fine winter layers and coarser summer layers are called varves (Fig. 4.6).

Any of the glacial deposits may subsequently be reworked. Wind, for example, produces loess, a silt deposit derived from glacial river bars and flood plains

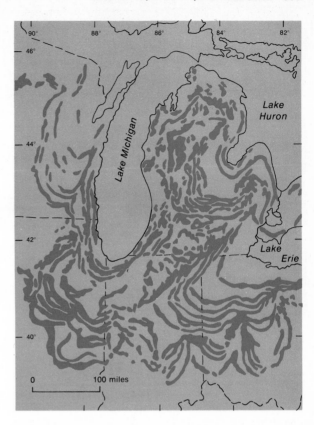

FIGURE 22.39   *Map of terminal and recessional moraines of the north-central United States. The festooned pattern outlines the former ice lobes. (After* Glacial Map of North America, *Geological Society of America.)*

FIGURE 22.40   *Outwash plain in foreground extends from the end moraine at the barn, near Bloomer, Wisconsin. (R. F. Black photo.)*

FIGURE 22.41 *Ice-walled lake deposit in Chippewa County, Wisconsin. A lake surrounded by ice during deglaciation partly filled with sediment washed into it. The sediment now stands high above the countryside. (R. F. Black photo.)*

FIGURE 22.42 *Kames in Ninemile Valley, New York. These are dirty gravel produced under the ice in part by streams dropping through moulins or openings and in part by streams flowing off the edge of the ice. (G. K. Gilbert, U.S. Geological Survey.)*

FIGURE 22.43 *An esker, or "serpent ridge," near Fort Ripley, Minnesota, as seen from the air. Eskers are believed to represent former streams that flowed beneath the glacier. (W. S. Cooper.)*

FIGURE 22.44 *Crevasse fill of till forms opposite wall of kettle near Bloomer, Wisconsin. (R. F. Black photo.)*

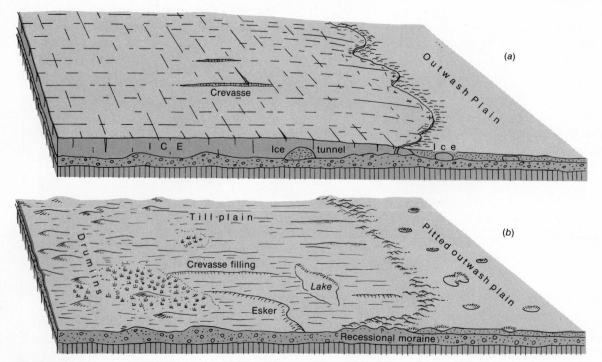

FIGURE 22.45 *Diagrams to illustrate the field relations of the various types of glacial deposit and their relations to the part of the glacier they formed from. (a) The margin of a retreating continental ice sheet; (b) the same area after the ice has melted. Compare the diagram with the actual photos of these features shown in Figs. 22.35, 22.39, 22.40, 22.43, and 22.44. (After Trewartha.)*

FIGURE 22.46 *Terminal moraine topography south of St. Paul, Minnesota. The low area in the middle is a typical iceblock kettle. It is surrounded entirely by high morainic knobs and hummocky ridges like that in the background. (Kenneth M. Wright.)*

(Fig. 22.47). It forms a blanket commonly tens of meters thick in the upper Mississippi Valley region.

Individual rock fragments, called erratics, foreign to their resting places, are carried by ice rafting long distances in water bodies. Some erratics from Antarctica have been recognized in oceanic deposits off South Africa. Many erratics were deposited directly from glaciers onto distinctly different bedrock. Some erratics are distinctive, and not only can they be traced to their source, but they show by their deployment the flow pattern of the ice that transported them (Fig. 22.48).

LANDSCAPE CHANGES

Glaciated landscapes in mountainous regions are characterized by U-shaped valleys, hanging valleys, faceted spurs, cirques, serrated ridges, saddle-shaped passes, round plucked bosses, and a variety of lakes and other features (Fig. 22.49). Glaciated landscapes inundated by continental ice sheets are commonly buried under tens of meters of debris (Fig. 22.45) and have had their former stream systems obliterated or strongly modified. Valley glaciers in the mountains accentuate relief by differential erosion (Fig. 22.50). Former ice sheets on the central plains, on the contrary, generally decreased relief by erosion of the highs and filling in of the lows. For example, southwestern Wisconsin has bedrock and topographic relief of several hundred meters with little drift, whereas southeastern Wisconsin has similar bedrock relief, but only tens of meters of topographic relief because of the thick drift in the valleys.

Perhaps the most striking change made by a valley glacier is the transformation of a V-shaped stream valley to a U-shaped glacial valley (Figs. 22.49 and 22.50). The form results from the glacier's efforts to conserve energy through establishment of a semicircular flow channel. A vigorous glacier immediately strips off the residual loose debris to bedrock and then abrades and plucks solid rock. Some valleys, like Yosemite Valley in the Sierra Nevada, have been deepened many hundreds of meters by glacial erosion. Under favorable circumstances, ice sheets may also accentuate pre-existing valleys, as was done in the Great Lakes basins, especially Superior, Michigan,

FIGURE 22.47 *Typical vertical cliff of loess in Illinois. (R. F. Black photo.)*

and Huron, and in the famous Finger Lakes of central New York State.

DRAINAGE CHANGES

Discharge of meltwater from valley glaciers typically carries abundant debris, glaciofluvial material, that blocks the down-valley area. The water coming from the ice has its gradient changed abruptly, as does a stream debouching from a mountain front. Sediment is dropped in the valley, gradually covering initial declivities until the stream is free to swing at will across the valley floor in a braided course. The initial channel inherited from preglacial days is obliterated by valley fill.

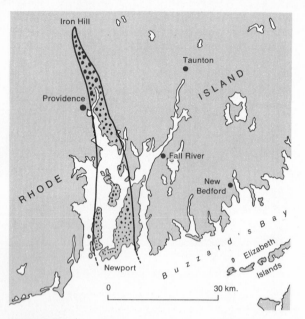

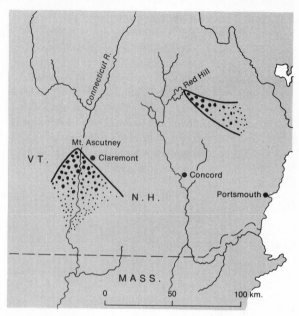

FIGURE 22.48 *Maps showing glacial boulder trains in New England. The black, iron-bearing rock of Iron Hill was distributed over a progressively wider area southward toward Newport. The rocks from Red Hill and Mount Ascutney were spread in a similar manner.*

Discharge of meltwater from ice sheets can accomplish a similar burial of former channels by huge alluvial outwash aprons. Of course, the ice that overruns former drainage channels can either erode them or fill them with drift. After destruction of the ice, a channel may be so modified as to be unavailable to surface drainage, or its drainage basin may have been taken over by another channel. Even the largest of the rivers in North America, like the St. Lawrence, Mississippi, Ohio, Missouri, and Columbia, have been moved at least locally from their previous channels by ice. The abandoned bedrock gorges in places along the upper Mississippi River provide mute testimony of drainage derangement (Fig. 22.51). Many deeply buried channels are found in the upper Mississippi Valley. The deeply buried gorges of some ancient streams, such as the Teays, that flowed westward across Ohio and Indiana, now provide reservoirs of groundwater.

Parts of some major rivers, notably the Ohio and

Missouri, owe their present positions to former courses marginal to ice sheets. Other channels are started as ice-walled canyons during the destruction of glaciers and lead to amazing canyon systems, such as those of Alberta, now abandoned (Fig. 22.52). Catastrophic flooding by water bursting from glacial lakes produced the "channeled scabland" of eastern Washington (Fig. 18.6).

Glaciation and Lakes

Lake basins formed by the activity of glaciers include those eroded by flowing ice, by the melting of buried stagnant ice within glacial meltwater streams (Fig. 22.45) or those dammed by glacial deposits or glaciers themselves. Still another type of lake is formed by melting in areas of permafrost.

Glacial abrasion excavated numerous lake basins in mountain valleys. Unlike a stream, no base level limits the extent of downcutting of a glacier, and so

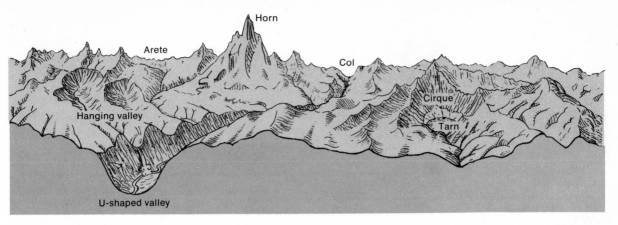

FIGURE 22.49   *Diagram showing the relations of landforms made by the mountain glaciation.*

many glacier-produced lake basins extend below sea level. Lake Chelan, Washington, for example, occupies a depression reaching more than 100 meters below sea level (Fig. 22.53).

Continental glaciation produced even more lake basins. Hundreds of thousands of lakes occur on the surface of the glacial deposits that cover large areas in northeastern North America and northwestern Europe. Melting of buried ice blocks (Fig. 22.54) or the blocking of drainage by moraines formed the thousands of lakes in Minnesota and Wisconsin (Fig. 22.46). Continental glaciers deepened and widened narrow stream valleys and deposited moraines across the southern end of the valleys, thus damming the

FIGURE 22.50   *Glaciated mountain valleys, now free of ice.This aerial photograph was taken looking southward from the northern border of the Canyon ranges, Mackenzie Mountains, Northwest Territories. The glacial sculpturing increases toward the interior, where excellent cirques appear. (National Air Photo Library Canada. Photo T4-116R.)*

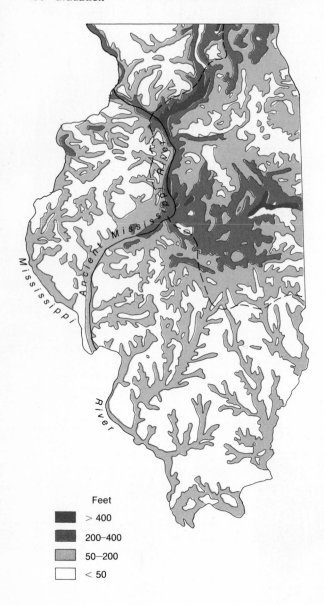

Feet

> 400

200–400

50–200

< 50

FIGURE 22.51 *Thickness of glacial drift, loess, and alluvium on bedrock in Illinois and the ancient channel of the Mississippi River from which it was diverted in late Wisconsin time. (After Frye et al., 1962.)*

drainage and producing the Finger Lakes of New York (Fig. 20.43). Some of the Finger Lake valleys were deepened below sea level. Continental glaciers also eroded many lake basins from the bedrock of the Canadian shield, with 250,000 lakes in Ontario alone.

Erosional resistance and structure of the bedrock, glacial erosion and deposition, and postglacial processes determined the form of the basins of the Great Lakes. Lakes Michigan and Huron occupy basins carved into shales and other weak rocks along the flanks of the Michigan structural basin by glaciers that followed preglacial river valleys in some places (Fig. 22.55). The resistant Niagara Dolomite forms ridges that separate Green Bay from Lake Michigan, Georgian Bay from Lake Huron, and Lake Erie from Lake Ontario. The breaching of the last-mentioned ridge by the headward retreat of Niagara Falls could drain Lake Erie within about 25,000 years at recent rates of retreat. Lake Superior, the deepest of the Great Lakes, is surrounded by more or less resistant rocks. Its 305-meter depth is due to glacial erosion of weaker rocks and downward crustal movements of the floor of the lake basin.

Old shorelines exist high above and below the levels of the present Great Lakes. Abandoned channels through which the excess water was discharged (Fig. 22.56) from the higher lake levels through the St. Croix, Illinois, Wabash, and Hudson Rivers show that there were higher and lower levels of the Great Lakes along the ice front during the later stages.

SEA-LEVEL CHANGES

Emerged interglacial shorelines around the world show that the oceans, at times in the Pleistocene epoch when glaciers were more restricted, rose 20 meters and more above their present level. When glaciers were more extensive than now, the water removed to make land ice lowered sea level as much as 120 meters below the present. Thus, as the volume of land ice fluctuated during the Pleistocene, sea level also fluctuated. Sea-level changes due to ice melt or ice formation are called eustatic changes. Organic matter that can be dated by radioactive carbon gives us precise ages of some of the later levels (Fig. 22.57).

Complicating the problem of eustatic changes in sea level are changes brought about by isostatic and

FIGURE 22.52 *Third Lake near Chitima, Alaska, is in the axis of a former glacial stream trench. (R. F. Black photo.)*

FIGURE 22.53 *Lake Chelan, Washington, occupies a depression reaching more than 100 meters below sea level. The preglacial Stehekin River Valley was deepened and its gradient reversed locally by erosion of a glacier that moved nearly 50 kilometers down valley. (Photo by Washington State Department of Commerce and Economic Development.)*

FIGURE 22.54   *Kettle Lake in northwest Wisconsin. (R. F. Black photo.)*

FIGURE 22.55   *Geologic cross section of the Michigan sedimentary rock basin, showing Lakes Michigan and Huron lying in zones occupied by relatively weak shales and salt-bearing rocks. Green Bay and Georgian Bay lie in zones of weak Ordovician rocks, separated from the lakes by the escarpment of Niagaran Dolomite. At the eastern end of Lake Erie, the Niagara River falls over this same dolomite formation. (From J. L. Hough,* Geology of the Great Lakes, *University of Illinois Press, 1958.)*

West

Green
Bay   Lake Michigan   Michigan

Precambrian rocks   Cambrian rocks   Ordovician rocks   Niagaran series   Pennsylvanian and Mississippian rocks, undifferentiated   Antrim Shale   Devonian rocks   Upper Silurian rocks   Salina Group

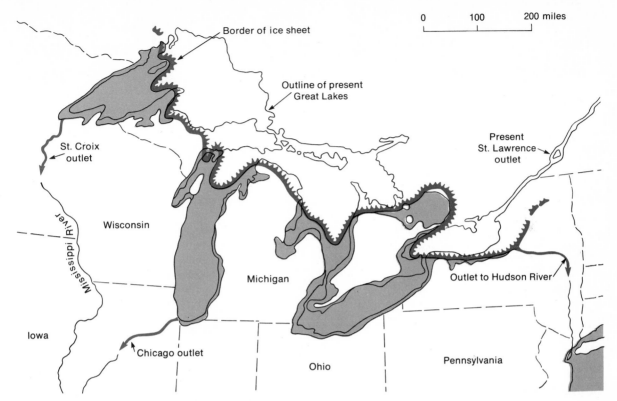

FIGURE 22.56 *Map showing overflow drainage outlets of the Great Lakes prior to the melting of the ice that blocked the St. Lawrence outlet. (After Taylor, 1927.)*

tectonic processes. Depression of the land may be caused by weight of ice or tectonism, or mountain-building processes. Despite the low specific gravity of ice, the weight of enormous ice sheets is far too great to be sustained as a simple load on the earth's crust. It causes the crust to bend or warp downward. By reasonable calculation, the Antarctic Continent is today depressed some 600 meters because of its load of ice. Were that load removed, the continent would rise to about its original position over a period of several thousand years. Rebounding, as the process is called, is still going on in Scandinavia, North America, and elsewhere since the disappearance of the ice sheets that covered them 6,000 to 10,000 years ago. The rate of rebound is fast at first, slowing gradually as equilibrium is approached. Marine beaches in those areas are now tilted and uplifted from 10 to over 100 meters.

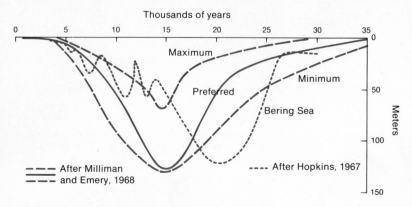

Thousands of years

FIGURE 22.57 *A preferred sea-level curve for the Atlantic and Pacific Oceans, with an envelope of maximum and minimum levels showing scatter of data as compared with a proposed curve for the Bering Sea.*

Maximum

Minimum

Preferred

Bering Sea

Meters

— — — After Milliman
— — — and Emery, 1968

- - - - After Hopkins, 1967

Glacial rebound is also recorded by deformed shore-lines of preglacial lakes, such as Lake Michigan, which tell something of the rates of retreat of the ice.

PLEISTOCENE GLACIATION

To a man best known as a poet, Johann Wolfgang von Goethe (1749–1832), goes the credit for first recognition of a former ice age. He was also one of the first to attribute the transport of erratic blocks to glaciers and to believe that an ice sheet formerly covered northern Germany.

It was recognized early that many erratics on the plains of northern Germany were not of rock types found in the Alps to the south, where the rivers originated, but were identical to rocks from Scandinavia. Because of the early influence of the church on geology, the Biblical flood was cited as a mechanism of transporting them there. Only through the study of the deposits associated with glaciers was it possible to reconstruct a history of glaciation—not just once, but several times. Widespread recognition of continental glaciation and of multiple glaciation has come only during the past century.

Detailed studies of some of the small glaciers in western United States indicate that they were regenerated only a few thousand years ago after having disappeared for a time. The fact that they occupy ancient cirques and U-shaped valleys indicates an older glaciation that was much more extensive than now. Ice today covers about 10 percent of the land area of the globe, but a variety of evidence indicates that more than 30 percent has been covered at times in the past.

Some large glaciers apparently have been in existence for tens of thousands of years, and glaciers existed in Antarctica throughout the Pleistocene epoch and well back into the Cenozoic era. Evidence of marked fluctuations are indisputable on land and in the sea. Abundant faunal and geochemical evidence of the Pleistocene is derived from studies of ocean-floor sediments that indicate numerous worldwide shifts in climate.

The last major Pleistocene ice sheet on North America seems to have originated in the northeastern part of the continent and to have spread out in all directions. It extended northwest into the Arctic Sea and southward to the valleys of the Missouri and Ohio Rivers. A smaller ice field formed in the mountains of western Canada and part of the Pacific Northwest. At first it consisted of scores of valley and piedmont glaciers that coalesced at many places into a nearly continuous sheet which met the one moving southwestward across the plains of Alberta.

Although the former ice sheets inundated most of the northern part of land in the Northern Hemisphere more than once, isolated areas were not covered during every advance. Some such "unglaciated" areas may have been refuges from which plants and animals later spread.

In the Great Lakes area, dating of a buried forest by means of radioactive carbon indicates that a continental ice sheet extended southward into Lake Michigan as recently as about 10,000 years ago. The deposits left by that glaciation are on top of lake deposits that in turn lie on an ancient soil and a buried forest dated about 12,000 years ago. These overlie other lake deposits that in turn rest on still older glacial deposits like a multiple layer cake. Throughout the area covered by continental glaciers, evidence of multiple glaciation is preserved only here and there. Correlation of one locality with another has been difficult. The evidence is fragmentary, and the magnitude of an ice retreat and readvance recorded at one point cannot always be determined. Thus, it is not always possible to distinguish local ice fluctuations from those that are regional. Scientists have not yet agreed on the number and extent of expansions of continental ice sheets in North America or in the world at large during the last 1 to 3 million years. Moreover, many older glaciations in various parts of the world are recorded on the surfaces of ancient rocks, well back into Precambrian time.

Causes of Glaciation

The cause of glaciation is for the most part climatic, but the connection is indirect and complex. Innumerable hypotheses have been offered to account for the drastic changes that evoked glaciation, but no theory is universally accepted. A cooling of the ocean waters during the late Tertiary is recorded by fossils and by isotopic geochemistry. This accompanied polar glaciation and prepared the way for the Pleistocene glaciation of the now temperate regions, but the cause of the cooling is not known. Hundreds of millions of years in geologic time have been characterized by an absence of glaciers. As long as the earth's poles of rotation were located in open oceans, no glaciation took place, except possibly an alpine type in extremely high mountains. Continental ice sheets during the Pleistocene epoch were favored by the South Pole in Antarctica and the North Pole in the Arctic Ocean. The Arctic Ocean was restricted from complete mixing with the Pacific and Atlantic Oceans by the configuration of the surrounding continents. Yet even during the Pleistocene, the climate at times was warmer than now.

Hypothetical explanations of glaciation commonly invoke continental uplift, modified atmospheric circulation, reduced solar heating, or combinations of these factors. It has been recognized that high, broad, uplifted land areas of the geologic past produced more climatic contrasts than did the expansive low-lying areas with widespread epicontinental seas. However,

we have no conclusive evidence that the altitude of northern and northeastern North America fluctuated up and down sufficiently prior to and during the Pleistocene epoch to have caused high-altitude glaciation that could have spread out from that region.

Changes in the meteorological circulation system of the globe, coupled with changes in topography, seem likely factors in increasing snowfall in the alpine areas of the cold regions and in triggering glaciation. A snow and ice cover tends to be self-perpetuating. It reflects solar radiation and insulates the earth from atmospheric heat. Thus, large areas of snow and ice tend to create their own cold regions and expand into ice sheets. In deglaciation of ice sheets, any reduction in nourishment causes a withdrawal of the ice sheets, followed by warming of the land and, later, the ocean.

The astronomical theory of glaciation in its various forms has many advocates. Times of radiation maxima and minima from the sun can be calculated from known periodicities of the tilt and wandering, or wobble, of the earth's axis and the earth's orbital path about the sun. Sunspot cycles are accompanied by minor variations in thermal energy. Such variations, when coupled with other factors, may be sufficient to induce glaciation.

The Pleistocene glaciation was not the first experienced on earth, and it probably will not be the last. The present pole positions in Antarctica and the Arctic Ocean favor another advance of ice. In fact, there is more ice on earth now than existed about 5,000 years ago and in still earlier interglacial times.

While a wide variety of conditions can be postulated for the causes of glaciation, and man has been searching for more than a century, no simple solution has emerged. The answer is likely to come from a complex intertwining of various disciplines, such as meteorology-climatology, oceanography, astrophysics-astronomy, botany, and geology. The search for the mechanism is taking on greater urgency. Many scientists believe that only slight variances from a norm of these parameters may trigger great changes in the earth's heat budget.

Quite recently in published geologic literature, there has been an upswing toward the theory that the earth will experience a new ice age. Some go as far as to say that not only is the earth ready, but it is overdue for an ice age, possibly within the next 100 years.

The thought that the earth is overdue for an ice age comes from sediment studies on climatic changes over the past 500,000 years. These studies have shown that ice ages do not occur every 100,000 years nor do they last that long. A time span of 10,000 to 15,000 years between glaciers seems more likely, and the present interglacial period is already 7,000 to 10,000 years old.

The sediment studies also show that minor climatic changes can be significant even though temperature variations are small. The temperature trends were on the rise from 1850 to 1960 (with exceptions); however, these trends have now reversed, and a cooling-off period seems to be present.

Man himself is now responsible for many unnatural changes in the atmosphere, stratosphere, hydrosphere, biosphere, and lithosphere. Which way will these changes throw the balance?

REFERENCES

Bird, J. B., 1967, *The Physiography of Arctic Canada,* The Johns Hopkins Press, Baltimore.

Black, R. F., 1954, *Permafrost—A review,* Geological Society of America Bulletin, vol. 65, pp. 839-856.

Dyson, J. L., 1962, *The World of Ice,* Alfred A. Knopf, Inc., New York.

Embleton, C., and King, C. A. M., 1968, *Glacial and Periglacial Geomorphology,* St. Martins Press, Inc., New York.

Flint, R. F., 1971, *Glacial and Quaternary Geology,* 2d ed., John Wiley & Sons, Inc., New York.

Hough, J. L., 1958, *Geology of the Great Lakes,* University of Illinois Press, Urbana.

Kamb, B., 1964, *Glacier geophysics,* Science, vol. 146, pp. 353-365.

Kingery, W. (ed.), 1963, *Ice and Snow—Properties, Processes and Applications,* The M. I. T. Press, Cambridge, Mass.

National Academy of Sciences-National Research Council, 1966, Proceedings, Permafrost International Conference, Publication 1287.

Paterson, W. S. B., 1969, *The Physics of Glaciers,* Pergamon Press, Ltd., Oxford.

Sharp, R. P., 1960, *Glaciers, Condon Lectures,* University of Oregon Press, Eugene, Ore.

Turekian, K. (ed.), 1971, *Late Cenozoic Glacial Ages,* Yale University Press, New Haven, Conn.

Wright, H. E., Jr., and Frey, D. G. (eds.), 1965, *The Quaternary of the United States,* Princeton University Press, Princeton, N. J.

CHAPTER 23

OCEANS AND SHORES

The ocean basins have an important place in geology. They represent an area of about 361 million square kilometers, or 70.8 percent of the earth's surface (Table 23.1). They are the ultimate resting place of land-derived sediments. In terms of relative abundance, the oceans hold about 86 percent of the earth's water. The ocean waters are the principal source of the precipitation that forms streams and sustains plants and animals. Much of this water is returned by streams and rivers to the ocean, where its load of sediment acquired enroute is deposited and eventually forms sedimentary rock. In addition, the ocean is the home of a host of organisms that play an important role in forming sedimentary rock.

This chapter will be concerned with the relief features of the ocean basins, the characteristics and motions of the ocean water, and the geologic work of waves and currents upon the shores and coasts.

OCEAN-BASIN PHYSIOGRAPHY

Methods of Investigation

The depth of the oceanic basin is determined mainly by sonic sounding. An echo sounder measures the

TABLE 23.1   Data On The Oceans

Area:	361×10^6 square kilometers		
Pacific:	180×10^6 square kilometers		
Atlantic:	107×10^6 square kilometers		
Indian:	74×10^6 square kilometers		
Mean depth:		3,795 meters	
Pacific average depth:		4,028 meters	
Atlantic average depth:		3,332 meters	
Indian average depth:		3,897 meters	
Maximum known depth:		11,034 meters (Pacific)	

Extreme depths in the Pacific Ocean:

Date	Place	Ship	Depth (meters)
1923	Mindanao Trench	German *Emden*	10,497
1952	Tonga Deep	U.S. *Horizon*	10,882
1959	Mariana Trench	U.S.S.R. *Vityaz*	11,028
1960	Mariana Trench	U.S. *Trieste*	11,034

Volume: $1,370 \times 10^6$ cubic kilometers
Average salinity: 34.482 parts per thousand
Total dissolved material: 5×10^{22} grams
Mean seawater density: 1.025 grams per cubic centimeter

time required for a sound to travel to the bottom and its echo to return to the ship. Based on the known velocity of sound in water, the instrument automatically registers the depth as the ship continues along on its course. A recording device is used to make a continuous record from which maps of the ocean basins are compiled. The determination of the exact position of the ship at sea at the time of sounding has greatly improved in recent decades by the use of electronic navigational equipment and artificial satellites.

Results of Sounding

The deepest parts of the sea are in the western Pacific Ocean, where several soundings have exceeded 10,000 meters (Table 23.1). The maximum known depth recorded is 11,034 meters (6.85 miles) in the Mariana Trench. Depths of 6 to 9 kilometers occur in about 4 percent of the area of the sea floor. The depth of about 20 percent of the ocean basins is 4.5 kilometers, and approximately two-fifths of the basin area is at depths of between 2.0 and 5.0 kilometers. The mean depth of the ocean is about 3,800 meters (Fig. 23.1).

Other studies of the ocean floor are carried out with the aid of seismic reflection and refraction surveys, gravity meters, magnetometers, heat-flow probes, current meters, undersea photography, and sonic sounding of sediments (Fig. 23.2). Sediment-sampling devices include coring and drilling instruments capable of obtaining cores of the bottom sediments up to 1,300 meters below the seabed.

A cooperative oceanographic investigation called the Deep Sea Drilling Project, using the drill-ship *Glomar Challenger* (Fig. 23.3), began a series of cruises in 1968. As of September 1973 a total of more than 400 holes had been drilled at 314 sites in the Atlantic, Pacific, and Indian Oceans, the Gulf of Mexico, and the Mediterranean Sea in water as much as 6,245 meters deep. The deepest penetration was 1,300 meters into the ocean floor. The oldest rocks obtained were lower Upper Jurassic in age (about 160 million years).

In August 1972, the Deep Sea Drilling Project began phase III, which will continue the drilling and coring of the ocean floors until August 1975. Following phase III, the Project will continue on a more international scope in terms of both scientists and financial support.

Major Relief Features

The principal relief divisions of the ocean floor are the continental shelves, continental slopes, continental rise, and deep-sea abyssal floor (Figs. 17.3 and 17.4).

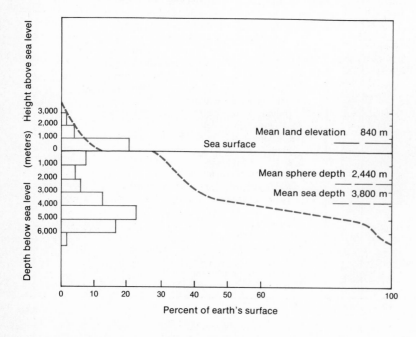

FIGURE 23.1 *Hyposographic curve of the oceans, showing the percentage of the earth's surface above any given elevation or below any given depth. (After Sverdrup et al., 1942.)*

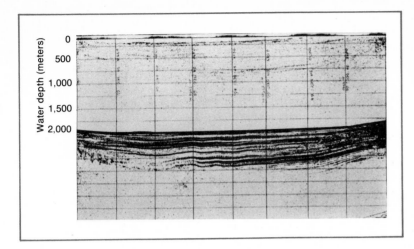

FIGURE 23.2   *Sparker reflection profile of layered materials on the sea floor in the Solomon Sea, western Pacific Ocean. Near the middle of the profile the deposits are about 600 meters thick. (George P. Woollard photo.)*

FIGURE 23.3   *Deep Sea Explorer. This is a port side view of the Deep Sea Drilling Project drilling vessel,* Glomar Challenger, *which is drilling and coring for ocean sediment in all the oceans of the world. Scripps Institution of Oceanography, of the University of California at San Diego, is the managing institution for DSDP under contracts with the National Science Foundation. The drilling vessel is owned and operated by Global Marine Inc., of Los Angeles, which holds a subcontract with Scripps to do actual drilling and coring work. The* Glomar Challenger *weighs 10,400 tons, is 400 feet long, and the 1-million-pound-hook-load-capacity drilling derrick stands 194 feet above the waterline. She is the first of a new generation of heavy drilling ships capable of conducting drilling operations in open ocean, using dynamic positioning to maintain position over the borehole. A re-entry capability was established on June 14, 1970, which will enable the changing of drill bits and re-entering the same borehole in the deep ocean. Forward is the automatic pipe racker, designed by Global Marine Inc., which holds 24,000 feet of 5-inch drill pipe. (Scripps Institution of Oceanography-University of California, San Diego.)*

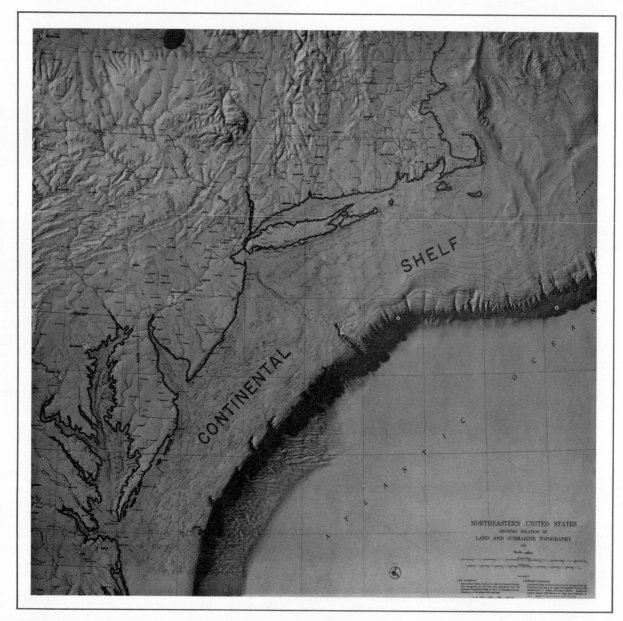

FIGURE 23.4 *Photograph of a relief model of the continental shelf off the northeastern coast of the United States. Note the submarine canyon that seems to extend from the Hudson River southeast across the continental shelf and down the continental slope toward the deep sea. (Aero Service Corporation: Division of Litton Industries.)*

Continental Shelves The continental shelves are the submerged margins of the continents to a depth of about 200 meters, where the sea floor begins to descend more abruptly. Their total area is about 7 million square kilometers, or about 2 percent of the area of the sea floor.

The surface of the shelves is varied. In places it is broken by channels (Fig. 23.4), ridges, enclosed depressions, submarine benches, and coral reefs. At some places the bottom material is solid rock; in others, it is mud, sand, or gravel. Along the coast of New England and eastern Canada, the shallow floor has many glacially formed features, among which are drumlins and deep troughs.

The continental shelves are especially sensitive to relative changes in sea level brought about by earth movements, erosion of the land, and deposition of sediments in the sea. Other changes are caused by loss of water to glaciers on land and the return of water to the sea by melting of the ice. Sinking of the coastline or a rise in sea level may cause the sea to spread farther upon coastal lowlands. Conversely, a rise of the coast or a lowering of sea level may convert part of the continental shelves into land coastal plains.

Continental Slopes At the outer edge of the continental shelves or the shelf break, the bottom slopes toward deeper depths, forming the continental slope. Its floor has a width of 15 to 30 kilometers (Figs. 17.4, 17.5, and 23.4). The continental slope gradually merges with the continental rise at a depth of 2,000 meters.

Continental Rise The continental rise is a gently sloping sedimentary apron present at the base of the continental slope. It is lacking along parts of the California coast and along areas of deep-sea trenches. It consists of sediments deposited on the margin of the deep-sea floor, and so it is analogous to a piedmont alluvial plain deposited by streams at the base of a mountain range. It is thought to be the product of turbidity currents and gravity transfer of sediments off the continental slope.

The combined area of the continental shelf and continental slope is about 3 times that of the continental rises. The deep-sea basin beyond the continental rise occupies about six-tenths of the earth's surface.

Found on the continental slopes, and often continuing as valleys across the continental rise, are the submarine canyons. The canyons that indent the continental slopes at many places are deep (1,200 meters), V-shaped, steep-walled, valleylike depressions (Fig. 23.5). Some of them have tributaries with a branched or dendritic pattern similar to that made by stream erosion on land. Some lie off the mouths of rivers, such as the Congo, Indus, Hudson, Delaware, and Columbia (Fig. 23.5), whereas others have little or no counterpart on land nearby. Their lower reaches extend to depths of 2,000 to 3,000 meters below sea level. Their walls in many places are rocky, and the deposits of sediment on the sea floor at their mouths on the continental rise illustrate that the canyons act as sediment chutes. Within the canyons, fine and even coarse sediments are carried into deep water.

The origin of these canyons is uncertain. They have been attributed to erosion by turbidity currents, mudflows and submarine slumping, erosion by rivers flowing across the continental margin during periods of lowered sea level, erosion by tidal currents, and faulting.

Turbidity currents are known to have broken trans-Atlantic submarine cables at sea, but their ability to cut deeply into solid rock is questioned by many geologists. A number of submarine canyons have been studied in great detail, such as the Scripps Canyon system off the California coast at La Jolla. These canyons occasionally show a sudden deepening following storms that seems to indicate submarine slumping or downslope movement of sand. It may be that these processes operate only in canyons already formed. The close resemblance of submarine canyons to land valleys suggests that subaerial erosion may be a partial explanation. Erosion by streams during the lowering of sea level in the Pleistocene epoch may have extended to depths of 100 to 300 meters but hardly to depths of thousands of meters where some canyons are found.

The opinions of geologists are divided over the various possible causes, or combinations of causes, that have been proposed. Two marine geologists, F. P. Shepard and R. Dill, studied over 93 canyons, and in a book published in 1966, they presented 28 questions that need answers before an acceptable theory can be agreed upon. It is very likely that one theory will not explain the origin of all submarine canyons.

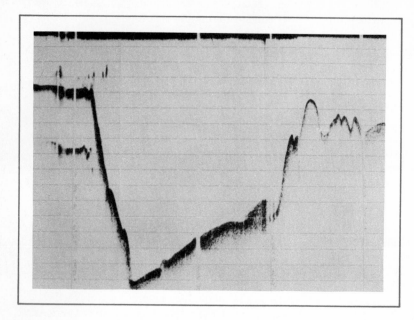

FIGURE 23.5   *Fathogram across As-
toria Canyon off the mouth of the Columbia River. The maximum ocean depth
shown at this crossing is about 600
meters. (Oregon State University School
of Oceanography.)*

Deep-Sea Abyssal Floor   Before the development of sonic sounding, the topography of the deep-
sea floor was imagined to be a monotonous plain. In
truth it is highly diversified by major and minor relief
features.

(*a*)   *The Abyssal Plains*   These generally represent
the flattest and deepest (3,000 to 6,000 meters) portions of the ocean floor. In all likelihood, the plains
represent depositional plains formed by sediments
brought by turbidity currents off the continental slope
and rise, as well as pelagic sediments derived from the
overlying ocean waters. The plains cover 30 percent of
the area of the ocean floor, but they vary considerably
in their appearance, shape, and areal coverage from
ocean to ocean (Fig. 17.4). Irregularities of a pre-exist-

ing rough topography of the oceanic crust create such
relief features as abyssal hills and seamounts. Some of
the more extreme topographic features are volcanoes,
volcanic islands and atolls, volcanic island ridges
(Hawaii), and guyots.

A guyot (pronounced gē-yō) is a flat-topped
seamount (Fig. 23.6). Numerous guyots have been discovered in the central and western Pacific ocean (Fig.
23.7). Some are very broad. The Cape Johnson Guyot
is 11 by 16 kilometers across the top, while the Horizon Guyot is up to 66 kilometers wide and 280 kilometers long (Fig. 23.6). Their summits, now found at
depths of 1,300 to 2,000 meters, apparently represent
volcanic peaks that were truncated by wave erosion in
the geologic past. The summits contain wave-rounded

FIGURE 23.6   *Profile of Horizon Guyot in the western Pacific. No vertical exaggeration. (After Hamilton, 1956.)*

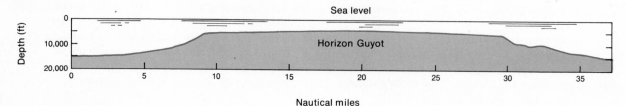

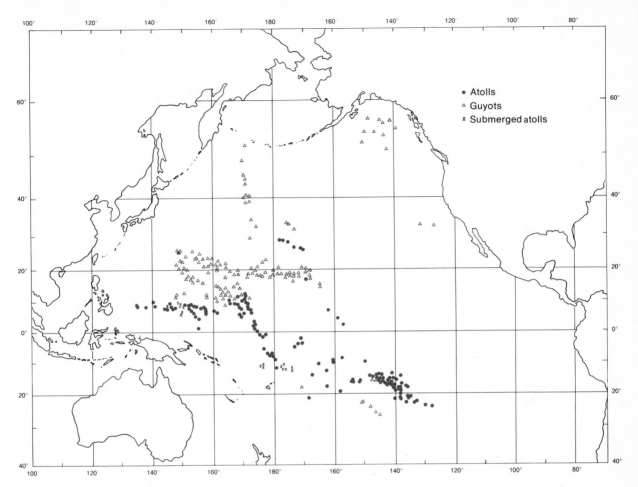

FIGURE 23.7   *Drowned ancient islands in the Pacific Basin that became guyots or atolls. (From Menard, 1964.)*

cobbles and shell fragments of shallow-water organisms. Thus, the guyots indicate a subsidence of the sea floor or a rather drastic rise in sea level. The fossils of shallow-water reef corals and mollusks dredged off the top of the Hess and Cape Johnson Guyots are middle Cretaceous in age (110 million years), indicating that the subsidence of submergence originated after this time.

Atolls are nearly circular to oval coral or algal reefs that surround a lagoon (Fig. 23.8). Darwin in 1842, Dana in 1885, and Davis in 1928 suggested that an atoll is formed at a certain stage in the history of a

coral reef growing on a sinking volcano. A reef may begin as a shallow bench on a shoal or as a fringing reef around an island. If it is a fringing reef that grows chiefly seaward, it changes to a barrier reef, separated from land by a lagoon, as the island sinks. Then, as the island sinks farther and finally disappears below sea level, the corals continue to build up the reef until it becomes an atoll (Fig. 23.9).

Drilling on Funafuti, Bikini, and Eniwetok Atolls (Fig. 23.10) penetrated coral limestone at depths of 338 to 1,400 meters. Inasmuch as reef-forming corals today live only in shallow water, substantial sub-

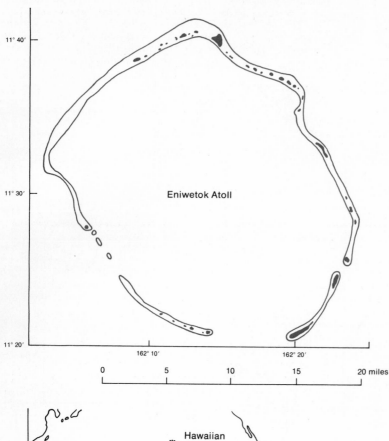

FIGURE 23.8 *Map of Eniwetok Atoll.
(After Ladd and Schlanger, 1960, U.S.
Geological Survey.)*

11° 40'

11° 30'

Eniwetok Atoll

11° 20'

162° 10'   162° 20'

0   5   10   15   20 miles

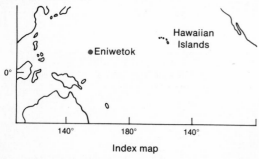

Hawaiian
Islands

●Eniwetok

0°

140°   180°   140°

Index map

sidence of the sea floor seems apparent. Foraminifera
of Upper Eocene age, also found in the deep borings,
indicate subsidence over a period of at least 50 million
years. The basement rock on Eniwetok was found to
be basalt. The foundations of the atolls evidently are
former volcanic islands. Many were truncated like
guyots but were maintained near sea level by corals
and associated organisms that grew upward as fast as

the regional sea floor subsided. On other platforms
where the corals could not keep up an upward growth,
guyots resulted instead of atolls.

(b) The Mid-Ocean Rises and Ridges These are a
nearly worldwide undersea mountain system, and they
constitute 23 percent of the area of the ocean floor
(Figs. 17.4, 17.6, 17.14, and 23.11).

A central fracture zone, or rift, similar to a graben,

and other large cross-fracture zones are found along the system. Some of these fracture zones are linear, wide, and deep and thus resemble trenches.

(*c*) *The Island Arc System*   This system is another major relief feature of the ocean basins. It was discussed in Chap. 17.

PHYSICAL AND CHEMICAL PROPERTIES OF THE OCEAN WATERS

Salinity

Seawater contains a vast quantity of dissolved substances. Seawater is an ionic solution with an average composition of 34.5 grams per 1,000 grams of water (Figs. 23.12 and 23.13).

The term salinity is used to refer to the composition of this solution and is expressed in parts per thousand ($^0/_{00}$). The solids dissolved in seawater are dispersed as ions. About 67 different chemical elements are found in seawater; however, as shown in Table 23.2, the main ions are chloride (Cl^-), sodium (Na^+), sulfate (SO_4^{--}), magnesium (Mg^{++}), calcium (Ca^{++}), potassium (K^+), bicarbonate (HCO_3^-), and bromide (Br^-). One of many hypothetical combinations to form salts is given in Table 23.3.

TABLE 23.2   Ionic Constituents of Seawater

Ion	Parts per thousand ($^0/_{00}$)	Total dissolved solids (wt. %)
Cl^-	18.98	55.07
SO_4^{--}	2.65	7.72
HCO_3^-	0.14	0.40
Br^-	0.07	0.19
$H_3BO_3^-$	0.02	0.01
F^-	0.00	0.10
Na^+	10.55	30.62
Mg^{++}	1.27	3.69
Ca^{++}	0.40	1.18
K^+	0.38	1.10
Sr^{++}	0.01	0.02
Totals	34.47	100.00

The salinity of ocean water may range between the extremes of 33 and 37 parts per thousand (33 $-37 \, ^0/_{00}$) but normally contains between 34 and 35 ($34-35 \, ^0/_{00}$). Salinity is decreased at the surface by precipitation at sea, by surface fresh-water runoff, or by icemelt. It increases by evaporation of water from the surface (Fig. 23.13), or by the formation of sea ice.

The possible sources of sea salt are several: the initial waters may have contained some dissolved salts in solution; some salts may have been dissolved from the rocks over which the water spread; some may have been supplied by degassing the earth during its

FIGURE 23.9 *Diagrams showing how an atoll may be developed. (a) An island with a fringing reef. (b) After subsidence the fringing reef has grown upward and become a barrier reef. (c) After further subsidence of the original island, the barrier reef has become an atoll.*

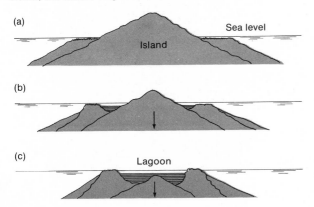

TABLE 23.3   Dissolved Solids in Seawater Stated as Salts

Salt	Parts per thousand ($^0/_{00}$)	Wt. %
Sodium chloride, $NaCl$	23.477	68.085
Magnesium chloride, $MgCl_2$	4.981	14.445
Sodium sulfate, Na_2SO_4	3.917	11.359
Calcium chloride, $CaCl_2$	1.102	3.196
Potassium chloride, KCl	0.664	1.926
Sodium bicarbonate, $NaHCO_3$	0.192	0.557
Potassium bromide, KBr	0.096	0.278
Boric acid, H_3BO_3	0.026	0.075
Strontium chloride, $SrCl_2$	0.024	0.070
Sodium fluoride, NaF	0.003	0.009
Totals	34.482	100.000

SOURCE: Gross, M. G., 1972.

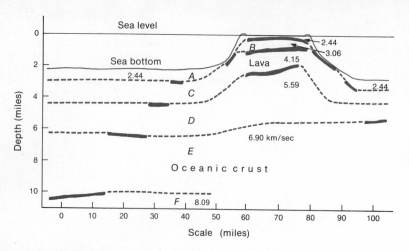

FIGURE 23.10 Inferred structure of the earth's crust at Eniwetok Atoll. The numbers are seismic velocities in kilometers per second. The vertical exaggeration is about 5 to 1. The two top layers, (A) and (B), are reef and lagoon deposits, (C) is basaltic lava (found by drilling), (D) is either denser lava or an igneous intrusion, (E) is normal oceanic crust, and (F) is the mantle. (After Raitt, 1957, U.S. Geological Survey.)

FIGURE 23.11 Mid-oceanic rises, ridges, and fracture zones. (From Menard, 1964.)

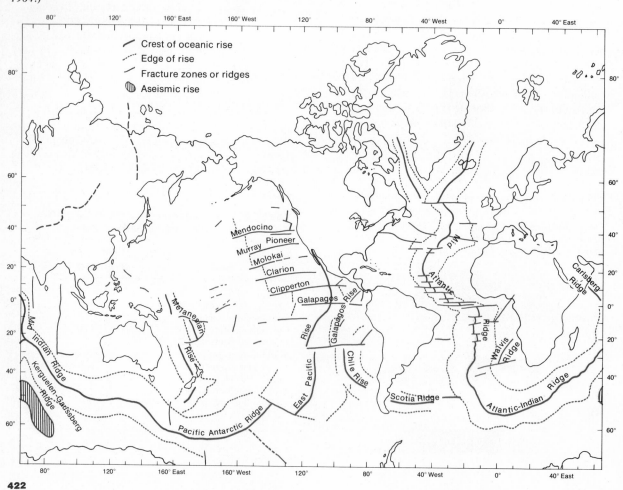

cooling; some derive from submarine vulcanism and hot springs on the sea floor; and some may have been contributed by river runoff that obtained it from groundwater after its solution from land rocks. In reference to the last-mentioned source, it should be noted that the dissolved-salt content of river water differs considerably from that of seawater. In river water, $HCO_3^- > SO_4^{--} > Cl^-$ and $Ca^{++} > Mg^{++} > Na^+$, but in the oceans, $Cl^- > SO_4^{--} > HCO_3^-$ and $Na^+ > Mg^{++} > Ca^{++}$. Silica, iron, and aluminum have a relatively low concentration in river water and have a concentration of less than 1 part per million in seawater.

The origin of the ocean's salinity still remains an intriguing question for chemical oceanographers. The general thinking today is that seawater has a remarkable constancy of composition in today's oceans. Studies of past oceanic life seem to indicate that the composition has changed little since 600 million years ago (the Cambrian period).

Temperature

The temperature of the ocean varies from equator to poles and from the surface to the abyssal depths. At the equator the surface temperature averages about 36°C, whereas in polar regions it is about −2°C, or near the freezing point of seawater. Water at the bottom of the deep sea varies from −2°C in polar regions to about +2°C in low latitudes. The temperature of seawater decreases rapidly from the surface down-

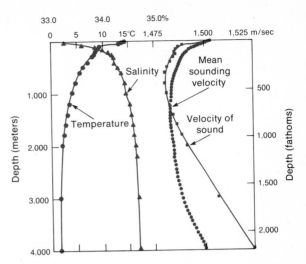

FIGURE 23.12 *Changes in temperature, salinity, and computed velocity of sound with increasing depth off southern California. The mean sounding velocity is the mean velocity from the surface to any specified depth. (After Sverdrup et al., 1942.)*

ward to the 400-meter level, where it is about 4°C. It decreases less rapidly to the 2,000-meter level, where it becomes relatively constant at or near 2°C (Fig. 23.12). The deeper portions of the oceans contain a high percentage of cold, dense water, which moves toward the equator from the polar-subpolar areas and thus forms part of the circulation of the deep sea.

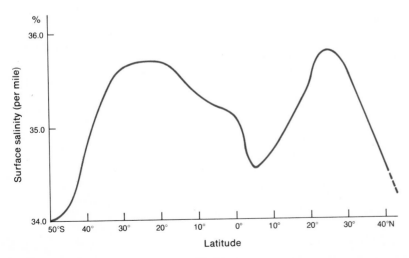

FIGURE 23.13 *Graph showing the average salinity of seawater in relation to latitude. Salinity is highest near 25°N and 20°S because of high evaporation and low precipitation there. It decreases toward the equator and toward high latitudes. (After Wüst, 1958.)*

The difference in surface temperatures at different latitudes reflects differences in solar heating. The slight heat received from the earth's interior, radioactivity on the sea bottom, and oxidation of organic matter are negligible in their contribution to the oceanic heat budget.

Light

Sunlight disappears rapidly with depth in the sea, with the blue wavelengths showing the deepest penetration. More than half of the visible wavelengths of light are absorbed in the first 10 meters of average seawater, 99 percent of it in the first 100 meters. Most of the energy absorbed is converted to heat, but a vital part is used in photosynthesis by plants.

Density and Pressure

At an average salinity and temperature, seawater has a density of about 1.025. In cold polar seas this increases to about 1.028, whereas in warm tropical seas it decreases to about 1.022. Changes in its density, due to changes in temperature, pressure, and salinity, are responsible for certain currents in the ocean.

The hydrostatic pressure at any particular depth in the sea is equal to the weight of the water above it. Although water is nearly incompressible, the pressure at great depths is enough to cause a very slight increase in its density. If this slight increase in density did not occur, sea level would be about 30 meters higher than it is now.

OCEANIC CIRCULATION AND MOVEMENTS

The circulation and movements of the sea extend from the surface to the deepest portions. Waves of varying height cross the sea in a complex endless succession. Generated primarily by steady winds over a wide expanse of open water, waves once started are little restricted by internal friction, and so they move through a distance of hundreds or even thousands of kilometers. As they strike the shore, they expend part of their energy in erosion and transportation of rock materials.

In addition to waves with their associated breakers and surf, sea movements include rip currents, tides, the surface wind-driven ocean currents, subsurface oceanic circulation and other density currents operating at depth, and large waves or tsunami caused by earthquakes. Tides were briefly treated in Chap. 3.

Causes of Movement

The principal causes of movement are wind, storms, density changes, and earthquakes.

Wind Friction The most common cause of movement is the drag, or friction, of the wind as it passes over the surface of the water. It is this frictional force that generates the spectrum of wave motion and impels the wind-driven surface currents in the world oceans.

Storms Very severe storms sometimes cause extreme wave and associated current motion. On October 5, 1864, a violent storm off the coast of India raised sea level about 8 meters along the shore and inundated a large land area. Similar events occurred along the Washington, Oregon, and California coastlines in 1972. Swells at sea reached 6 meters in height, and intense waves killed six persons along the beaches. The storm originated in the Gulf of Alaska and generated severe wave action 48 hours after it dissipated.

Density Changes Evaporation removes great quantities of water from the ocean, increasing the salinity and hence the density of the surface waters, which then tend to sink. The effect of cooling temperatures also increases the density. Precipitation and warming have the opposite effect. Cold, dense water from the polar regions sinks and moves toward the equator. It later rises as upwellings or divergences in other areas of the oceans.

Changes in water density, whether due to evaporation, temperature changes, or suspended sediment, are the prime cause of deep-ocean circulation.

Earthquakes Seismic sea waves of great destructive capabilities have been known to sweep inland for several kilometers and to have reached heights tens of meters above normal sea level. Waves of this sort, as were mentioned earlier (Chap. 13), are caused by earthquakes. They are called tsunami.

Surface Ocean Currents

In the tropical trade-wind belts, where the winds are steady and fairly constant in direction, the surface

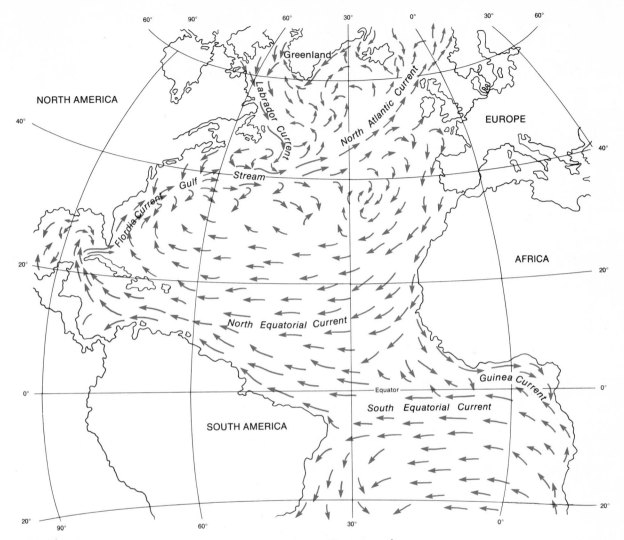

FIGURE 23.14 *Surface ocean currents in the Atlantic Ocean. The westward equatorial currents, the Gulf Stream, and the large gyral in the North Atlantic are outstanding.(After Schott, 1935.)*

waters of the sea are dragged along with the wind (Fig. 23.14). The trade winds blow from the northeast north of the equator and from the southeast south of the equator. The water is pushed toward the equator from both sides; there it drifts westward as an equatorial current, which would encircle the earth if continents did not block its path. This equatorial drift is the origin of the main currents in equatorial oceans. In the North Atlantic, the westward-drifting equatorial waters approach the east coast of Brazil, and the current is deflected along the north coast of South America. It moves toward the eastern end of the Caribbean Islands, where it divides again, part of it crossing the Caribbean Sea and entering the Gulf of Mexico. This branch emerges through the Straits of Florida as the Gulf Stream. In the North Atlantic, the Gulf Stream can be recognized as a distinct water mass by its color, temperature, and salinity. Similar currents are formed in the South Atlantic and Pacific Oceans.

The shapes of the continental shelves, the configuration of the coastlines, and the Coriolis effect due to the earth's rotation modify ocean currents and to some extent control their movements; but the winds are the primary agent in maintaining the currents.

Surface currents carry little sediment; their chief effect is climatic. The Gulf Stream, for example, supplies great quantities of heat and moisture to the British Isles and vicinity. Even where these currents

impinge on shorelines, their velocity is too small to be significant in erosion compared to the much greater effect of waves.

Subsurface Ocean Currents

Certain subsurface ocean currents are density currents. Dense, cold, polar surface water sinks in the North Atlantic off the coast of Greenland and moves southward as a deep current as far as 60°S (Fig. 23.15). In the southern latitudes, cold Antarctic water sinks and moves below it as a bottom current. Another current, the intermediate current, flows northward from 50 to 60°S to about 20°N at a depth of about 1,000 meters. Other near-surface currents sink near 40°N and 40°S

FIGURE 23.15   *North-south profile of subsurface density currents in the Atlantic Ocean, based on (a) differences in salinity, and (b) differences in temperature. (After Wüst, 1958.)*

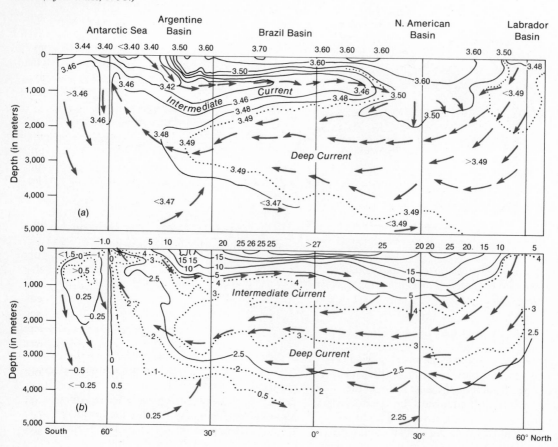

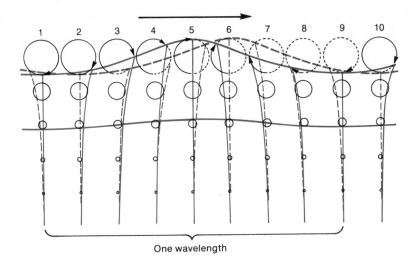

FIGURE 23.16 *Sketch of a wave to show the relative movements of the water particles at different parts of the wave form and at different depths. The orbits are drawn to scale. The dashed line shows the progression of the wave a short time later. (After Kuenen, 1950.)*

One wavelength

and flow along a depth of about 600 meters toward the equator, where they move up and begin to return near the surface. A complicated circulatory system prevails in the Atlantic, but somewhat similar systems exist in all oceans. This system is delicately balanced in dynamic equilibrium between heating, cooling, dilution by fresh water, and concentration by evaporation. It is also modified by the shape of the basin, the configuration of the ocean floor, and the earth's rotation. This deep circulation requires thousands of years to complete one cycle.

In partially enclosed basins such as the Mediterranean, special considerations must be taken into account to explain the subsurface circulation system.

Wave Motion and Wave Characteristics

Wave motion in deep water is the composite effect of the movement of adjoining water particles in nearly circular orbits. The orbits form one after another in a close overlapping succession, like a chain reaction, in the direction of wave propagation (Fig. 23.16). An object resting on the surface of the water may be seen to rise and move forward as the wave rises and then to sink and move backward as the wave subsides, without any appreciable net change in position. The same type of motion takes place within the water itself: the wave form moves forward, but except for a little water blown off the wave crest and a very slow mass transport as the orbits shift laterally, most of the

water itself does not progress. At the surface of the water the orbit's diameter represents the wave height.

Height   Wave height is measured vertically from trough to crest. Most waves on the ocean are less than 3 meters high, with some 3 to 6 meters high. Storm waves often reach heights of about 10 meters near the British Isles and 16 meters in Antarctic waters.

Length   The length of waves, measured horizontally from crest to crest, ranges from less than 1 meter on a relatively quiet lake to several hundred meters in the area of generation on a stormy sea. As the waves leave the storm area, the high storm waves gradually flatten and lengthen to become eventually the long, low waves of the swell. The swell may have wavelengths of 300 to 600 meters.

The depth affected by wave motion depends upon the size of the orbits or the wave height and the relationship between the depth of the water and the wavelength. In general, where the depth of water exceeds one-half the wavelength, the wave motion decreases with increasing depth and is absent or weak at a depth equal to one-half the wavelength (Fig. 23.16); where the depth of water is less than one-twentieth of the wavelength, the wave motion will affect the sea floor.

Period and Speed   The period of waves, i.e., the time interval between successive crests or troughs, is generally small. The ordinary, small, closely spaced waves come ashore only a few seconds apart, but in a train of long waves the crests may be 10, 12, or occasionally, as much as 20 seconds apart. The period between them is directly related to the wavelength.

The speed of deep-water waves is directly related to the period in seconds. Hence, a wave with a period of 6 seconds travels about 35 kilometers per hour. An exceptional wave, with a period of 20 seconds, will travel about 100 kilometers per hour.

Interference   Several superimposed sets of waves of different lengths, heights, and directions of travel frequently are in progress simultaneously, and so the resulting wave analysis shows irregularity rather than a rhythmic pattern. Different sets of waves interfere with one another; they nullify each other where they are completely out of phase, and they amplify each other wherever their rising and falling phases coincide.

Wave Force   Waves exhibit tremendous power according to their height and length. An average wave 2 meters high exerts a pressure of about 15 metric tons per square meter. Wave pressures of storm waves are much higher. Extreme pressures last only a fraction of a second, but the hammering of waves is repetitive.

A few examples will illustrate their power. At Wick, Scotland, a 1,200-metric-ton piece of a breakwater was shifted bodily off its foundation. Years later, a 2,350-metric-ton block of concrete, set to repair the previous damage, was torn away. At Cherbourg, France, a 3.2-metric-ton boulder was tossed over a 6-meter wall. Wave-thrown rocks have broken glass on the lighthouse at Tillamook Rock, Oregon, 45 meters above sea level, and once a 61-kilogram rock fell through the roof of the keeper's house 30 meters above sea level. Windows on the lighthouse at Dunnet Head, Scotland, 100 meters above sea level, have also been broken by rocks thrown by storm waves.

Other Motions Resulting From Waves

Breakers   Waves entering shallow water are slowed by the friction of touching bottom in water with a depth less than half their wavelength. They become progressively shorter, higher, and steeper, until their tops curl and spill over, and they finally break or collapse into a turbulent surf (Fig. 23.17). The undulating wave form is destroyed, and the wave energy carries the water forward (Fig. 23.18). Gravity then returns the excess water on the beach as backwash into the sea. Waves striking steep gravelly shores generally form plunging breakers instead of the spilling type seen on gently sloping sandy shores.

Since waves of different heights and lengths break at different depths, the position of the breakers shifts back and forth over time according to tidal heights and the state of the sea. The surf zone lies between the shore and the line of breakers.

FIGURE 23.17   *Numbered stages in the origin of breakers and surf. Long low waves at sea (1) begin to drag bottom (2) at depths equal to about one-half the wavelength and become shorter and higher on a sloping bottom (3), until the depth becomes too shallow to allow the water particles to complete their orbits. Hence the waves curl over the collapse at the line of breakers (4). The water then sloshes forward as a turbulent sheet (5) that flattens out in an uprush on the beach (6). After a brief backwash on shore, the entire sequence is repeated by the next wave. Sand on the bottom is stirred into suspension in the surf.*

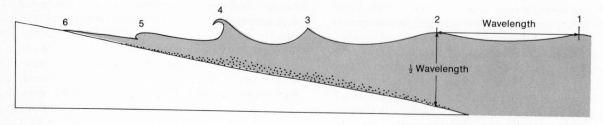

FIGURE 23.18 *Beach at Ecola State Park, Oregon. Waves break on a gently sloping bottom, water sloshes forward onto a sandy beach, and the excess water then returns to the sea as backwash. Lobate marks left by the waves can be seen on the sand. The sea shown here is relatively quiet and near low tide. (Oregon State Highway Commission.)*

Rip Currents Rip currents are narrow channels of water flowing seaward through the breakers from the surf zone. Fed by longshore currents in the surf zone, they pass through the breaker zone in narrow bands (Fig. 23.19). They may attain a speed of about 1 meter per second. They serve to discharge water and sediment brought to the beach by wave-drift currents.

Wave Refraction, Diffraction, and Reflection Waves approaching shore obliquely are bent by refraction (Fig. 23.20). While that part of the wave nearest shore touches bottom and slows down, its offshore counterpart continues forward at its regular speed. The wave front therefore pivots in shallow water and attempts to approach parallel to shore. All parts of the shore, then, whether bays or headlands, are subjected to nearly head-on wave attack. Because of refraction of waves, the attack tends to spread out and weaken in bays and to concentrate and strengthen against the sides and ends of headlands. Refraction patterns appear on many photographs of irregular, rocky shorelines (Fig. 23.20).

Waves also undergo diffraction; that is, they transfer energy parallel to their wave crests. As the waves pass the edge of an obstruction, such as the end of a jetty, the transfer of energy creates waves of a smaller height behind the obstruction.

Waves breaking against cliffs or artificial sea walls may be reflected, just as light is reflected by a mirror. Both diffraction and reflection must be considered in the design of harbors and yacht basins to minimize possible damage to small vessels and to avoid blocking the entrance with sand.

Beach Drifting Where oblique waves are incompletely refracted, the oblique push of water onshore and the return of the water almost directly downslope on the beach combine to transport rock particles in zigzag saw-tooth paths along the beach as beach drift (Fig. 23.21). This type of movement may carry pebbles many kilometers from their sources and transport sand hundreds of kilometers.

Shore Currents Oblique waves also generate currents along the shore. Some of the water near shore is driven forward as a current that parallels the general trend of the coastline. This current may have a velocity of a few meters per second and may be capable of

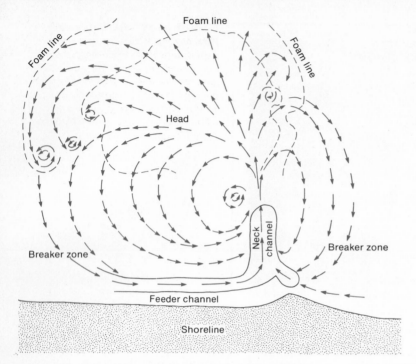

Foam line

Foam line

Foam line

Foam line

Head

Breaker zone

Neck channel

Breaker zone

Feeder channel

Shoreline

FIGURE 23.19 *Sketch of a rip current, showing a feeder channel, neck, head, and foam lines in relation to the breaker zone and shoreline. The lengths of the arrows indicate relative velocities. (After Shepard, 1972.)*

FIGURE 23.20 *Curved wave fronts resulting from refraction, Curry County, Oregon. Obstructions in the irregular shoreline cause the arcuate wave fronts. A small bay at the left is cut off by a bay bar (behind tree), and sand also has accumulated in the lee of the small sea stacks (resistant rock masses isolated by wave erosion). (Oregon State Highway Commission.)*

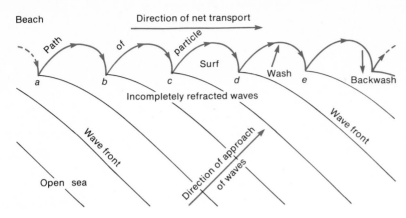

Beach

Direction of net transport

Path of particle

Surf

Wash

Backwash

a   b   c   d   e

Incompletely refracted waves

Wave front

Open sea

Direction of approach of waves

Wave front

FIGURE 23.21 *Beach drift caused by oblique approach of waves. The inclined uprush of each wave carries particles of sediment diagonally up the beach, and the backwash returns them nearly directly downslope. Thus, the particles are caught by the next swash and hurled up the beach again. The net movement is from (a) to (b), (b) to (c), and so on, but many trips from many waves are required to move more than a wavelength, instead of the single trips shown.*

dragging the bottom and transporting sand and gravel by traction. It also transports particles raised from the bottom in the surf zones, particularly sand suspended in the turbulent water.

Shore currents and beach drifting transport large quantities of sand (Table 23.4) not only along continuous beaches but also around headlands, across inlets, and into submarine canyons.

Shore Processes in Lakes   All shoreline processes of seacoasts are active in lakes, but where the waves are smaller, some processes may be less effective. Some seacoasts experience less vigorous wave attacks and smaller waves than some lake shorelines, but usually lakeshore processes are less severe.

Shoreline recession on some parts of the Great Lakes has been as much as 2 to 3 meters per year in poorly consolidated materials. Shorelines are particularly susceptible to attack during periods of high water produced by seasonal variations in water level.

Wave attack may be enhanced or reduced by a seiche, which is a periodic tilting of the entire surface of the lake, similar to the movement of a single wave in a sloshing tub of water. Wind and passage of atmospheric high- and low-pressure areas are the most common causes of seiches, although crustal movements and earthquakes also cause them. A seiche was set in motion in Kenai Lake, Alaska, by the March 1964 earthquake. A seiche started by a shift in the bottom of

Hebgen Lake, Montana, during the 1959 Madison Canyon earthquake overtopped Hebgen Dam several times. The surface of a lake is horizontal only when the wind and pressure differences permit it to come to rest. Atmospheric pressure differences at opposite ends of large lakes will cause the water to go down in the area of high pressure and up in the area of low pressure. An oscillation of the water may continue for days. Wave erosion is enhanced during the high of a seiche because the deeper water offshore reduces bottom friction. The same is true at high tide on a seacoast. Wave erosion is reduced by the increased bottom friction during a low tide.

TABLE 23.4   Quantities of Sand Moved Along-shore

Place	m³/year*	Date
Sandy Hook, New Jersey	377,145	1885–1933
Sandy Hook, New Jersey	333,540	1933–1951
Absecon Inlet, New Jersey	306,000	1935–1946
Galveston, Texas	334,687	1919–1934
Oxnard, California	765,000	1938–1948
Port Hueneme, California	382,500	1938–1948

SOURCE: Johnson, J. W., 1957, *The littoral drift problem at shoreline harbors,* Proceedings of the American Society of Civil Engineers, Waterways and Harbors Division, vol. 83, paper 1011.
*Each cubic meter weighs about 1.8 metric tons.

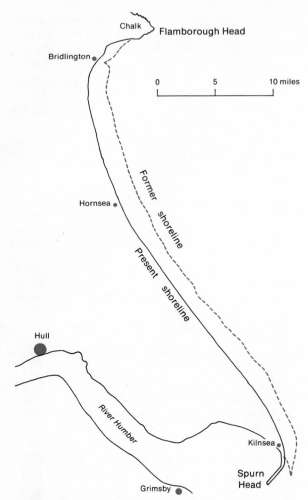

FIGURE 23.22 *Shore erosion on the coast of Yorkshire, England, where about 30 townsites have been eroded away since Roman times. Flamborough Head on the north is composed of resistant chalk. Spurn Head on the south is a spit. The intervening smooth shore line is a cliff of easily eroded glacial till, sand, and gravel. (After Steers.)*

SHORE EROSION

Methods

Waves erode the shores in several ways. In unconsolidated rock, such as clay, silt, sand, gravel, or glacial till, "washing" action of waves is enough to loosen and remove the rock. In more resistant solid rock, considerable force is necessary to bring about erosion. Solid rock is commonly eroded from coasts by pounding waves that create hydraulic wedges in the cracks until the rock yields block by block. This is a type of quarrying action. Another method by which waves erode coasts is abrasion. Waves with suspended sand and gravel abrade a rocky shore and wear it away. The action of waves cuts notches or slits and undermines the rock standing above the direct reach of waves. Seawater also attacks limestone shores by solution. In constricted passages the ebb and flow of tides may scour the bottom.

Shore Retreat

The general effect of wave erosion is an encroachment of the sea upon the land (Fig. 23.22). In loose or closely fractured solid rock, the rate of retreat locally may be several meters a year.

An example of rapid wave cutting is furnished by the island of Helgoland in the North Sea. Its perimeter was reduced by wave erosion from 200 kilometers in the year 800 to only 5 kilometers in 1900. Its area decreased to about 0.6 square kilometer before protective steps were taken to preserve it for a German naval base before World War I.

The Holderness coast of Yorkshire, England, was eroded along a 60-kilometer front between 1852 and 1952 at an average rate of 1 meter per year and locally as much as 3 meters. The sea cliff is composed of relatively weak glacial sediments and averages 15 meters high. The loss since Roman times is estimated to have been equal to a strip of land 4 kilometers wide (Fig. 23.22).

Differential Erosion

Most solid rocks erode unevenly. As waves incessantly pound the shore, they reveal differences in rock resistance owing to differences in degree of induration, fracturing, and abrasive hardness. As a result of this unequal erosion, the shoreline becomes notched or minutely jagged in plan view on a map. Conversely, where the shore is composed of rocks having equal resistance, it is eroded away evenly, creating a generally smooth coastline (Fig. 23.22).

Wave erosion produces a characteristic array of shore features. One of the most typical is a wave-cut

FIGURE 23.23 *Sea cliff and caves at La Jolla, California. The cliff has been undercut by waves. (Arnold, U.S. Geological Survey.)*

The boulders, cobbles, pebbles, and sand used as erosional tools by waves are also eroded (Fig. 23.28). They become well rounded and reduced in size by attrition. They generally undergo sorting by grain size and specific gravity during their repeated movements. This sorting may concentrate magnetite, chromite, gold, platinum, tin ore, tungsten ore, zircon, garnet, and rare-earth minerals in sufficient quantity to produce mineable deposits in many places.

SHORE DEPOSITION

The materials broken by waves and shore currents may be moved ashore by strong waves or tranported into deeper water beyond wave reach offshore. Often, the material may be moved along the coast into a bay

FIGURE 23.24 *The wave-notched base of a sea cliff near Nanaimo, British Columbia. (Canada Department of Interior.)*

cliff (Fig. 23.23). A cliff cut in solid rock commonly has a large, horizontal, basal notch along the general plane of direct wave attack (Fig. 23.24). By eroding a weak spot in the lower part of a wave-cut cliff, waves may make a wave-cut cave or arch. A rupture through the roof of such a cave allows water to surge up through the opening, especially at times of storm or high tide, as a blowhole (Fig. 23.25). Eroded fracture zones become joint chasms (Fig. 23.25).

Any resistant mass isolated from the mainland by erosion of the intervening rock becomes a sea stack or chimney (Fig. 23.26). The enlargement of a joint and the collapse of sea arches help to isolate sea stacks. Indentations larger than joint chasms eroded into areas of weak rock become coves and their counterparts, the seaward projections of resistant rock, remain as headlands.

As erosion advances upon the land by disintegrating the rocks along the shore, the area of beveling widens into a plane surface called a wave-cut terrace, which is partly exposed at low tide (Fig. 23.27). This surface generally shows numerous irregularities due to differential erosion.

FIGURE 23.25 *Joint chasm being eroded along a fracture in basaltic lava. (Oregon State Highway Commission.)*

FIGURE 23.26 *Group of sea stacks, Bandon, Oregon. Wave erosion has removed the less resistant rock that formerly surrounded them. (Oregon State Highway Commission.)*

FIGURE 23.27 *Elevated sea terrace, sea cliff, and wave-cut terrace exposed at low tide, near Devils Punch Bowl State Park, Oregon. Both terraces occur on tilted layers of sandstone and mudstone. The continued advances of the sea formed both terraces, the upper one before uplift of the coast and the lower one (still forming) later. (Oregon State Highway Commission.)*

FIGURE 23.28 *Beach deposit of wave-rounded cobbles near Newport, Oregon.*

435

or quietly allowed to settle as the waves and currents slacken.

Forms of Deposits

The principal shore deposits are beaches, spits, bars, barrier beaches, and wave-built terraces (Fig. 23.29). Not only are these features common along the present coastlines and lake shores, but apparently, they were also common along ancient shores. Some of these ancient shore deposits, now preserved as a part of the rock record, are important sources of petroleum and natural gas. Having an open texture, these fossil shore

FIGURE 23.29 *Forms of shore deposits. (1) Spit; (2) baymouth bar; (3) recurved spit, or hook; (4) loop; (5) cuspate spit; (6) tie-bar, or tombolo.*

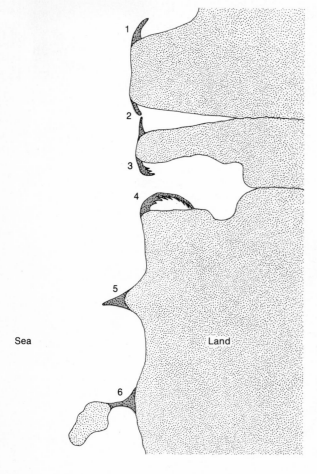

Sea

Land

deposits serve as excellent reservoir rocks for the accumulation of such fluids.

Beach A typical beach deposit is a ridge or embankment of gravel or sand that slopes seaward (Figs. 23.30 and 23.31). Beach deposits may be removed and later replaced by seasonal changes in the state of the sea. Deposition on the seaward side may widen a beach area to hundreds of meters or even kilometers, usually by deposition of multiple beach ridges.

Spit A typical spit is an offshore continuation of a beach from a point of land or into the mouth of a bay. Sedimentary particles repeatedly washed onto a beach by the diagonal swash of oblique waves are carried directly down the shore face by the backwash (Fig. 23.21). In a series of such downbeach motions, the sediment particles are forced to travel along the beach until they come to the edge of a land area. Here they develop, first as a shoal and later as an accumulation of new land. The accumulation grows into a long spit by continued beach drifting and by the drag of currents parallel to shore (Fig. 23.29). The rate and extent of shore drifting can be measured by tracing the movement of distinctive materials, such as a particular rock type, radioactive sands, or introduced fluorescent sands (Table 23.4).

Some spits are blunted on their growing ends by refracted waves or by cross currents. As a result, their tips are bent into recurved spits or hooks (Fig. 23.29). Occasionally, a hook is doubled back completely, becoming attached to land at both ends as a loop. Spits connecting islands to the mainland or to each other are tombolos (Figs. 23.29 and 23.32). Those that extend almost or entirely across the outer end of a bay become baymouth bars (Figs. 23.29 and 23.33). Spits commonly develop on both sides of a headland truncated by wave erosion.

Bar The term bar is applied loosely to almost any offshore embankment of loose sand and gravel deposited by waves or currents. In addition to specific landforms such as spits, tombolos, and baymouth bars, many bars are merely submerged shoals of varying configuration.

Some bars and spits have crescentic outlines. Deposits of this nature can be seen projecting offshore

FIGURE 23.30 *A shingle beach at Three Island Cove, Cape Breton Island, Nova Scotia. A shingle beach consists of disk-shaped, flattish beach pebbles that overlap one another like shingles on a roof. (Geological Survey of Canada)*

FIGURE 23.31 *A bayhead beach, Emerald Bay. Laguna Beach, California. The highway is on a sea terrace. (Spence Air Photos.)*

FIGURE 23.32 *Land-tied islands, Spruce Head Island, Maine. The sand bar tying the island to the shore is a tombolo. (Bastin, U.S. Geological Survey.)*

FIGURE 23.33 *Baymouth bars and coastal lagoons near Eureka, California. The lagoon in the center is cut off from the Pacific Ocean by a sand bar. A similar lagoon occurs above (south). This type of sand bar has been formed from a spit that was elongated to cut off a bay. Baymouth and other sand bars have the effect of simplifying a shoreline. (Fairchild Aerial Surveys, Inc.)*

from Cape Hatteras, North Carolina, and Cape Canaveral, Florida (Fig. 23.34).

A barrier beach is a long offshore bar trending parallel with the coast. It is separated from the mainland by a lagoon (Fig. 23.35) and is usually capped by hills of wind-blown sand. It is built up by waves that strike broadside against a gently sloping, sandy sea floor. A barrier beach may be attached to the mainland at one end or it may be an island. Inlets through it serve as tidal channels into and out of the lagoon. Tidal currents flowing through such channels may build tidal deltas in the quiet water behind a barrier beach. In the long course of time, waves, longshore currents, and winds may slowly move a barrier beach landward or along shore. Many popular beaches, like the one at Atlantic City, New Jersey, are barrier beaches.

The deposition of spits, baymouth bars, and barrier beaches tends to smooth or simplify an irregular coastline.

Wave-Built Terrace   The deposits washed seaward off the beach often accumulate below the level of wave agitation. If dumped into deep water on a steeply sloping sea bottom, they come to rest in an inclined position. If spread over a flat sea floor, they tend to build up a wave-built terrace below the level of wave base. Because of sea-level changes caused by glaciation during the Pleistocene epoch, wave-built terraces are generally lacking adjacent to the present coastlines. Good examples can be seen, however, along the exposed shores of former lakes.

Lakes of the Oceans and Shores

The combined action of wind and waves along the shores of the oceans, seas, and large lakes sometimes creates other lakes, ponds, and lagoons (Fig. 23.36). This is accomplished by building spits and bars across the entrances of bays or from a headland across a recess of the original shoreline. Such lakes occur along the south shore of Lake Ontario and along the Atlantic coast from New England to Panama.

When uplift or shore processes cut off an estuary or embayment of the sea, the water commonly is freshened by inflow from streams until a fresh-water lake is formed. Lake Maricaibo, Venezuela, and Lake Nicaragua have fresh-water sharks, suggesting a former connection with the sea.

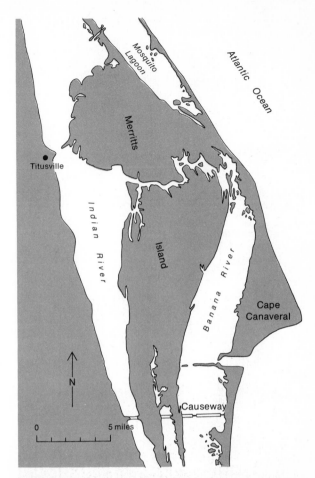

FIGURE 23.34   *Cape Canaveral, Florida, a cuspate shore deposit.*

SHORE CONTROL

The artificial control of shoreline erosion and deposition presents many problems. To prevent undue erosion of shores and the destruction of valuable seaside property, man builds various artificial structures to block wave and current action. These structures afford only limited protection against a stormy sea, however, and their maintenance is generally expensive. In some places they upset the previous behavior of waves and shore currents. In some localities unforeseen erosion has set in against a previously stable coast. In other places, undesirable deposition of sand

FIGURE 23.35 *Model of the New Jersey coast at Barnegat Bay, showing a long barrier beach, an offshore sand ridge built up by waves and currents. (Aero Service Corporation: Division of Litton Industries.)*

has occurred. Some structures are designed to induce deposition. They are used to catch additional sand where beach prograding is wanted.

The building of jetties at Lake Worth Inlet, Florida, caused deposition on the north (upcurrent) side and starved the beaches farther down the coast on the lee side so that erosion set in near Palm Beach (Fig. 23.37). The problem has been met by pumping sand past the inlet.

Construction of a breakwater in 1929 to protect the harbor at Santa Barbara, California, caused 600,000 tons of sand per year to collect in front of the breakwater. By 1934, the capacity of the site was reached, and sand began to go around the end of the breakwater to fill in the harbor (Fig. 23.38). By 1952, more than 9 million metric tons of sand had accumulated in the harbor behind the breakwater. Sand now is pumped from the harbor continuously and fed to the beach farther south, where beach erosion had become serious.

FIGURE 23.36 *Fresh-water pond formed in a drowned river mouth blocked by shoreline deposits along the California coast. (Wm. C. Bradley photo.)*

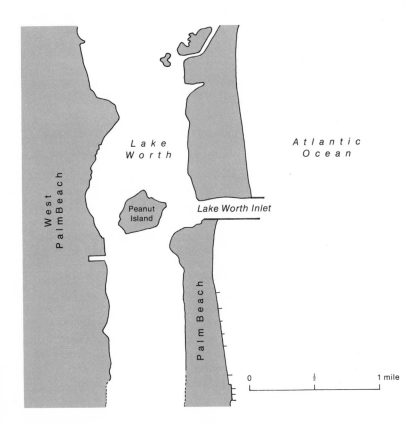

FIGURE 23.37 *Lake Worth Inlet, Florida, where artificial bypassing of sand is practiced in order to maintain the beach at Palm Beach with the aid of groins.*

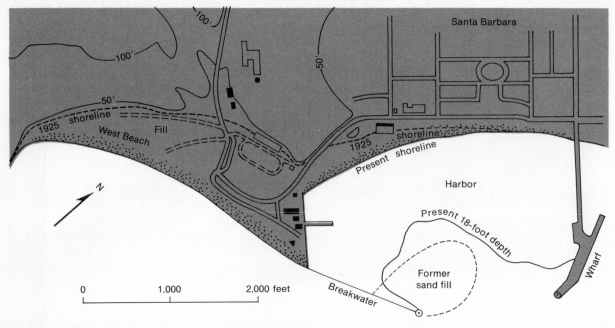

FIGURE 23.38 *Accumulation of beach-drift sand at Santa Barbara, California, mainly west of the breakwater and around its end where dredging is necessary.*

At Santa Monica, California, a 600-meter breakwater was built about 600 meters offshore to provide shelter for small ships. However, the breakwater created a wave shadow behind it, and this resulted in an accumulation of sand onshore. The sand first formed a point on the beach, and after 30 years the fill grew to be several hundred meters wide.

Despite the difficulties cited, engineers usually can select a design appropriate to a particular setting with reasonable success and take steps to counteract any undesirable changes in the regimen elsewhere. Scale-model studies made before construction help to avoid costly mistakes.

Artificial addition of sand to beaches is easily managed. Structures built perpendicular to shores across the zone of beach drift have been used successfully to cause accretion of sand alongshore (Fig. 23.39.)

The remarkable beach of the Mississippi coast at Biloxi and Gulfport is artificial. Sandy mud is pumped onshore, where gentle waves remove the fine particles and leave the sand component to form a beautiful beach. This beach, which is 30 kilometers long, is the longest artificial beach in the world. Beyond 30 meters offshore in water about 1 meter deep, waves have not winnowed out the fine components, and so the sea floor there is soft and muddy.

Sand has been dredged or pumped from nearby sources to supply artificial nourishment of beaches on the coasts of Long Island, New Jersey, Florida, Lake Michigan, and California. Bypassing of sand past jetties and inlets is now carried out routinely at many places.

FIGURE 23.39   *Beach-drift groins on California coast near Santa Monica. (Spence Air Photos.)*

REFERENCES

Bascom, W., 1964, *Waves and Beaches,* Anchor Books, Doubleday & Company, Inc., New York.

Carson, R.L., 1951, *The Sea Around Us,* Oxford University Press, New York.

Fairbridge, R. W. (ed.), 1966, *The Encyclopedia of Oceanography,* vol. 1, Reinhold Book Corporation, New York.

Groen, P., 1967, *The Waters of the Sea,* D. Van Nostrand Company, Inc., Princeton, N.J.

Gross, M.G., 1972, *Oceanography,* Prentice-Hall, Inc., Englewood Cliffs, N.J.

Guilcher, A., 1958, *Coastal and Submarine Morphology,* Methuen & Co., Ltd., London.

Keen, M. J., 1968, *An Introduction to Marine Geology,* Pergamon Press, New York.

King. C.A.M., 1962, *Oceanography for Geographers,* McGraw-Hill Book Company, New York.

Kuenen, Ph.H., 1950, *Marine Geology,* John Wiley & Sons, Inc., New York.

Menard, H.W., 1964, *Marine Geology of the Pacific,* McGraw-Hill Book Company, New York.

Moore, J.R. (ed.), 1971, *Oceanography* (Readings from Scientific American), W.H. Freeman and Company, San Francisco.

Ross, D.A., 1970, *Introduction to Oceanography,* Appleton Century Crofts, New York.

Shepard, F. P., 1959, *The Earth Beneath the Sea,* The Johns Hopkins Press, Baltimore.

———, 1972, *Submarine Geology,* (3d ed.), Harper & Row, Publishers, Incorporated, New York.

————and Wanless, H.R., 1971, *Our Changing Coastlines,* McGraw-Hill Book Company, New York.

————and Dill, R.F., 1966, *Submarine Canyons and Other Sea Valleys,* Rand McNally & Company, Chicago.

Turekian, K. K., 1968, *Oceans,* Prentice-Hall, Inc., Englewood Cliffs, N.J.

Weyl, P. K., 1969, *Oceanography, An Introduction to the Marine Environment,* John Wiley & Sons, Inc., New York.

CHAPTER 24

THE WORK
OF WIND

Previous chapters have noted the effects of wind in accelerating evaporation, distributing rain clouds over the land, modifying temperatures, and generating waves and currents at sea. The wind also acts as an erosional agent.

The gradational results of wind are based not so much upon the strength of the wind as upon the great volume of air involved. The geologically active or turbulent part of the atmosphere, the troposphere, is about 6 kilometers thick at the poles, 11 kilometers in middle latitudes, and 18 kilometers at the equator. Ordinarily, dust is confined to the lower levels of this volume of atmosphere, but atomic explosions and explosive volcanic eruptions can carry dust to the uppermost levels of the atmosphere and even into the stratosphere beyond.

A thin film of dust transported by the wind does not seem significant until a sample is weighed from a measured area and the total tonnage for the earth is computed. One gram, or $\frac{1}{28}$ ounce, of dust (about the quantity one can heap on a dime) per 0.1 square meter, adds up to about 22 metric tons per square kilometer. At Lincoln, Nebraska, in March 1935, 5 times this quantity (5 grams per 0.1 square meter) accumulated in 4 days. An additional 9 grams per 0.1 square meter fell in the area in a 2-week period in April. In a severe dust storm, visibility may be considerably reduced (Fig. 24.1).

A dust storm beyond earth comparisons occurred on the planet Mars during the 2-month period September to November 1971. This storm was recorded in photographs by Mariner 9. The dust storm reached thicknesses of 6 kilometers and encompassed the entire globe for nearly the entire 2-month period.

The Martian dust, which covers vast areas of the planet is estimated to be 10 microns (0.010 millimeter) in diameter, comparable to finely powdered talc. The atmosphere of Mars is $\frac{1}{100}$ as dense as that of earth; therefore, winds of 160 kilometers per hour (100 miles per hour) would be needed to begin a dust storm, even in the fine Martian dust. The dust storms are most common and severe when the planet is closest to the sun, indicating that intense convectional heating may cause the winds. Large dune fields, over 135 kilometers long, form as the dust is deposited. It must be apparent from the above examples that wind as a geologic agent is most effective where moisture and precipitation are at a minimum.

FIGURE 24.1 *An approaching dust storm in western Oklahoma. Such a storm may carry dust and sand a great distance from their origin. (Pictures, Inc.)*

FIGURE 24.2 *Cave rocks near Sierra La Sal, Dry Valley, Utah. In this wind-swept plain and mesa area, wind is a powerful eroding agent. Loose material is carried away by the wind as fast as it is weathered. Note the scoured and grooved section of bedrock. (Jackson, U.S. Geological Survey.)*

WIND EROSION

Deflation

The wind erodes loose dry materials by deflation. The impact of the wind removes small particles of suitable clay and silt size according to the competence of the wind. These particles are lifted from the ground by the eddies, whirlwinds, and updrafts of the wind and are kept suspended by the turbulence of the air.

The extent of deflation is not easily measured. Generally, only a small amount of material is removed at any one time, and that small amount comes from a large surface area. Locally, however, deflation can be readily apparent. In the American southwest, loose fine material is swept off great expanses of bare rock, with the assistance of rainwash, often as fast as weathering loosens it (Fig. 24.2). During the "dust-bowl" years of the 1930s, the soils of large areas of marginal farmland on the Great Plains were severely damaged by deflation.

Deflation Basins   Basins eroded by the wind are especially impressive evidence of deflation. Dried-up lake floors, underlain by silt, volcanic ash, and dia-tomite, in south-central Oregon and eastern California, have had basins carved into them by the wind (Fig. 24.3). These basins are generally broad and shallow, but some reach depths of 14 meters. They range in size from less than an acre to a square kilometer or more. Lake sediments in Danby Playa, California, have been eroded to a relief of 4 to 5 meters over an area of 26 square kilometers.

Hundreds of similar basins occur in fine-grained bedrock on the western Great Plains. One basin near Laramie, Wyoming, called Big Hollow, is 15 kilometers long, 5 kilometers wide, and 30 to 50 meters deep.

Still larger and deeper deflation basins are found in the Gobi Desert of Asia, where undrained depressions up to 50 kilometers long and 30 meters deep occur in disintegrated granite. Fine particles of feldspar and quartz loosened by weathering were blown out at a time when the climate was drier than it is now.

Similar depressions in the Libyan Desert of northwestern Egypt have also been attributed to deflation, although some faulting may have been involved also. Seven of them have depths between 200 and 500 meters, The Qattara Basin, about 18,000 square kilometers in area, has an average surface level of 67

FIGURE 24.3   *Deflation of soft lake beds near Fossil Lake, Lake County, Oregon. Wind erosion of loose material has hollowed out a broad basin, leaving remnants. The highest of these remnants, protected by hard layers or shrubs, is about 2 meters high. In the foreground, fragments of hard rock, too heavy to be blown away, remain as lag gravels, or residuals.*

meters below sea level and an extreme depth of 134 meters below sea level (Fig. 24.4). Whether or not faulting contributed to this relief, wind erosion has at least maintained and enlarged the basin.

Another feature of deflation occurs when particles too large or too heavy for the wind to remove by deflation lag behind, and so they are called lag gravels (Fig. 24.3). The continued accumulation of granules, pebbles, and larger fragments may in time virtually cover the ground with a close-set pattern of loose fragments that constitute a desert pavement. Such a cover prevents further deflation of the underlying finer material.

Sandblasting

The wind erodes solid rocks by abrasion, i.e., the use of a natural sandblast mechanism comparable to the technique man uses to clean metal castings or stone buildings. Damage to automobile windshields and paint by a single sandstorm, the "frosting" and ultimate destruction of glass windows exposed along a sandy shore, and the marring of the pyramids in Egypt may be familiar to the reader. Similar sandblasting intermittently over a long period of time may etch parallel grooves or long shallow grottoes around the base of sandstone hills or along inclined bedding (Fig. 24.5), thereby adding more sand to the wind.

Some sand- or siltblasted rocks develop a polish, or more commonly, a somewhat dull surface. On certain fine-grained massive rocks, such as quartzite, the polish may be highly lustrous. Wind-abraded sand grains develop a mat or frosted surface as a result of chipping, pitting, and minute cracking.

Pebbles subjected to natural sandblasting are polished (if uniformly hard) and pitted and grooved (if uneven in resistance). Given time enough, they become faceted with one or more flat faces to form ventifacts or dreikanter (Fig. 24.6).

Many odd-shaped landforms are evolved by the combined effects of weathering, other erosional processes, and perhaps to some degree, sandblasting. These landforms may include table rocks, pedestals, or similar sculpturings. In open areas, a complex of small fluted and undercut ridges, called yardangs, and intervening grooves several meters wide may be made on a bare rock surface. Except for aiding in the development or intensifying of these features, the quantitative effects of wind abrasion are relatively minor, as compared with deflation.

WIND TRANSPORTATION

Generally, dust particles are carried in suspension, and sand grains are carried along the surface of the ground. Exceptionally, very fine sand also may be lifted hundreds of meters off the ground.

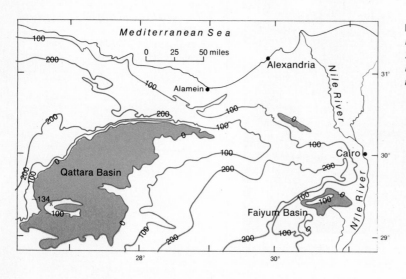

FIGURE 24.4 *Wind-eroded basins of northern Egypt. The huge Qattara Basin, 282 kilometers long, reaches 134 meters below sea level. The contours are in meters. (After Ball, 1933.)*

FIGURE 24.5   *Wind-eroded sandstone. The bedding planes of the sandstone have been etched into relief by wind scouring. (Gregory, U.S. Geological Survey.)*

FIGURE 24.6   *Wind-polished and faceted pebbles, or ventifacts, from the Big Horn Basin, Wyoming.*

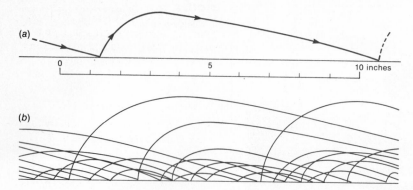

FIGURE 24.7 *(a) Path of an individual sand grain in saltation. (b) Paths in a cloud of saltating sand grains. (After Bagnold, 1941.)*

Wind-blown sand grains carried along the surface may bounce at high angles a few centimeters to a few meters off the ground and are impelled forward in curved trajectories by the higher wind velocities off the surface so that in falling back they land forcefully at angles of 10 to 16°. They not only ricochet upon landing, but they also bump into other grains in a continuous chain reaction of saltation (Fig. 24.7). Thus the sand moves forward jerkily in an irregular fashion. In a strong wind the moving cloud of sand grains may reach as high as 2 meters, but ordinarily it is less than a meter.

The wind's ability to transport particles is conditioned by its velocity and turbulence. A light wind can carry dust in suspension, and a gentle breeze can roll fine sand. However, a strong breeze with a velocity of 22 meters per second can move grains as large as a millimeter in diameter. Gales and hurricanes can carry sand in suspension to heights of hundreds of meters and can roll pebbles 5 to 7 centimeters in diameter. The load carried by the wind is a direct function of its velocity.

Sources of Load

The sources of the wind's load are several. The principal source is probably the rock waste formed by weathering, disintegrating sandstones, the flood-plain and sand-bar deposits of rivers, glacial moraines, and beach sands also contribute (Fig. 24.8). The deposits of dried-up lakes commonly serve as immediate sources. Volcanic explosions supply tremendous quantities of light-weight, highly angular rock material that is well suited to the capacity of the wind.

Suspended Load

The quantity of material carried in suspension in dust storms is very large, as shown by the measurements at Lincoln, Nebraska, cited previously.

Changes Enroute

Sand grains undergo notable changes during their transportation by the wind, in addition to the frosting already mentioned. Rolling or bouncing along the surface of the ground, they are subjected to mutual wear (corrasion), which reduces their sizes and rounds their shapes, even in the very fine sizes. Repeated winnowing also sorts the sand so that in time a large portion of a wind-blown sand deposit falls within a few size grades. Dust is removed, and coarse particles are left behind.

By contrast, dust particles carried in suspension, where collisions are ineffectual, retain their sizes and angularity.

Range of Transportation

The distances traveled by wind-borne particles vary over a wide range. The distant spread of volcanic dust from Krakatoa, Vesuvius, and Katmai was noted in an earlier chapter. Ordinary rock dust, lifted by dust storms, has been carried in a few days from Australia to New Zealand, from the Gobi Desert in Asia to ships off the coast of Japan, from the Sahara Desert to northern Europe, and from the southwestern United States to the Atlantic seaboard.

Sand moving on the ground ordinarily travels only a few kilometers. In France, however, sand has been blown 8 kilometers inland from the shore, and the great sand deposits of the Sahara Desert lie on a limestone plateau 160 kilometers from the outcrops of sandstone that apparently disintegrated to supply the sand. Wind-blown material from the Sahara has been found in deep-sea sediments off the Caribbean Islands area, over 4,500 kilometers away.

WIND DEPOSITION

The principal causes of deposition by the wind are slackening or cessation of the wind, precipitation of rain or snow, and the presence of obstructions.

Slackening of the wind allows deposition of both dust and sand. The particles settle out because of gravity, as decreased turbulent-air velocity causes a loss of the particles' kinetic energy.

Precipitation clears the air by washing. Occasionally, the moisture-laden dust is sufficient to make the rain muddy or to discolor snow.

Obstructions apply mainly to drifting sand. The obstructions may be of any sort, single boulders, rock pinnacles, rock hills, ridges, fences, buildings, bushes, a tree, or a forest. Sand accumulates on both the windward and lee sides of many obstructions. A

hummock of sand once started acts as a self-made obstruction that thereafter tends to grow in size by encouraging additional deposition upon it. Locally, sand may be blown out of reach into a pond or lake or drifted over the edge of a cliff.

Forms of Eolian Deposits

Because dust is carried in suspension and sand by movement along the surface, their deposits are segregated into widespread sheets of deposited dust or local piles of sand. Both are eolian deposits. Where the wind-deposited dust has an appreciable thickness, it is called loess. The natural piles of sand are dunes.

Loess Loess deposits are widespread in the central Mississippi River basin and the Great Plains. Substantial deposits of loess also occur in the Palouse Hills of eastern Washington, along the Ohio River, in north-central and eastern Europe, the Sudan, Argentina, and China. The fine-grained, friable, and fertile deposits of loess coincide with the areas of the world's wheat production.

The Mississippi Valley loess ranges in thickness from less than a meter to many tens of meters. It consists mostly of silt-sized particles of a great variety of

FIGURE 24.8 *Sand dunes on the Oregon coast encroaching upon a forest. (National Park Service.)*

FIGURE 24.9 *Beach sand blown inland by wind along coast of California south of Point Sal. A beach is a common source of wind-borne sand. (Spence Air Photos.)*

minerals. These include quartz, feldspar, calcite, hornblende, and mica, all of which are generally in a fairly fresh, little-weathered state. It overlies glacial drift or various older rocks where drift is lacking. Although generally uncemented, the loose grains of loess are angular enough to interlock so that sides of excavations dug into loess may stand in nearly vertical cliffs for years. Tubules left by the deep roots of grasses and commonly coated with calcium carbonate give the loess a vague columnar jointing.

Loess is deposited under either cold-dry or warm-dry conditions and from either glacial or desert sources. Glacial loess, as in Iowa, apparently was derived from the dried-up flood-plain streams that carried glacial meltwaters or was blown directly off the Pleistocene ice sheets themselves. The thickest deposits lie downwind on uplands adjacent to the Missouri and Mississippi Rivers. They thin toward the east. The glacial flood plain must have been the major source. The dust may well have been caught and held in place by vegetation. Loess is now being deposited in a similar setting along most braided glacial streams in Alaska.

Except along the Missouri River, the loess of Nebraska and Kansas thickens toward the Sand Hills area of northwestern Nebraska. It evidently was de-

posited in part by ancient dust storms of the desert or nonglacial dust-bowl type.

The loess of northwestern China, said to be hundreds of meters thick, presumably was derived from the desert of central Asia under warm-dry conditions. Having been dissected by later stream erosion, it was subject to disastrous landslides triggered by a severe earthquake in December 1920. About 100,000 persons lost their lives in the collapsing hills of silt. The Yellow River in China acquired its name because of the coloration of the fine-grained loess the river carries in suspension.

Dunes Dunes are hills or ridges of wind-blown sand, generally in motion. Some sand deposits are stationary because of their relation to some fixed obstruction or because they have become fixed by growth of vegetation.

The sand in dunes comes from sandy beaches (Fig. 24.9) or other shore features, sandy flood plains or terrace deposits of streams, plains of sandy glacial outwash, sandy portions of alluvial fans (Fig. 24.10), dry lake floors, or the disintegration and winnowing of previous sandy sediments.

The most abundant mineral in dune sands ordinarily is quartz, but mixtures of minerals are common.

Dunes covering about 647 square kilometers in the White Sands area of New Mexico are composed of gypsum (Fig. 24.11), and calcite dunes (derived from coral) occur on Bermuda.

The migration of a dune results from the transfer of sand from the windward side to the dune crest, where the sand rolls or slides down the slip face in the wind shadow on the leeward side (Figs. 24.12, 24.13, and 24.14). The windward side usually slopes less than 10° and the leeward side about 25 to 34°. Many dunes shift downwind only a few meters a year. Dunes along the Bay of Biscay near Bayonne in southwestern France have been advancing landward at a steady rate of 1 to 2 meters per year for decades. On the southeast coast of the Baltic Sea, a dune advanced 13 kilometers in about 70 years. A few small dunes on bare rock floors in areas of strong winds travel hundreds of meters in a year. Isolated dunes in Asia have traveled as much as 20 meters on an extremely windy day. Dunes in Imperial Valley, California, migrated an average of 27 meters per year over one 7-year period.

Dunes generally range in height from less than 1 meter to more than 10 meters. Locally, they reach hundreds of meters, as in Great Sand Dunes National Monument in Colorado and in the Sahara Desert.

Dunes develop diverse shapes, according to the supply of sand, the lay of the land, the restricting vegetation, and the steadiness of the direction of the winds. The principal types are lee dunes, or sand drifts, other longitudinal dunes, barchans, seifs, transverse dunes, and complex dunes.

(*a*)   *Lee Dunes*   Lee dunes are longitudinal dunes that develop as long narrow sand ridges in the shape of a wind shadow behind a rocky ledge or behind clumps of vegetation or as sloping embankments blown over a cliff.

All these lee dunes are constant in position, but with the addition of great quantities of sand, typical migratory dunes may form and move away from their tail ends.

(*b*)   *Other Longitudinal Dunes*   Dunes are also elongated in the direction of the effective winds, where strong winds blow across areas of scanty sand, or where winds have to contest with the holding effects of

FIGURE 24.10   *Death Valley near Stove Pipe Wells, California. Alluvial fans have accumulated on the left, and seiflike sand dunes occur in the foreground. (Spence Air Photos.)*

FIGURE 24.11 *Dunes made of gypsum at White Sands National Monument, New Mexico. Although most dunes consist chiefly of quartz grains, there are exceptions, such as these gypsum dunes. (New Mexico State Tourist Bureau.)*

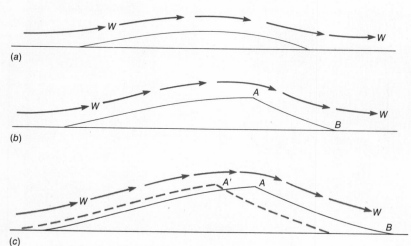

FIGURE 24.12 *(a) Longitudinal profile of an early sand-pile stage of a dune, streamlined in the direction of the wind (W) but with little or no distinctive plan. (b) Longitudinal profile of a sand dune with a definite crest (A) and a slip face (A to B) in the wind shadow. (c) Longitudinal profile of a sand dune migrating with the wind (W) from A' to A by transfer of sand from the windward side to the lee slope.*

FIGURE 24.13   *Sand dunes on the east side of Imperial Valley, California. They are mostly crowded barchans (crescent-shaped dunes). Note the ripple-marked, gentle, windward slopes and the steep slip face on the leeward side. (Frashers, Inc.)*

FIGURE 24.14   *Sand dunes, with well-marked ripples, on the east side of Imperial Valley, California. The steep, advancing front of a dune is marked by sliding sand on its slip face (on right). (Frashers, Inc.)*

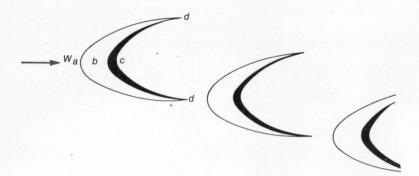

FIGURE 24.15 *Outline sketch of barchans, such as are found in the Arabian Desert, showing wind direction and the echelon arrangement of dunes. (a) Windward margin; (b) crest; (a-b) windward slope; (b-c) slip face; (d) horns, or points of wings; (W) wind direction.*

grass or small shrubs. Oblique winds are most effective to move the sand into ridges. These fluted dunes grade into other forms. Some are blown up the slopes of valley walls.

(c) Barchans   Barchans (Fig. 24.15) form in open areas, unrestricted by topographic or vegetational barriers, where the wind direction is fairly constant and the sand supply is limited, particularly over a solid rock base. A typical barchan is crescent shaped, with the gentle windward slope on the outer curve and the points, or wings, of the crescent drawn out with the wind on the leeward side in a streamlined form.

(d) Seifs   A seif is a longitudinal dune considerably modified by quartering winds. One simple variety resembles, in plan outline, an Arabian sword (Fig. 24.16). It is associated with barchans, and conversions of barchans to seifs, and vice versa, are said to occur. Another variety is essentially a lee dune with a knife-edge crest between one rounded windward side and a steeper slip face on the other. A third variety is a Saharan sand ridge with a sharp, broadly wavy crest. The forms of seifs, especially their crests, frequently change when the wind shifts from its prevailing direction. Seifs are especially well developed in the Arabian, Libyan, Saharan, and Australian deserts.

(e) Transverse Dunes   Transverse dunes form

where the sand supply is copious and the dry-weather wind direction is constant, as along many seacoasts. The great bulk of sand in such a setting piles up in a ridge or a series of ridges across the wind. Their crests are generally sinuous and uneven, partly from unequal accumulation and partly from wind-swept channels between high points on their summits. Large patches of transverse dunes seem to advance in multiple ranks.

(f) Complex Dunes   Complex dunes, lacking clean-cut forms, develop where wind directions are variable, sand is abundant, and perhaps, vegetation interferes. Barchans, seifs, or transverse dunes locally become crowded and hence overlap, losing their distinctive shapes in a confused welter of diverse slopes (Fig. 24.17).

The dunes in the vast tracts of thick sand in Arabia cover almost one-third of the 2,500,000-square-kilometer area of the country. Northeast trade winds prevail there, but seasonal monsoons and other wind variations occur. Some of these winds reach gale velocities. The resulting dunes take on a great variety of shapes not readily classified.

Other Dune Features

(a) Blowouts   A blowout is a wind-excavated basin in a sand dune. It is usually spoon or horseshoe

FIGURE 24.16 *Outline sketch of seifs, such as are found in the Libyan Desert, showing prevailing winds (solid arrow) and strong secondary winds (dashed arrow), with the tendency of small dunes to unite into a longitudinal ridge.*

shaped, open toward the oncoming wind. The sand dug from it piles up around the margins of the horseshoe. On the basis of its new shape, such a modified dune is sometimes classified as a parabolic dune. Blowouts in fixed dunes may be started by a rent in a sod-covered dune or by the reduction of vegetation, following fire or reduced rainfall.

(b) *Internal Structure*   The deposition of sand on ever-changing slopes develops cross bedding within a dune. This consists of diversely oriented laminae of coarse and fine sand in wedge-shaped or curved units that reflect the changes in wind direction and intensity. If the migration of a dune is consistently in one direction, most of the laminae conform to the advancing slip face and are inclined in one direction. Conversely, the creation and later filling of blowouts, the upward mounding of sand, and the shifting of the orientation of slip faces in accordance with varying wind directions tend to develop a complex unsystematic arrangement of concave and convex units. Eolian cross bedding of either type is so distinctive in form that it marks certain sandstones as of dune origin (Fig. 24.18).

(c) *Ripple Marks*   Ripple marks are common on sand dunes (Figs. 24.11, 24.13, and 24.14) and on other sand surfaces associated with dunes. Their size and spacing are controlled by grain size and wind ve-locity. They are formed at the interface between wind and sand under the sorting action of surface creep. The coarsest grains collect at the crests of the ripples and the finest in the troughs. Ripple marks are asymmetrical, like small dunes. Variations in wind intensity modify them, continually adjusting and readjusting their crests. Strong winds may obliterate them entirely.

(d) *Association of Lakes*   In dry climates, deflation basins are likely to contain water only during wet seasons, and in deserts they may remain dry. But dunes and dune topography also occur in humid regions, where the basins among the dunes fill with water to form lakes or pools. Some excellent examples occur along the southern and eastern shores of Lake Michigan. Migrating dunes, propelled by the wind, may block a stream channel to form a lake, or along the shore they may block off a small bay or estuary to form a lagoon (Fig. 24.19).

CONTROL OF DUNES

Wind-blown sand locally requires control in order to protect man-made structures. It encroaches upon roads, railroads, airfields, buildings, and cultivated fields. Various methods of control are in use. A

FIGURE 24.17   *Sand dunes beside the All-American Irrigation Canal, eastern Imperial County, California. They range in form from simple barchans (middle foreground) to crowded complexes where sand is abundant. (Spence Air Photos.)*

FIGURE 24.18   *Sandstone, showing eolian cross bedding, or dune structure. (Lee, U.S. Geological Survey.)*

FIGURE 24.19   *Lake Chad, Africa, in part is blocked by dunes that have encroached upon its basin during an arid period. (NASA photo.)*

roadway can be built above the surrounding level to promote sweeping rather than deposition by the wind. Trenching with a bulldozer interrupts the sand movement temporarily. Planting sand-tolerant grasses, shrubs, and trees is widely practiced. Wire-and-slat structures similar to snow fences may serve for a while. Oiling the sandy ground with an asphaltic or waxy oil or covering the sand with gravel, cinders, or crushed rock, or other coarse material also prevents further sand movement. The method of control and the material to be used are usually determined by a cost-benefit ratio.

Occasionally, dune development is encouraged in order to protect natural land features. In the early 1930s, sand accumulation was promoted to protect and stabilize the Outer Banks of North Carolina. The purpose was to develop barrier dunes that would confine the upper limit of wave uprush. In doing so, overwash and channel development could be prevented.

This program has had a detrimental effect. Wave energy has tended to become more concentrated, resulting in a narrowing of the distance from dunes to shoreline. In a 13-year period, the dune to shoreline distance was narrowed 30 to 40 meters from the original distance of 100 to 125 meters.

REFERENCES

Bagnold, R.A., 1965, *The Physics of Blown Sand and Desert Dunes*, Methuen & Co., Ltd., London.

Blackwelder, E., 1931, *The lowering of playas by deflation*, American Journal of Science, 5th Series, vol. 21, pp. 140–144.

———, 1934, *Yardangs*, Geological Society of America Bulletin, vol. 45, pp. 159–166.

Bryan, K., 1931, *Wind-worn Stones or Ventifacts—A Discussion and Bibliography*, pp. 29–50, National Research Council Reprint and Circular Series 98, Report of the Committee on Sedimentation, Washington, D.C.

Frye, J.C., 1950, *The origin of Kansas Great Plains depressions*, pp. 1–20, Kansas Geological Survey Bulletin 86, Part 1.

Long, J.T., and Sharp, R.P., 1964, *Barchan-dune movement in Imperial Valley, California*, Geological Society of America Bulletin, vol. 75, pp. 149–156.

Lugn, A.L., 1962, *The Origin and Sources of Loess*, University of Nebraska Studies, New Series no. 26, Lincoln, Nebraska.

Sears, P.B., 1947, *Deserts on the March*, 2d ed., University of Oklahoma Press, Norman, Oklahoma.

Sharp, R.P., 1949, *Pleistocene ventifacts east of the Bighorn Mountains, Wyoming*, Journal of Geology, vol. 57, pp. 175–195.

Thorp, J., Smith, H.T.U., et al., 1952, *Pleistocene Eolian Deposits of the United States, Alaska, and Parts of Canada* (map, scale 1:2,500,000), Geological Society of America, Boulder, Colorado.

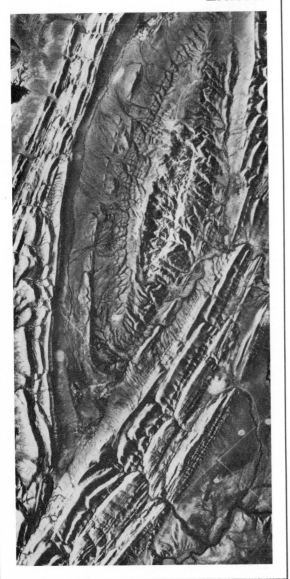

CONCLUSION
GEOLOGY: THE SCIENCE OF A CHANGING EARTH

The survey of The Work of Wind concludes this study of Geology: The Science of a Changing Earth. But does the study truly end? With change presented as the leitmotiv, can the study *ever* end?

All processes and events considered in this text reflect the variable, relentless, and eternal force that is change: from the finite formation of rocks and minerals to the infinite evolvement of geologic time; from the local effects of an abrupt earthquake to the shaping of continents over millions of years by plate tectonics; from mass gradation by gravity transfer to small-particle gradation by the wind. Certainly, the statement "no vestiges of a beginning, no prospects of an end" has more meaning now than it has ever had, for this is the essence of change.

And since the study of change is science, not only should the science of geology now be more apparent, so too should the concept of geology as a science. This concept is no different from that of any other science. As is so aptly expressed by Warren Weaver, a researcher, in the preface of *Investigating the Earth* (a textbook of the Earth Science Curriculum Project): "Science is an adventure. . . . It is an artistic enterprise stimulated by curiosity . . . served by disciplined imagination . . . and is a means of winning knowledge from ignorance. . . . It is a way of life and an unfinished human enterprise."

Geology is indeed The Science of a Changing Earth.

Look abroad thro' Nature's range.
Nature's mighty law is change.
Robert Burns

APPENDIXES

PROPERTIES OF COMMON MINERALS

Name	Composition	Color	Luster	Cleavage and fracture
Alabaster	(See gypsum)			
Albite	(See plagioclase)			
Amphibole	(See hornblende)			
Andesine	(See plagioclase)			
Anhydrite	$CaSO_4$ Calcium sulfate	White, gray, or tinted	Vitreous to pearly	3 rectangular cleavages
Anorthite	(See plagioclase)			
Asbestos	(See serpentine)			
Augite	$Ca(Mg,Fe,Al)$ $(Al,Si)_2O_6$ Complex aluminous ferromagnesian silicate	Dark green to nearly black	Vitreous to dull	Prismatic of 87° and 93°; basal parting
Biotite	$K(Mg,Fe)_3 AlSi_3O_{10}$ $(OH)_2$ Complex ferromagnesian silicate	Dark green, brown, black	Vitreous, pearly, or submetallic	Highly perfect in 1 plane
Bytownite	(See plagioclase)			
Calcite	$CaCO_3$ Calcium carbonate	Colorless, white, gray, or any color	Vitreous to earthy	Rhombohedral, perfect in 3 inclined planes
Chromite	$FeCr_2O_4$ Iron chromite	Brownish black to iron black	Pitchy submetallic to metallic	Fracture uneven
Corundum	Al_2O_3 Aluminum oxide	Brown, gray, red, blue, yellow, colorless	Vitreous	No cleavage; parting basal
Diamond	C Carbon	Colorless, yellow, red, blue, green gray, black	Adamantine	Perfect octahedral, 4 planes
Dolomite	$CaMg(CO_3)_2$ Calcium magnesium carbonate	White, pink, gray, yellow, brown black	Vitreous to pearly	Rhombohedral; 3 inclined planes
Fluorite	CaF_2 Calcium fluoride	Green, violet, white, yellow, colorless, blue brown	Vitreous	Perfect octahedral; 4 planes
Galena	PbS Lead sulfide	Lead gray	Metallic to dull	Perfect cubic; 3 rectangular planes
Garnet	$(Ca,Fe,Mg,Mn)_3$ $(Al,Fe,Cr)_2(SiO_4)_3$ Complex silicate	Red, brown, yellow, green	Vitreous to resinous	Conchoidal fracture

Hardness	Specific gravity	Crystal system	Other properties
3–3½	2.7–3.0	Orthorhombic	Usually in granular masses
5–6	3.2–3.6	Monoclinic	Crystals 4- or 8-sided in cross section; opaque or translucent
2½–3	2.7–3.2	Monoclinic	Readily splits into flexible flakes; transparent to translucent; disseminated scales common
3	2.72	Hexagonal	Effervesces freely in dilute HCl; transparent to opaque; crystals common
5½	4.3–4.6	Isometric	Streak dark brown; granular masses or disseminated grains in basic rocks; translucent to opaque
9	3.9–4.1	Hexagonal	Hexagonal crystals often barrel shaped; transparent to translucent; ruby and sapphire gems
10	3.2–3.5	Isometric	Transparent to opaque; occurs in octahedral crystals or as pebbles in placer gravels
3½–4	2.85	Hexagonal	Effervesces weakly in strong HCl; crystal faces often curved; transparent to translucent; occurs in fine- to coarse-grained masses
4	3.0–3.2	Isometric	Cubic crystals common; masses fine to coarse granular; transparent to translucent
2½	7.3–7.6	Isometric	Heavy to heft; opaque; cubes common; streak grayish black; ore of lead
6½–7½	3.4–4.3	Isometric	Occurs in well-formed equidimensional crystals or granular masses; transparent to translucent; used as an abrasive

PROPERTIES OF COMMON MINERALS *(Continued)*

Name	Composition	Color	Luster	Cleavage and fracture
Graphite	C Carbon	Steel gray to black	Dull earthy or metallic	Perfect, 1 way; flakes
Gypsum	$CaSO_4 \cdot 2H_2O$ Hydrous calcium sulfate	Colorless, white, gray, yellow, red, blue	Vitreous, pearly, or satiny	1 poor
Halite	$NaCl$ Sodium chloride	Colorless, white, or any color	Vitreous	Excellent, cubic; 3 rectangular planes
Hematite	Fe_2O_3 Iron oxide	Reddish brown to steel gray and iron black; red earth	Metallic or dull	Fracture uneven
Hornblende	$Ca_2(Mg,Fe,Al)$ $(Al,Si)_8O_{22}(OH)_2$ Complex ferromag-nesian silicate	Light to dark green, brown, or black	Vitreous to silky	Perfect prismatic at angles of 56° and 124°
Kaolinite	$Al_2Si_2O_5(OH)_4$ Hydrous aluminum silicate	White if pure; yellow, red, brown	Dull earthy in mass; scales pearly	Scales (rare); earthy fracture
Labradorite	(See plagioclase)			
Limonite	$Fe_2O_3 \cdot nH_2)$ (varies) Hydrous iron oxide	Yellow, brown, black	Dull earthy to submetallic	No cleavage; frac-ture conchoidal or earthy
Magnetite	Fe_3O_4 Iron oxide	Iron black	Metallic, sub-metallic, or dull	Fracture uneven
Microcline	$KAlSi_3O_8$ Potassium aluminum silicate	White, pink, red, yellow, green, gray	Vitreous or pearly	2 cleavages at nearly 90°
Oligoclase	(See plagioclase)			
Olivine	$(Mg,Fe)_2SiO_4$ Magnesium iron silicate	Pale green, olive green, to brownish green, according to iron content	Vitreous	Conchoidal fracture
Opal	$SiO_2 \cdot nH_2O$ Hydrous silica	Colorless, white, or any color	Vitreous, dull, or greasy	Conchoidal fracture
Orthoclase	$KAlSi_3O_8$ Potassium alumi-num silicate	White, pink, red, gray, colorless	Vitreous to pearly	2 perfect at 90°
Plagioclase feldspars	$NaAlSi_3O_8$ to $CaAl_2Si_2O_8$ Sodium-calcium aluminum silicates	Colorless, white, gray, bluish, greenish	Vitreous to pearly	2 perfect at 86°; fracture uneven

Hardness	Specific gravity	Crystal system	Other properties
1–2	1.9–2.3	Hexagonal	Opaque; greasy feel; marks paper; streak lead gray; in dense masses or flexible leaves
2	2.2–2.4	Monoclinic	Clear, often twinned crystals are *selenite;* fibers, *satin spar;* and massive form, *alabaster;* transparent to opaque
2–2½	2.1–2.3	Isometric	Salty to taste; cubic crystals or granular masses; transparent to translucent
5½–6½ (or soft)	4.9–5.3	Hexagonal	Streak red or red-brown; opaque; many varieties; steel gray, micaceous, and foliated form is *specular hematite;* main ore of iron
5–6	2.9–3.3	Monoclinic	Crystals often prismatic; granular masses of fibers; opaque or translucent
1–2½	2.2–2.6	Monoclinic	Smooth or somewhat greasy feel; earthy odor when moist; plastic when wet; opaque or translucent
1–5½	3.4–4.0	Amorphous	Streak always yellow brown; opaque; massive, concretionary, and other varieties; ore of iron
5½–6½	4.9–5.2	Isometric	Strongly magnetic; streak black; usually granular-massive or loose in black sand
6–6½	2.55	Triclinic	Crystals and cleavable masses
6½–7	3.2–3.6	Orthorhombic	Usually in granular masses or scattered crystals; transparent to translucent
1–6½	2.1–2.3	Amorphous	May show play of colors; structural forms varied
6	2.5–2.6	Monoclinic	Transparent to opaque; rock constituent or coarsely cleavable to granular masses
6–6½	2.6–2.7	Triclinic	Isomorphous series include albite, oligoclase, andesine, labradorite, bytownite, and anorthite; crystals striated; some labradorite shows blue opalescence

PROPERTIES OF COMMON MINERALS *(Continued)*

Name	Composition	Color	Luster	Cleavage and fracture
Pyrite	FeS_2 Iron sulfide	Pale brass yellow; tarnishes brown	Metallic	Fracture conchoidal or uneven
Pyroxene	(See augite)			
Quartz	SiO_2 Silica	Generally colorless or white; may be any color	Crystals vitreous; fractures oily	No cleavage; conchoidal
Serpentine	$H_4Mg_3Si_2O_9$ Hydrous magnesium silicate	Olive green, yellow green, brownish green, brown; often mottled	Greasy, waxy, dull; silky if fibrous	Conchoidal to splintery fracture; may be fibrous
Sphalerite	ZnS Zinc sulfide	White (pure), yellow, brown, black (ferrous)	Resinous to adamantine	6-direction cleavage
Talc	$H_2Mg_3(SiO_3)_4$ Hydrous magnesium silicate	White, gray, pale green	Pearly, dull or greasy	1 perfect, flaky

Hardness	Specific gravity	Crystal system	Other properties
6–6½	4.9–5.2	Isometric	Brittle; opaque; streak greenish to brownish black; striated cubes common
7	2.65	Hexagonal	Transparent to translucent; hexagonal prisms common; many varieties; cryptocrystalline form is *chalcedony*
2½–5	2.5–2.8	Monoclinic	Smooth to greasy feel; translucent; massive, platy, and fibrous (asbestos) varieties
3½–4	3.9–4.2	Isometric	Streak pale yellow; transparent to opaque; main ore of zinc
1	2.6–2.8	Monoclinic	Greasy feel; massive or foliated; flakes are flexible but not elastic; forms soapstone

THE METRIC SYSTEM
AND CONVERSION FACTORS

LENGTH

1 kilometer (km) = 10^3 meters
$\qquad\qquad$ = 0.621 statute miles
$\qquad\qquad$ = 0.540 nautical miles
1 meter (m) = 10^2 centimeters
$\qquad\qquad$ = 39.4 inches
$\qquad\qquad$ = 3.28 feet
$\qquad\qquad$ = 1.09 yards
1 centimeter (cm) = 10 millimeters (mm)
$\qquad\qquad$ = 0.394 inch
$\qquad\qquad$ = 10^4 microns (μ)
1 micron (μ) = 10^{-3} millimeter
$\qquad\qquad$ = 0.000394 inch

AREA

1 square centimeter (cm²) = 0.155 square inch
1 square meter (m²) = 10.8 square feet
1 square kilometer (km²) = 0.386 square statute mile

VOLUME

1 cubic kilometer (km³) = 10^9 cubic meters
$\qquad\qquad$ = 10^{15} cubic centimeters
$\qquad\qquad$ = 0.24 cubic statute mile
1 cubic meter (m³) = 10^6 cubic centimeters
$\qquad\qquad$ = 10^3 liters
$\qquad\qquad$ = 35.3 cubic feet
$\qquad\qquad$ = 264 U.S. gallons
1 liter (1) = 10^3 cubic centimeters
$\qquad\qquad$ = 1.06 quarts
$\qquad\qquad$ = 0.264 U.S. gallons
1 cubic centimeter (cm³) = 0.061 cubic inch

MASS

1 metric ton (1 tonne) = 10^6 grams
$\qquad\qquad$ = 2,205 pounds
1 kilogram (kg) = 10^3 grams
$\qquad\qquad$ = 2.205 pounds
1 gram (g) = 0.035 ounce

SPEED

1 meter per second (m/sec) = 2.24 statute miles per hour
$\qquad\qquad$ = 1.94 knots
1 centimeter per second (cm/sec) = 0.033 feet per second

TEMPERATURE

Conversion formulas:
$$°C = \frac{°F - 32}{1.8}$$
$$°F = 1.8\,(°C) + 32$$

ENERGY

1 gram-calorie (usually calorie, cal) = $\frac{1}{860}$ watt-hour
$\qquad\qquad$ = $\frac{1}{252}$ British thermal unit (Btu)

INDEX